Springer Texts in Statistics

Springer Texts in Statistics

Alfred	Elements of Statistics for the Life and Social Sciences
Christensen	Plane Answers to Complex Questions: The Theory of Linear Models
du Toit, Steyn, and Strumpf	Graphical Exploratory Data Analysis
Kalbfleisch	Probability and Statistical Inference: Volume 1: Probability. Second Edition
Kalbfleisch	Probability and Statistical Inference: Volume 2: Statistical Inference. Second Edition
Keyfitz	Applied Mathematical Demography. Second Edition
Kiefer	Introduction to Statistical Inference
Peters	Counting for Something: Statistical Principles and Personalities

Jack Carl Kiefer

Introduction to Statistical Inference

Edited by
Gary Lorden

With 60 Illustrations

Springer-Verlag
New York Berlin Heidelberg
London Paris Tokyo

Gary Lorden
California Institute of
Technology
Pasadena, CA 91125

AMS Classification: 62–01, 62FXX

Library of Congress Cataloging in Publication Data
Kiefer, Jack
Introduction to statistical inference.
(Springer texts in statistics)
Includes index.
1. Mathematical statistics. I. Lorden, Gary.
II. Title. III. Series.
QA276.16.K54 1987 519.5 86-31638

Typeset by Asco Trade Typesetting Ltd., Hong Kong.
Printed and bound by R.R. Donnelley and Sons, Harrisonburg, Virginia.
Printed in the United States of America.

9 8 7 6 5 4 3 2 1

ISBN 0-387-96420-7 Springer-Verlag New York Berlin Heidelberg
ISBN 3-540-96420-7 Springer-Verlag Berlin Heidelberg New York

Preface

This book is based upon lecture notes developed by Jack Kiefer for a course in statistical inference he taught at Cornell University. The notes were distributed to the class in lieu of a textbook, and the problems were used for homework assignments. Relying only on modest prerequisites of probability theory and calculus, Kiefer's approach to a first course in statistics is to present the central ideas of the modern mathematical theory with a minimum of fuss and formality. He is able to do this by using a rich mixture of examples, pictures, and mathematical derivations to complement a clear and logical discussion of the important ideas in plain English.

The straightforwardness of Kiefer's presentation is remarkable in view of the sophistication and depth of his examination of the major theme: How should an intelligent person formulate a statistical problem and choose a statistical procedure to apply to it? Kiefer's view, in the same spirit as Neyman and Wald, is that one should try to assess the consequences of a statistical choice in some quantitative (frequentist) formulation and ought to choose a course of action that is verifiably optimal (or nearly so) without regard to the perceived "attractiveness" of certain dogmas and methods.

Besides the centrally important decision-theoretic concepts such as admissibility and Bayes and minimax procedures, Kiefer is concerned with providing a rational perspective on classical approaches, such as unbiasedness, likelihood-based methods, and linear models. He also gives thorough consideration to useful ways of reducing and simplifying statistical problems, including sufficiency, invariance (equivariance), and asymptotics. Many of these topics are taken up in Chapter 4, which provides both an overview and a foundation for developments in later chapters, especially Chapters 6 and 7 on sufficiency and estimation, respectively. Chapter 5, on linear unbiased estimation, as well as Chapters 8 and 9, on hypothesis testing and confidence intervals, cover quite a bit of "standard statistical method''; although many insights and helpful ideas are provided, beginners who want to learn the usual techniques of Student's t-tests, F-tests, and analysis of variance will likely find the treatment too brief. Supplementary use of a standard text is recommended for this purpose. The first three chapters are short and introductory; Chapter 1, in particular, gives a clear and challenging view of what Kiefer has in mind. Appendix A lists some notational conventions.

Following Jack Kiefer's death in 1981, many of his colleagues, friends, and former students were eager to see his course notes published as a book. The statistical community owes its thanks to the late Walter Kaufmann-Bühler, of Springer-Verlag, for pursuing the project, and to Jack's wife, Dooley, for agreeing to it. In editing the notes, I have made only a few minor changes in the narrative, concentrating instead on correcting small errors and ensuring the internal consistency of the language, notation, and the many cross-references. The result, I believe, fully preserves the style and intentions of Jack Kiefer, who communicated his persistent idealism about the theory and practice of statistics with characteristic warmth and enthusiasm to generations of students and colleagues. May this book inspire future generations in the same way that Jack's teaching inspired all of us.

Gary Lorden
Pasadena, California

Contents

CHAPTER 1

Introduction to Statistical Inference

A typical problem in probability theory is of the following form: A sample space and underlying probability function are specified, and we are asked to compute the probability of a given chance event. For example, if $X_1, \ldots, X_n$ are independent Bernoulli random variables with $P\{X_i = 1\} = 1 - P\{X_i = 0\} = p$, we compute that the probability of the chance event $\left\{\sum_{i=1}^{n} X_i = r\right\}$, where r is an integer with $0 \leq r \leq n$, is $\binom{n}{r} p^r (1-p)^{n-r}$.

In a typical problem of statistics it is not a single underlying probability law which is specified, but rather a *class* of laws, any of which may *possibly* be the one which actually governs the chance device or experiment whose outcome we shall observe. We know that the underlying probability law is a member of this class, but we do not know which one it is. The object might then be to determine a "good" way of guessing, on the basis of the outcome of the experiment, which of the *possible* underlying probability laws is the *one* which actually governs the experiment whose outcome we are to observe.

EXAMPLE 1.1. We are given a coin about which we know nothing. We are allowed to perform 10 independent flips with the coin, on each of which the probability of getting a head is p. We do not know p, but only know that $0 \leq p \leq 1$. In fact, our object might be, on the basis of what we observe on the 10 flips, to guess what value of p actually applies to this particular coin. We might be professional gamblers who intend to use this coin in our professional capacity and who would clearly benefit from knowing whether the coin is fair or is strongly biased toward heads or toward tails. The outcome of the experiment cannot tell us with complete sureness the exact value of p. But we can try to make an accurate guess on the basis of the 10 flips we observe.

There are many ways to form such a guess. For example, we might compute [number of heads obtained in 10 tosses]/10 and use this as our guess. Instead, we might observe how many flips it took to get the first head, guessing p to be 0 if no head comes up in 10 tosses, and to be 1/[number of toss on which first head appears] if at least one head occurs. A third possibility is to guess p to be $\pi/8$ or 2/3, depending on whether the number of heads in 10 tosses is odd or even. Fourth, we might ignore the experiment entirely and always guess p to be 1/2. You can think of many other ways to make a guess based on the outcome of the experiment.

Which of these ways of forming a guess from the experimental data should we actually employ? The first method suggested may seem reasonable to you. Perhaps the second one does not seem very unreasonable either, since one would intuitively expect to get the first head soon if p were near 1 and later (or not at all) if p were near 0. The third method probably seems unreasonable to you; a *large* or *small* number of heads might indicate that p is near 1 or 0, respectively, but the knowledge only that the number of heads is *even* seems to say very little about p. The fourth method, which makes no use of the data, may also seem foolish.

Having formed some intuitive judgment as to the advisability of using any of these four guessing schemes, you may be asked to *prove that method* 1 *is better than method* 3. You will probably be unable even to get started on a proof, no matter how many of the classical works and elementary textbooks on statistics you consult. Yet, all of these books will tell you to use method 1, and in fact you will probably not find a reference to *any* other method! At best you will find some mention of a "principle of unbiased estimation" or a "principle of maximum likelihood" (don't worry now about what these mean) to justify the use of method 1, but this only transfers the inadequacy of the discussion to another place, since you will not find any real justification of the use of these "principles" in obtaining guessing methods. In fact, you will not even find any satisfactory discussion of what is meant by a "good" method, let alone a proof that method 1 really *is* good.

Surely it should be possible to make precise what is meant by a "good" guessing method and to determine which methods are good. It is possible. Such a rational approach to the subject of statistical inference (that is, to the subject of obtaining good guessing methods) came into being with the work of Neyman in the 1930s and of Wald in the 1940s. Almost all of the development of guessing methods before that time, and even a large amount of it in recent years, proceeded on an intuitive basis. People suggested the use of methods in many probabilistic settings because of some intuitive appeal of the methods. We shall see that such intuitive appeal can easily lead to the use of very bad guessing methods.

It is unfortunate that the modern approach to statistical inference is generally ignored in elementary textbooks. Books and articles developed in the modern spirit are ordinarily written at a slightly more advanced mathematical level. Nevertheless, it is possible to discuss all of the important ideas of modern statistical inference at an accessible mathematical level, and we shall try to do this. First we must describe the way in which a statistics problem is specified. This description can be made precise in a way that resembles the precise formulation of probability problems in terms of an underlying sample space, probability function, and chance events whose probabilities are to be computed.

CHAPTER 2

Specification of a Statistical Problem

We shall begin by giving a formulation of statistics problems which is general enough to cover many cases of practical importance. Modifications will be introduced into this formulation later, as they are needed.

We suppose that there is an experiment whose outcome the statistician can observe. This outcome is described by a random variable or random vector X taking on values in a space S. The probability law of X is unknown to the statistician, but it is known that the distribution function (df) F of X is a member of a specified class Ω of df's. Sometimes we shall find it convenient to think of Ω not as a collection of df's F but rather as a collection of real or vector parameter values θ, where the form of each possible F is specified completely in terms of θ, so that a knowledge of the value taken on by the label θ is equivalent to a knowledge of F. The members of Ω are sometimes referred to as the *possible states of nature*. To emphasize which of the *possible* values of F or θ is the unknown one which actually *does* govern the experiment at hand, this value is sometimes referred to as the "*true* value of F or θ," or the "*true* state of nature."

EXAMPLE 2.1(a). In Example 1.1 (Chapter 1), we can let $X = (X_1, X_2, \ldots, X_{10})$, where X_i is 1 or 0 according to whether the ith toss is a head or a tail. S consists of the 2^{10} possible values of X. The class Ω of possible df's (or probability laws) of X consists of all df's for which the X_i are independently and identically distributed with $P(X_i = 1) = \theta = 1 - P(X_i = 0)$, where $0 \leq \theta \leq 1$ (θ was called p in Example 1.1). It is convenient to think of these df's as being labeled by θ. It is easier to say "$\theta = 1/3$" than to say "the X_1 are independently and identically distributed with $P\{X_1 = 1\} = 1 - P\{X_1 = 0\} = 1/3$." Of course, the description "$\theta = 1/3$" can be used *only* because we have carefully specified the meaning of θ and the precise way in which it determines F.

It is implied in our discussion of specifying a statistical problem that by conducting an experiment the statistician can obtain some information about the "actual state of nature" F in Ω. It can often be shown by using the law of large numbers or other asymptotic methods (like those discussed later in Chapter 7) that if one could repeat the experiment an infinite number of times (independent replications), then one could discover F exactly. In reality the statistician performs only a finite number of experiments. The more experiments made, the better is the approximation to performing an infinite number of experiments.

It is frequently the case that statisticians or experimenters proceed as follows. We decide on an experiment and specify the statistical problem (as described later) in terms of the experiment. We then decide whether the results that would be obtained would be accurate enough. If not, we decide on a new experiment (often involving only more replications, like flipping the coin more) and work out the specification of a new statistical problem.

The behavior, just described, of a statistician, is often called *designing an experiment*. It is an aspect of statistics which we will not consider further. We will in the sequel be mostly concerned with questions of how best to use the results of experiments already decided upon.

Returning to our general considerations, we must introduce some notation. We will want to compute various probabilities and expectations involving X and must somehow make precise which member of Ω is to be used in this computation. Thus, in our Example 1.1, if we say, "the probability that $\Sigma X_i = 4$ is .15," under which of the possible underlying probability laws of X has this probability been computed? To eliminate confusion, we shall in general write

$$P_{F_0}\{A\} \quad \text{or} \quad P_{\theta_0}\{A\}$$

to mean "the probability of the chance event A, computed when the underlying probability law of X is given by $F = F_0$ (or by $\theta = \theta_0$)". Thus, in our Example 1.1 we would write

$$P_\theta\left\{\sum_{i=1}^{10} X_i = 4\right\} = \binom{10}{4}\theta^4(1-\theta)^6.$$

We shall extend the use of this notation to write

$$p_{F_0;Y} \quad \text{or} \quad p_{\theta_0;Y}$$

or

$$f_{F_0;Y} \quad \text{or} \quad f_{\theta_0;Y}$$

for the probability mass or density function of a random variable Y (which has been defined as a function of X) when the probability law of X is given by $F = F_0$ (or $\theta = \theta_0$). Thus, in the preceding example, if $Y = \sum_{i=1}^{10} X_i$, we would write

$$p_{\theta;Y}(4) = \binom{10}{4}\theta^4(1-\theta)^6.$$

We shall carry this notation over to expectations, writing

$$E_{F_0}g(X) \quad \text{or} \quad E_{\theta_0}g(X)$$

and

$$\text{var}_{F_0}\, g(X) \quad \text{or} \quad \text{var}_{\theta_0}\, g(X)$$

for the expectation and variance of $g(X)$ when the probability law of X is given by $F = F_0$ (or $\theta = \theta_0$). In our example, we would thus write

$$E_\theta Y = 10\theta, \qquad \text{var}_\theta\, Y = 10\theta(1-\theta).$$

We now describe the next aspect of a statistics problem which must be specified. This is the collection of possible actions which the statistician can take, or of possible statements which can be made, at the conclusion of the experiment. We denote this collection by D, the *decision space*, its elements being called *decisions*. At the conclusion of the experiment the statistician actually only chooses *one* decision (takes one action, or makes one statement) out of the possible choices in D. The statistician must make such a decision.

EXAMPLE 2.1(a) (continued). In our previous example, the statistician was required to guess the value of θ. In this case, we can think of D as the set of real numbers d satisfying $0 \le d \le 1$:

$$D = \{d: 0 \le d \le 1\}.$$

Thus, the decision ".37" stands for the statement "my guess is that θ is .37."

EXAMPLE 2.1(b). Suppose our gambler does not want to have a numerical guess as to the value of θ but only requires to know whether the coin is fair, is biased toward heads, or is biased toward tails. In this case the space D consists of three elements,

$$D = \{d_1, d_2, d_3\},$$

where d_1 stands for "the coin is fair," d_2 stands for "the coin is biased toward heads," and d_3 stands for "the coin is biased toward tails."

Note that in both of these examples, D can be viewed as the collection of possible answers to a question that is asked of the statistician. ("What do you guess θ to be?" "Is the coin fair, biased toward heads, or biased toward tails?")

EXAMPLE 2.1(c). An even simpler question would be "Is the coin fair?" We would have

$$D = \{d_1, d_2\},$$

where d_1 means "yes (the coin is fair)" and d_2 means "no (the coin is biased)."

To see how D might be regarded as the collection of possible *actions* rather than *statements*, suppose the U.S. Mint attempted to discourage gamblers from unfairly using its coins, by issuing only coins which were judged to be fair. In this case, the Mint's statistician, after experimenting with a given coin, would either throw it into a barrel to be shipped to a bank (d_1) or throw it back into the melting pot (d_2). Actually, we shall see that, in the general theoretical approach to statistics, the practical meaning of the various elements of D as statements or actions does not affect the development at all: once S, Ω, D, and the function W (to be described later) have been specified, we would reach the same conclusions for any interpretation of the physical meaning of the elements of D. In two different applications for both of which a given specification of S, Ω, D, and W is appropriate, we will reach the same conclusion regarding what statistical procedure ("guessing method") to use, regardless of what the applications are. The same statistical technique would be used in conjunction with agronomical or astronomical work, provided the possible outcomes of the experiment (S), possible probability laws (Ω), mathematical representation of the space of possible decisions (D), and loss function (W) are the same in the two settings. The fact that d_1 stands for one physical statement in the agronomical application and for another in the astronomical setting does not matter.

Example 2.1(d). Suppose that our gambler does not merely want a guess as to the value of θ which governs the coin at hand, but rather a statement of an interval of values which is thought to include θ. In this case we can think of each element d of D as being an interval of real numbers from d' to d'' inclusive, where $0 \leq d' \leq d'' \leq 1$, so that D is the set of all such intervals; or, equivalently, we can think of each d as being the ordered pair (d', d'') of endpoints of an interval. In either case, the decision $d = (d', d'')$ stands for the statement "My guess is that $d' \leq \theta \leq d''$." A guess in this form, such as "I guess that $.35 \leq \theta \leq .4$," may not seem to be so precise as the statement "I guess that $\theta = .37$," the latter form being that encountered in Example 2.1(a). We shall see later that in practice it is often more desirable to give a guess in the form of the present example than in the form of Example 2.1(a), because of the possibility of attaching to an interval of values a simple measure of confidence in the statement that the underlying value θ does fall in the stated interval.

In order to specify a statistical problem completely, we must state precisely how right or wrong the various possible decisions are for each possible underlying probability law F (or θ) of X. (We repeat that, in order to emphasize that we are speaking of that unknown value of θ, among all possible values of θ, which describes the distribution of X in the *particular experiment now at hand*, books and articles often refer to this value as the *true value of* θ. This true value may of course change when we move on to another experiment for which S and Ω are exactly the same—for example, when we look at another coin in Example 2.1.)

For example, in Example 2.1(c) it may be that, if the true value of θ is fairly close to .5, say $.495 < \theta < .505$, then we deem it correct to make decision d_1 ("the coin is fair") and incorrect to make decision d_2, since if $|\theta - .5| < .005$ the coin is judged to be close enough to fair for us to call it such. On the other hand, if $|\theta - .5| \geq .005$, we may feel that d_2 is to be thought of as correct and d_1 as incorrect. If we judge that making a correct decision causes us to incur no loss, whereas any incorrect decision causes us to incur the same positive loss L as any other incorrect decision, we can write down analytically the loss encountered if the true distribution of X is given by θ and if we make decision d, as follows:

$$W(\theta, d_1) = \begin{cases} 0 & \text{if } |\theta - .5| < .005, \\ L & \text{otherwise;} \end{cases}$$

$$W(\theta, d_2) = \begin{cases} L & \text{if } |\theta - .5| < .005, \\ 0 & \text{otherwise,} \end{cases}$$

Similarly, in any statistics problem we define $W(F, d)$ (or $W(\theta, d)$) to be the *loss incurred* if the *true distribution of X is given by F (or θ) and the statistician makes decision d*. The function W is one of the data of the problem; the loss $W(F, d)$ (or $W(\theta, d)$) which will be encountered if F (or θ) is the true distribution and decision d is made must be stated precisely for every possible F (or θ) in Ω and every possible decision d in D. The reason for this is that, as we shall see later, the choice of a good statistical procedure (guessing rule) depends on W. A procedure which is good for one loss function W may be very poor for another loss function which might be appropriate in other circumstances.

How do we obtain the loss function? In practice it is often difficult to judge the relative seriousness of the various possible incorrect decisions one might make for each possible true situation, so as to make W precise. In some settings W can be written down in units of dollars on the basis of economic considerations, but in other settings, particularly in basic scientific research, the losses (which might be in units of "utility") incurred by making various decisions are difficult to make precise. For example, if an astronomer makes several measurements (which include chance errors) on the basis of which he wants to publish a guess of the distance to a certain galaxy, what is the seriousness of misguessing the distance by 5 light-years, as compared to misguessing the distance by only 1 light-year?

Practically speaking, the essential fact is that there are many important practical settings where we can find *one* statistical procedure which is fairly good for any of a variety of Ws whose forms are somewhat similar. From this we can conclude that it may not be so serious if the astronomer's judgment as to the values of losses due to making various possible decisions is not exactly what it should be (that is, what it would be if the astronomer could foresee all possible future developments which might result from publishing this distance, such as winning the Nobel Prize, or getting fired, or the United States sending the first dog to a star, and could determine from these the disutilities of all bad guesses). For as long as the loss function the astronomer writes down has roughly the same shape and tendencies as the loss function he or she should be using, a statistical procedure which we determine to be good using the astronomer's loss function will also be good for the actual loss function he or she should be using. We will indicate what this means in an example:

EXAMPLE 2.2. Suppose the astronomer's measurements are independent random variables, each with normal $\mathcal{N}(\mu, \sigma^2)$ distribution, mean μ and variance σ^2, where μ is the actual distance to the star. If this distance is known to be at least h light-years but nothing else is known, we can write $\theta = (\mu, \sigma^2)$ and

$$\Omega = \{(\mu, \sigma^2) : \mu \geq h, \sigma > 0\}.$$

The decision, as in Example 1.1(a), is a real number:

$$D = \{d : d \geq h\},$$

the decision $d = 100$, meaning "I guess the distance μ to be 100 light-years." The loss function would probably reflect the fact that the larger the amount by which the astronomer misguesses μ, the larger the loss incurred. For example, we might have

$$W((\mu, \sigma^2), d) = |\mu - d|$$

or

$$W((\mu, \sigma^2), d) = (\mu - d)^2,$$

and you can think of many other forms of W which exhibit the same tendencies. The important practical fact is that *a certain statistical procedure (guessing method) which is good assuming one of these forms of* W *will also be good for many other* W's *of similar form.* We shall discuss this fact in detail later on.

Thus, in practice it will often suffice to make a rough judgment as to the form of W. *Such a rough judgment, and the resulting determination of a good statistical procedure, is to be preferred greatly to the classical approach to statistics wherein no attempt is made to determine* W, *but instead an "intuitively appealing" procedure is used.* We shall see that such intuition can easily lead to the use of very poor procedures. The use of procedures which do turn out to be good ones in certain settings is not logically justifiable on intuitive grounds, but only on the basis of a careful mathematical argument.

EXAMPLE 2.2 (continued). Although the same statistical procedure may be good for many different W's of similar form, such a procedure may be a poor one if we are faced with a W which is quite different, and this is one reason why even a rough judgment as to the form of W is better than the intuitive approach which does not consider W at all. For example, suppose that underguesses of the values of μ are more serious than overguesses (they may result in giving the dog too little fuel). This may be reflected in a loss function such as

$$W((\mu, \sigma), d) = \begin{cases} 2(\mu - d) & \text{if } \mu \geq d, \\ d - \mu & \text{if } d \geq \mu. \end{cases}$$

For such a loss function, we would use a different statistical procedure than we would for the loss functions exhibited previously, which were symmetric functions of $d - \mu$.

EXAMPLE 2.3. We shall now give an example to illustrate how W might be determined on economic grounds. A middleman buys barrels of bolts from a manufacturer at \$15 per barrel. He can sell them to customers at \$25 per barrel, with a guarantee that a barrel contains at most 2 percent defective bolts, the customer receiving his money back while keeping the barrel of bolts in the event that he finds more than 2 percent defective bolts in the barrel. (We assume the customer always discovers how many defective bolts there are.) The middleman also has the option of selling the barrels with no guarantee, at \$18 per barrel. He removes a random sample of 200 bolts from each barrel, inspecting the 200 bolts from each barrel for defectives and deciding on the basis of this inspection whether to make the decision d_1 to sell that barrel at \$25 with a guarantee or the decision d_2 to sell it at \$18 without a guarantee. (The selling prices are for the barrels with 200 bolts removed from each; the cost of inspection is included in the \$15; the inspection procedure destroys the inspected bolts, which therefore cannot be returned to the barrels.) Let θ be the (unknown) proportion of defectives in a given barrel after the sample of 200 has been removed. If $\theta \leq .02$, it would be best to ship the barrel out under a guarantee, for \$25 (that is, to make decision d_1); to ship it out at \$18 without a guarantee would result in making \$7 less profit. On the other hand, if $\theta > .02$ and the middleman makes decision d_1, he will have to refund the \$25 as guaranteed and will receive nothing for the bolts; by making decision d_2, he will receive \$18 in this circumstance, $\theta > .02$, too. His profit (revenue minus cost) on a barrel with proportion θ of defectives and on which he makes decision d is thus given by the following table:

		Proportion θ of defectives in barrel	
		$\leq .02$	$> .02$
Decision made	d_1	\$10	$-$\$15
on barrel	d_2	\$3	\$3

This being profit, we could take the values of the loss function $W(\theta, d)$ to be the *negative* of those just tabled. For example, $W(.05, d_1) = \$15$, and $W(.01, d_2) = -\$3$. Note that a negative loss is the same as a positive gain. Sometimes statisticians work not with the absolute loss as exemplified here, but rather with a quantity called the *regret*. For each F, the excess of the loss $W(F, d')$ over the *minimum possible loss* for this F (incurred by making the most favorable possible decision) is the regret $W^*(F, d')$ when F is true and decision d' is made. Thus,

$$W^*(F, d') = W(F, d') - \min_d W(F, d),$$

where $\min_d W(F, d)$ means the minimum over all d of $W(F, d)$. In our example, the regret function W^* is thus given by

$$W^*(\theta, d_1) = \begin{cases} 0 & \text{if } \theta \leq .02; \\ \$18 & \text{if } \theta > .02; \end{cases}$$

$$W^*(\theta, d_2) = \begin{cases} \$7 & \text{if } \theta \leq .02; \\ 0 & \text{if } \theta > .02. \end{cases}$$

The regret is always nonnegative, and zero regret corresponds to making a decision which is most favorable for the F under consideration.

In future considerations we shall not concern ourselves with the question of whether the W being used is the primitive loss function or is actually the regret function derived from such a loss function. Usually the selection of a good statistical procedure, which is our eventual aim, will not depend much on which of these is being used. Whenever a loss function W has the property that $\min_d W(F, d) = 0$ for each F (that is, for each F there is a "correct" decision for which the loss is zero, and no better decision), we see that W^* and W are identical. The reader should check that the examples of W's displayed in Examples 2.1 and 2.2 have this property.

To summarize, then, a *statistical problem* is specified in most settings by stating S, Ω, D, and W. You should check the previous examples and note which ones have been completely specified. As we have mentioned, there will be modifications to this statement of a statistical problem (that is, there will be additional specifications) in some settings, which we will introduce later.

Once a statistical problem has been specified, the problem of statistical inference is to select a *statistical procedure* (which we earlier called a "guessing method," and which is sometimes also called a *decision function*) which will describe the way in which a decision is to be made on the basis of the outcome observed to occur in the experiment. (You will recall that such a decision must be made.) A *statistical procedure* t is a function from S into D. Once a statistical procedure t has been selected, it is used as follows: The experiment is conducted, the chance variable X being observed to take on a value x_0, an element of S. The function t then assigns an element d_0 of D, given by $d_0 = t(x_0)$, corresponding to this value x_0 in S. The statistician makes decision d_0. There are many statistical procedures from which to choose in any problem, and we must take care to distinguish between different procedures under discussion by using different symbols for them.

Example 2.1(a) (continued). In the first mention of the example, as Example 1.1, four statistical procedures were described for the case $D = \{d : 0 \leq d \leq 1\}$

of Example 1.1(a). We can write out the definitions of these procedures analytically as follows:

$$t_1(x_1,x_2,\ldots,x_{10}) = \sum_{i=1}^{10} x_i/10;$$

$$t_2(x_1,x_2,\ldots,x_{10}) = \begin{cases} 1/j & \text{if } x_1 = \cdots = x_{j-1} = 0 \text{ and } x_j = 1, \\ 0 & \text{if } x_1 = \cdots = x_{10} = 0; \end{cases}$$

$$t_3(x_1,x_2,\ldots,x_{10}) = \begin{cases} \pi/8 & \text{if } \sum_{i=1}^{10} x_i \text{ is odd}, \\ 2/3 & \text{if } \sum_{i=1}^{10} x_i \text{ is even}; \end{cases}$$

$$t_4(x_1,x_2,\ldots,x_{10}) = 1/2.$$

Make sure that you understand these analytic descriptions of the procedures which were previously described verbally.

EXAMPLE 2.1(b) (continued). The statistical procedures t_i, $i = 1, 2, 3, 4$, in the previous example were real-valued functions because D was an interval of real numbers. In Example 2.1(b) a procedure t does not take on real values, but rather the values d_1, d_2, d_3. For example, the procedure t' which tells us to state that the coin is fair if exactly half the tosses are heads and to state that the coin is biased toward heads or tails, respectively, if more or less than half the tosses are heads, is described as follows:

$$t'(x_1,\ldots,x_{10}) = \begin{cases} d_1 & \text{if } \sum_{i=1}^{10} x_i = 5, \\ d_2 & \text{if } \sum_{i=1}^{10} x_i > 5, \\ d_3 & \text{if } \sum_{i=1}^{10} x_i < 5. \end{cases}$$

The procedure t'' which says to ignore the data and state that the coin is fair is described by

$$t''(x_1,\ldots,x_{10}) = d_1.$$

How are we to compare various statistical procedures in a given problem, to select that procedure which we will actually use? In any practical problem, *any* statistical procedure we use can *possibly* lead to an unfortunate decision. For example, in Example 2.1(a), if we use the procedure t_1, it *might* happen that $\theta = .02$ but that we observe all heads ($x_1 = x_2 = \cdots = x_{10} = 1$) and therefore guess θ to be $t_1(1,1,\ldots,1) = 1$ (that is, we make the decision $d = 1$). This is certainly an inaccurate guess, and if we are using t_1 and the circumstances are that $\theta = .02$ and all heads come up in the experiment, then we will incur loss $W(.02, 1)$, which will be relatively large for any of the possible W's we have listed as illustrations in Example 2.1(a). The essential point we must keep in mind, however, is that such incorrect decisions, although *possible*, will not necessarily be very *probable*. Intuitively speaking, a "good" statistical procedure is one for which the probability is large that a favorable decision (one for which small loss is incurred) will be made, whatever the true F may be; we must make this statement precise.

Whatever the (unknown) true F may be, the decision $t(X)$ which will be made if the statistical procedure t is used is a random variable, since X, the result to be obtained in the experiment, is a random variable. The values taken on by this random variable are elements of D (not necessarily real numbers). Thus, if S is a discrete sample space, we can compute, for each fixed d,

$$P_F\{\text{decision } d \text{ will be made when procedure } t \text{ is used}\}$$
$$= \sum_{x \text{ such that } t(x)=d} p_{F;X}(x).$$

This quantity, which we shall hereafter write as

$$q_t(F, d),$$

is called the operating characteristic of the procedure t, considered as a function of F and d. We can compute this function q_t for any procedure t, and it tells us the probability, when F is true and we use t, of making any decision d. Note the importance of keeping straight what the probability of an event means: P_F means that the probability is to be computed when the underlying true probability law of X is F.

EXAMPLE 2.1(a) (continued). Since $\sum_{i=1}^{10} X_i\ (=10t_1(X))$ has a binomial probability function with parameters 10 and θ when the label value of the true probability law is θ, we have

$$q_{t_1}(\theta, d) = \begin{cases} \binom{10}{10d}\theta^{10d}(1-\theta)^{10(1-d)} & \text{if} \quad d = 0, 1/10, 2/10, \ldots, 1; \\ 0 \quad \text{otherwise.} \end{cases}$$

Since the probability that the first head comes on the rth toss is $\theta(1-\theta)^{r-1}$ and the probability of obtaining no heads in 10 tosses is $(1-\theta)^{10}$, we also obtain

$$q_{t_2}(\theta, d) = \begin{cases} \theta(1-\theta)^{d^{-1}-1} & \text{if} \quad d = 1, 1/2, 1/3, \ldots, 1/10; \\ (1-\theta)^{10} & \text{if} \quad d = 0; \\ 0 \quad \text{otherwise.} \end{cases}$$

We also have

$$q_{t_4}(\theta, 1/2) = 1.$$

Note in each case that, for fixed t and θ, the function $q_t(\theta, d)$, considered as a function of d, is a probability function.

EXAMPLE 2.1(b) (continued). For the procedure t' defined previously, we have

$$q_{t'}(\theta, d_1) = \binom{10}{5}\theta^5(1-\theta)^5;$$

$$q_{t'}(\theta, d_2) = \sum_{i=6}^{10}\binom{10}{i}\theta^i(1-\theta)^{10-i};$$

$$q_{t'}(\theta, d_3) = \sum_{i=0}^{4}\binom{10}{i}\theta^i(1-\theta)^{10-i}.$$

EXAMPLE 2.4. In problems of "acceptance sampling," there are two decisions: the decision d_1 to accept the lot being inspected, and the decision d_2 to reject the lot. In this context what is usually called the *operating characteristic* of a procedure t is actually the function $q_t(\theta, d_1)$. When there are only two decisions, as in this example, the rest of what we have called the operating characteristic is immediately obtainable, since

$$q_t(\theta, d_2) = 1 - q_t(\theta, d_1).$$

The operating characteristic of a given statistical procedure tells us the probabilities that it will result in making various decisions (due to the chance nature of the experiment) for each possible true probability law F. Suppose the true F is F_0. By chance, the procedure t may produce a favorable decision, or a slightly bad one, or a very bad one. If the experiment results in an outcome $X = x'$, the decision $d' = t(x')$ will be made, and the loss incurred will be $W(F_0, d')$. The probability of the event that decision d' will be made if we use procedure t and if the true F is F_0 is $q_t(F_0, d')$. Thus, the *expected* loss which will be incurred when t is used and F_0 is the true law of X is

$$\sum_d W(F_0, d)q_t(F_0, d).$$

We denote this quantity by

$$r_t(F_0).$$

We computed this quantity in two steps, since we wanted to define the operating characteristic q_t on the way. We could also have written down a formula for $r_t(F_0)$ directly by noting that the event $X = x'$, which has probability $p_{F_0;X}(x')$ in the discrete case if F_0 is the underlying true probability law, entails making the decision $t(x')$ if procedure t is used and thus entails incurring the loss $W(F_0, t(x'))$. Hence, the *expected* loss from using procedure t when F_0 is true is

$$r_t(F_0) = E_{F_0}W(F_0, t(X)) = \sum_x W(F_0, t(x))p_{F_0;X}(x).$$

For each procedure t, the function r_t is a real-valued function, defined on the space Ω of possible underlying probability laws of X. This function r_t is called the *risk function of the procedure t*.

We shall base our comparison of the various possible procedures, and our choice of the one of them which is to be used in a given statistical problem, on the risk functions of the procedures. It is very important to acquire a feeling for the meaning of the risk function.

Throughout this book we shall write $\bar{X}_n = n^{-1} \sum_{i=1}^{n} X_i$.

EXAMPLE 2.1(a) (continued). Suppose the loss function is $W(\theta, d) = (\theta - d)^2$, which is often called "squared error." We then have (since $10\bar{X}_{10}$ is binomial with mean 10θ and variance $10\theta(1-\theta)$)

$$\begin{aligned} r_{t_1}(\theta) &= E_\theta(t_1(X) - \theta)^2 \\ &= E_\theta(\bar{X}_{10} - \theta)^2 = \theta(1-\theta)/10. \end{aligned}$$

Also,

$$\begin{aligned} r_{t_4}(\theta) &= E_\theta(t_4(X) - \theta)^2 \\ &= E_\theta(1/2 - \theta)^2 \\ &= (1/2 - \theta)^2. \end{aligned}$$

Suppose $t_5(x_1, \ldots, x_{10}) = \sum_{i=1}^{6} x_i/6$. (Thus, t_5 is actually based on only the first six tosses of the coin, ignoring the outcomes on the other tosses.) We then have

$$\begin{aligned} r_{t_5}(\theta) &= E_\theta(t_5(X) - \theta)^2 \\ &= E_\theta(\bar{X}_6 - \theta)^2 = \theta(1-\theta)/6. \end{aligned}$$

One can compute the risk functions of t_2 and t_3, but we shall omit these slightly messier computations.

It is always instructive to draw graphs of the risk functions of procedures under consideration:

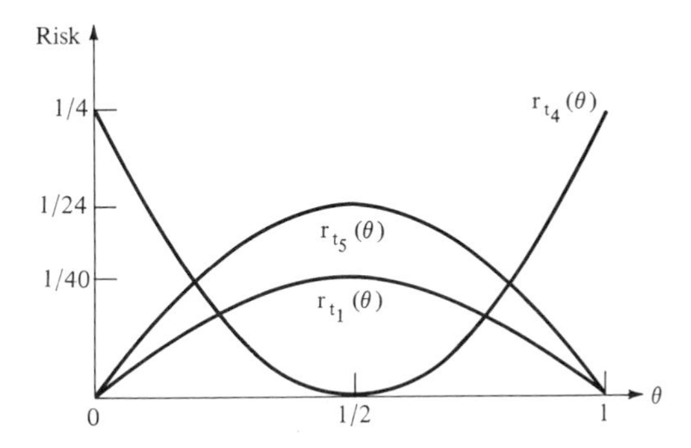

One can think of these graphs in the following way: If the true (unknown) value of θ is θ', we look at the ordinates of the three graphs corresponding to the abscissa value $\theta = \theta'$. These three ordinate values tell us the risk (expected loss) when the true θ value is θ' if we use t_1, t_4, and t_5. The smaller the risk, the better is the performance of a procedure when $\theta = \theta'$. Thus, when $\theta = 1/2$, the procedure t_4 yields risk 0, and t_1 and t_5 perform better than t_4 when $\theta = 1$.

Of course, in any given experiment, once a decision has been made, there is no question of chance: the decision is good or bad, and a definite loss has been incurred (we will not know what it is, since we do not know the true value of F). The risk tells us the *expected* loss in advance of performing the chance experiment. Alternatively, you can think of the risk intuitively in terms of the law of large numbers: if the statistician or society uses a procedure t' repeatedly in a statistical problem which comes up often, then the average of the *actual* losses incurred in all those cases for which the true θ is θ' will be close to $r_t(\theta')$, if the number of such cases is large.

If there were a procedure t^* whose risk function were, for *all* F, no greater than that of any other procedure (that is, if $r_{t^*}(F) \leq r_t(F)$ for all F and all procedures t), that would clearly be the procedure to use. For no matter what the true F might be, less expected loss would be incurred from using t^* than from using any other procedure. Unfortunately, there are no practical statistical problems for which such a "uniformly best" procedure t^* exists. Rather, the situation is as illustrated in the preceding example, where for one true F (e.g., $\theta = 1/2$) one procedure (namely, t_4) is best, whereas for another true F ($\theta = 1$) another procedure (t_1) is better than the first. We do not, of course, know the value of θ which governs our experiment, so we cannot know whether we would be better off using t_1 or t_4 with the experiment at hand! In our example, how are we to choose between procedures t_1 and t_4? This will require quite a bit of additional discussion. However, one thing is clear already: we would certainly never use t_5, since

$$r_{t_1}(\theta) \leq r_{t_5}(\theta)$$

for all θ, with strict inequality except when $\theta = 0$ or 1. Thus, no matter what the true value of θ may be, we would expect to be at least as well off from using t_1 as from using t_5 (and usually *better* off, as indicated by the strict inequality). Thus, although we still have to answer the question of whether it is preferable to use t_1 or t_4, there will never be any question about using t_5: anyone who uses t_5 could only do better, and never worse, by using t_1; among t_1, t_4, and t_5, we could rule out t_5 from consideration at once.

In general, we formalize this notion by saying that a procedure t' is *better than* a procedure t'' in a given statistical problem if

$$r_{t'}(F) \leq r_{t''}(F) \quad \text{for all } F,$$

with strict inequality holding for some F. If $r_{t'}(F) = r_{t''}(F)$ for all F, we say that t' and t'' are *equivalent*; it is perfectly possible for two procedures to be different functions on S but to have identical risk functions. If neither of two procedures is better than the other and they are not equivalent, we say that they are *incomparable* (t_1 and t_4 in the example).

If, for a given procedure t, there is another procedure t' which is better than it, we say that t is *inadmissible*; otherwise, we say the t is *admissible*. Thus, t_5 is inadmissible in the preceding example. We have not proved that t_1 and t_4 are admissible (as they are, although t_2 and t_3 are not), since we have not compared them with all other possible procedures. (If t_1, t_4, and t_5 were the only procedures to be considered, then the graphs would show us that t_1 and t_4 are admissible.) Just as we saw that there is no reason to use t_5 in our example, so we can see that in general there is no need to consider inadmissible procedures in a statistical problem. It is enough to consider only the admissible procedures, since for any inadmissible procedure t there will exist an admissible procedure t' which is better than t. [A regularity condition which will almost always be satisfied in practice is actually needed to ensure the validity of this last statement.] If two procedures are equivalent, it is immaterial which of the two is used, from our point of view.

Thus, as one general aim in the study of statistical inference, we could hope to make a list, for each statistical problem, of the admissible procedures. We shall have occasion to describe such lists of all admissible procedures in certain problems, but this will not usually be our chief concern. The reason for this is that the practical user of statistics does not want a list of many admissible procedures, but rather a *single* procedure to use to make a decision based on the outcome of the experiment at hand. Thus, we will be more concerned with the selection of the procedure to be used, than with obtaining a list of all admissible procedures.

How are we to select one procedure out of the class of all admissible procedures? Nothing we have discussed up to this point gives us any guide. Indeed, although every mathematical statistician would agree that we need only consider admissible procedures in making this selection, in many problems there would be wide disagreement among statisticians as to which admissible procedure should actually be used. There is no entirely mathematical basis for making this selection. Rather, one must introduce some extramathematical criterion for selecting a procedure from among the admissible procedures. This criterion might be based on certain axioms for "rational" human behavior which are not the province of this course. Suffice it to say that there are many such criteria now in existence and that, rather than use one of them dogmatically, we shall mention and illustrate the use of several of them (not, however, without criticism) in the chapters that follow.

2.1. Additional Remarks on the Loss Function

1. *Scale of* W. You should keep in mind that the choice of units of loss will not affect our conclusions. Thus, if W is replaced by $100W$ because we decide to measure loss in cents instead of dollars, the same procedures will be admissible, or Bayes relative to a given ξ, or minimax, or invariant, etc.

(these last terms will be defined in Chapter 4), for every risk function is multiplied by the same constant, 100. The conclusions would similarly remain unchanged if W were replaced by $W + c$ where c is a constant, since this merely involves adding the same constant c to every risk function.

2. If the loss function W can only take on the values 0 and 1, and if, for each F, we term a decision d "correct" or "incorrect" for this F according to whether $W(F,d) = 0$ or $= 1$ (there may be 0, 1, or more decisions of each of these two kinds for any F), then the risk function becomes

$$r_t(F) = E_F W(F, t(X)) = P_F\{W(F, t(X)) = 1\}$$
$$= P_F\{t \text{ reaches an incorrect decision}\},$$

which gives a simple interpretation of the risk function in such settings.

A W which can only take on the values 0 and 1 is sometimes called a *simple loss function.*

3. Sometimes the intuitive objection is raised that expected loss "does not sufficiently weigh the catastrophic effect of large losses" and that the variance or dispersion of chance losses should also be taken into consideration. For example, one might consider $E_F\{[W(F, t(X))]^2\}$ instead of $r_t(F) = E_F W(F, t(X))$. An answer to this is that reasonable axioms of rational human behavior lead to the conclusion that there is a (correct) loss function such that expected loss is the quantity which the rational person should try to make small and should use as a basis for comparison of several procedures; hence, that the preceding objection indicates that the objector has not chosen the loss function so as to reflect correctly the great disutility of certain decisions under certain F's. (In the preceding example, one might be more correct in using not the given W, but rather its square, as the loss function.)

A similar remark applies to sequential settings (see Chapter 3), wherein the number of observations n is a chance variable. Some people point to $\text{var}(n)$ as something which is as important to consider as En, but the logical basis for such a consideration would be that the cost (in utility) of taking n observations is not merely proportional to n, but is perhaps quadratic in n.

In discussing Bayes procedures in Chapter 4 we shall have occasion to average $r_t(F)$ over various values of F. An answer similar to that given previously applies there to the objection that large values of $r_t(F)$ should be given heavier weight: the objector has not chosen W properly.

4. Sometimes, we shall see, statisticians (for good or bad reasons, usually the latter) restrict considerations to procedures in some specified class $\mathscr{D}'$ rather than considering the class $\mathscr{D}$ of *all* procedures t. We then use such phrases as "t^* is admissible relative to $\mathscr{D}'$" to mean that t^* is in $\mathscr{D}'$ and that no t' in $\mathscr{D}'$ is better than t^*. In this last circumstance, there may or may not exist a t'' in $\mathscr{D}$ that is better than t^*, so that t^* could be admissible in $\mathscr{D}'$ but either inadmissible or admissible in $\mathscr{D}$. (If it is inadmissible in $\mathscr{D}'$, the

existence of t' in $\mathscr{D}' \subset \mathscr{D}$ that is better than t^*, shows that t^* is inadmissible in $\mathscr{D}$.) Similarly, procedures may be Bayes or minimax relative to $\mathscr{D}'$, meaning the competing procedures are restricted to $\mathscr{D}'$.

A procedure may have an appealing property when compared only to procedures in $\mathscr{D}'$ but may be much worse than some other procedure in $\mathscr{D} - \mathscr{D}'$. This is one of the dangers one must be aware of in restricting consideration to procedures that possess some (perhaps intuitively appealing) property, as we shall see.

Problems

[Suggestion: Work one of these three similar problems.]

2.1. X is a Bernoulli random variable with $P_\theta\{X = 1\} = 1 - P_\theta\{X = 0\} = \theta$. It is known only that $0 \leq \theta \leq 1$, and it is desired to guess the value of θ on the basis of X. If the guessed value is d and the true value is θ, the loss is $(\theta - d)^2$.

(a) Specify S, Ω, D, and W.

(b) [You can save time by working (e) first but may find it easier to work (b) first.] Determine and plot on the same graph the risk functions of the procedures t_1, t_2, t_3, t_4, t_5, and t_6 that are defined as follows:

$$t_1(x) = x;$$

$$t_2(x) = (2x + 1)/4;$$

$$t_3(x) = (x + 1)/3;$$

$$t_4(x) = 1/2;$$

$$t_5(x) = 1 - x;$$

$$t_6(x) = 0.$$

[*Note:* Your calculations will be minimized if you first compute the risk function of a general procedure t of the form $t(x) = a + bx$. A check: $r_{t_3}(\theta) = \theta^2/3 - \theta/3 + 1/9$.]

(c) From these calculations, can you assert that any of these six procedures is inadmissible?

(d) On the basis of the risk functions, if one of these 6 procedures must be used, which procedure would you use, and why? (*Note:* Don't consult any references in answering this. Later you will find out the precise meaning of your present intuition.)

(e) Suppose X is replaced by the vector $(X_1, X_2, \ldots, X_n)$, where the X_i are independent, identically distributed random variables with $P_\theta\{X_i = 1\} = 1 - P_\theta\{X_i = 0\} = \theta$. The procedures t_1, t_2, and t_3 considered when $n = 1$ have counterparts for all values of n, given by

$$t_{1,n}(x_1, \ldots, x_n) = \sum_{i=1}^{n} x_i/n = \bar{x}_n,$$

$$t_{2,n}(x_1, \ldots, x_n) = \left[\sum_{i=1}^{n} x_i + \frac{n^{1/2}}{2}\right] \Big/ [n + n^{1/2}],$$

$$t_{3,n}(x_1,\ldots,x_n) = \left[\sum_{i=1}^{n} x_i + 1\right] \Big/ [n+2].$$

(We shall later describe three commonly used "optimality" criteria, one of which is satisfied by each of these procedures.) Show that

$$r_{t_{1,n}}(\theta) = \theta(1-\theta)/n, \qquad r_{t_{2,n}}(\theta) = 1/4(\sqrt{n}+1)^2,$$
$$r_{t_{3,n}}(\theta) = [1 + (n-4)\theta(1-\theta)]/(n+2)^2$$

and plot graphs of these risk functions (or, rather, of $nr_{t_{i,n}}$ to make the results comparable to those of part (b) for n large (e.g., for $n = 10{,}000$). Also plot $r_{t_{4,n}}$ where $t_{4,m} = 1/2$ for all $(x_1,\ldots,x_m)$. [Use the fact that $\sum_1^n x_i$ is binomially distributed, and, in particular, its first two moments; begin, as in part (b), by finding the risk of a general procedure of the form $t_{a,b,n}(x_1,x_2,\ldots,x_n) = a + b\sum_1^n x_i$.]

(f) If n is large, which of the four procedures of part (e) would you use, and why? (Your answer to this last may differ from the answer to part (d) for the case $n = 1$; does it?)

Note: The risk functions to be plotted in parts (c) and (e) are linear or quadratic functions. Be sure to plot accurately the values corresponding to $\theta = 0$, $1/2$, and 1.

(g) If the statistician decides to restrict the procedure to be of the form $t_{a,b,n}$ mentioned at the end of (e) (as has been done for arithmetical simplicity in this example!), what pairs of values (a, b) are permitted (so as to yield only decisions in D)?

(h) [*Optional*] Show that the procedure $t_{4,n}$, defined in part (e), is admissible for each n. [Hint: How can t' satisfy $r_{t'}(1/2) \le r_{t_{4,n}}(1/2)$?]

(i) [*Optional*] For $n = 1$, show that t_6 of (b) is admissible. [Hint: If t' is better than t_6, show that $r_{t'}(0) \le r_{t_6}(0)$ implies $t'(0) = 0$, and then compare the two risk functions for θ near 0, in terms of $t'(1)$.]

2.2. X is a Poisson random variable (rv) with $P_\lambda\{X = j\} = e^{-\lambda}\lambda^j/j!$ for $j = 0, 1, 2, \ldots$. It is known only that $\lambda \ge 0$, and it is desired to guess the value of λ. Since the experimenter feels the loss is roughly like squared error $(d-\lambda)^2$ when the true λ is small but is like squared *relative* error $(\lambda^{-1}d - 1)^2$ when λ is large, he or she chooses loss function $(\lambda - d)^2/(1+\lambda^2)$ to reflect this behavior.

(a), (c), (d), (f): Same statement as in Problem 2.1.

(b) [You can save time by working (e) first but may find it easier to work (b) first.] Determine and plot on the same graph the risk functions of the 6 procedures t_i defined by

$$t_1(x) = x;$$
$$t_2(x) = (x + \sqrt{1/2})/(1 + \sqrt{1/2});$$
$$t_3(x) = x/2;$$
$$t_4(x) = 2x;$$
$$t_5(x) = 0;$$
$$t_6(x) = 1.$$

[Your calculations will be made simpler if you first compute the risk function of a general procedure of the form $t(x) = a + bx$. A check: $r_{t_3}(\lambda) = (\lambda^2 + \lambda)/4(1 + \lambda^2)$.]

(e) Suppose X is replaced by the vector $(X_1, X_2, \ldots, X_n)$ of independent and identically distributed (iid) Poisson rv's with mean λ. The procedures corresponding to t_1, t_2, t_3, t_5 are

$$t_{1,n}(x_1, \ldots, x_n) = \bar{x}_n,$$

$$t_{2,n}(x_1, \ldots, x_n) = \left(\sum_1^n x_i + \sqrt{\frac{n}{2}}\right)\Big/\left(n + \sqrt{\frac{n}{2}}\right),$$

$$t_{3,n}(x_1, \ldots, x_n) = \bar{x}_n/2,$$

$$t_{5,n}(x_1, \ldots, x_n) = 0.$$

Compute the risk functions of these four procedures, and plot $nr_{t_{i,n}}$ for these four procedures (to make the scale comparable to that of (b)). $\Big[$Note that $\sum_1^n X_i$ has the Poisson distribution with mean $n\lambda$ when X_i is Poisson, mean λ. Again, you may find it easier first to find $(1 + \lambda^2)^{-1} E_\lambda \left(a + b \sum_1^n X_i - \lambda \right)^2$ for general $a, b.\Big]$

(g) Suppose the statistician decides to restrict consideration to procedures $t_{a,b,n} = a + n^{-1} b \sum_1^n X_i$ of the form mentioned at the end of (e). He or she is concerned about the behavior of the risk function of the procedure when λ is large. Show that the risk function approaches 0 as $\lambda \to \infty$ if and only if $b = 1$. *Among procedures with* $b = 1$, show that the choice $a = 0$ gives uniformly smallest risk function. [This justification of the procedure $t_{1,n} = t_{0,1,n}$, *under the restriction* to procedures of the form $t_{a,b,n}$, will seem more sensible to many people than a justification in terms of the "unbiasedness" criterion to be discussed in a later section.]

(h) Show that the procedure $t_{6,n}$, defined by $t_{6,n}(x_1, \ldots, x_n) = 1$ for all $x_1, \ldots, x_n$, is admissible for each n. [Hint: How can t' satisfy $r_{t'}(1) \le r_{t_{6,n}}(1)$?]

(i) (*Very optional*) When $n = 1$, show that $t_{0,0,1}$ (the t_5 of (b)) is admissible among *all* procedures by verifying the following steps: (i) If t' is better than $t_{0,0,1}$ for the given W, it is also better for the squared error loss function, for which its risk function minus λ^2 is $h_{t'}(\lambda) = E_\lambda\{t'(X)[t'(X) - 2\lambda]\}$. (ii) If t' is better than $t_{0,0,1}$, then $E_\lambda[t'(X)]^2 < \infty$ for all λ. (iii) Show that $h_{t_{0,0,1}}(\lambda) = 0$ for all λ, and (from (i) and (ii), after a change in variables in the series for $E_\lambda t'(X)$)

$$h_{t'}(\lambda) = e^{-\lambda}\left\{t'(0) + \sum_{j=1}^{\infty} \lambda^j [[t'(j)]^2/j! - 2t'(j-1)/(j-1)!]\right\}.$$

By considering $h_{t'}(\lambda)$ for λ near 0, show that, for t' to be better than $t_{0,0,1}$, it is necessary that $t'(0) = 0$; then, by considering successively the coefficient of λ^1, λ^2, etc., in the sum, conclude that $t'(1) = 0$, $t'(2) = 0$, etc. (iv) Conclude that no t' *better* than $t_{0,0,1}$ exists. [This argument is more complicated than in part (i) of Problem 2.1, where X could only take on two values.]

2.3. X is a normal rv with variance 1 and unknown mean θ, to be guessed. For the reasons given in Problem 2.2, the loss is $(\theta - d)^2/(1 + \theta^2)$.

(a), (c), (d), (f): Same statement as in Problem 2.1.

(b) [You can save time by working (e) first but may find it easier to work (b) first.] Determine and plot on the same graph the risk functions of the 6 procedures t_i defined by

$$t_1(x) = x;$$

$$t_2(x) = (1 + x)/2;$$

$$t_3(x) = x/2;$$

$$t_4(x) = 2x;$$

$$t_5(x) = 0;$$

$$t_6(x) = 1.$$

[Your calculations will be made simpler if you first compute the risk function of a general procedure of the form $t(x) = a + bx$. A check: $r_{t_4}(\theta) = (\theta^2 + 4)/(1 + \theta^2)$.]

(e) Suppose X is replaced by the vector $(X_1, \ldots, X_n)$ of iid normal $\mathcal{N}(\theta, 1)$ rv's. The procedures corresponding to t_1, t_2, t_3, t_6 are

$$t_{1,n}(x_1, \ldots, x_n) = \bar{x}_n;$$

$$t_{2,n}(x_1, \ldots, x_n) = (\bar{x}_n + n^{-1})/(1 + n^{-1});$$

$$t_{3,n}(x_1, \ldots, x_n) = n^{1/2}\bar{x}_n/(1 + n^{1/2});$$

$$t_{6,n}(x_1, \ldots, x_n) = 1.$$

Compute the risk functions of these four procedures, and plot $nr_{t_i,n}$ for these four procedures (to make the scale comparable to that of (b)). [Note that $\bar{X}_n$ has the $\mathcal{N}(\theta, n^{-1})$ law when X_1 has the $\mathcal{N}(\theta, 1)$ law. Again, you may find it easier first to find $(1 + \theta^2)^{-1}E_\theta(a + b\bar{X}_n - \theta)^2$ for general a, b.]

(g) Suppose the statistician decides to restrict consideration to procedures $t_{a,b,n} = a + b\bar{X}_n$ of the form mentioned at the end of (e). He or she is concerned about the behavior of the risk function when $|\theta|$ is large. Show that the risk function approaches 0 as $|\theta| \to \infty$ if and only if $b = 1$; and that, *among procedures with* $b = 1$, the choice $a = 0$ gives uniformly smallest risk function. [See comment at end of Problem 2.2(g).]

(h) Show that $t_{6,n}$ is admissible for all n. [Hint: If t' is better than $t_{6,n}$, what can you conclude from $r_{t'}(1) \leq r_{t_{6,n}}(1)$?]

CHAPTER 3

Classifications of Statistical Problems

When statisticians discuss statistical problems they naturally classify them in certain ways. In the sequel we give an informal discussion of several such classifications in terms of the nomenclature we have introduced. Many of the categories mentioned later correspond to important ideas. Discussion of statistical problems in a particular category therefore becomes an important topic. Thus the following is intended also to given an informal outline of some topics we shall cover in later chapters. You may not understand all aspects of this outline at a first reading, but you are urged to refer to this chapter as various topics are taken up in order to fit them into a proper perspective.

Statistical problems have been and are classified on any of several bases (the terminology is that used in standard statistics books), as for example the structure of Ω or of D. Accordingly we give several separate classifications:

(i) The structure of Ω

(a) Parametric case

(b) Nonparametric case

The *parametric cases* of statistical problems are all those in which the class of all df's F in Ω can be represented in terms of a vector θ consisting of a finite number of real components, in a natural way. (The exact meaning of the term *natural way* need not concern us here; essentially, it means that F and W depend on θ in a reasonably smooth fashion.) All other problems are called *nonparametric*. The reader should check that Examples 2.1 to 2.4 were all parametric in nature.

EXAMPLE 3.1. Suppose we take n measurements on the length of an object, with a given measuring instrument. The experiment is thus the observation of a vector of n components, $X = (X_1, \ldots, X_n)$, where X_i is the ith measurement.

The measurements are independent and identically distributed according to some df G with finite unknown mean μ_G and finite unknown variance σ_G^2. Here μ_G is the actual length of the object being measured, but we know nothing about the probability law of the errors of measurement, except that these errors have finite variance and mean 0. (That is, we assume we *know* that the errors of this instrument cancel out on the average in many repeated measurements, so that the expected value of each measurement is the actual length μ_G of the object.) We may know some bounds on the length, say $a < \mu_G < b$. (If we know only that the length is positive, we would have $a = 0$, $b = \infty$.) The observations X_i may have a normal df, or an exponential df, or a uniform df, or any of the countless forms which are analytically too intractable to be written down easily. Here S is n-dimensional Euclidean space (the collection of all n-vectors with real components), the possible df's F of X can be labeled in terms of the possible df's G of X_1 (since $F(x_1, \ldots, x_n) = G(x_1) \ldots G(x_n)$, according to our assumption that the X_i are independently and identically distributed), and the space D will be taken to be the set $\{d : a < d < b\}$, a and b being the known bounds on the length μ_G of the object whose length we want to estimate (guess at) on the basis of the observations. Thus, the decision d means "I guess the length of the object to be d." Finally, suppose $W(G, d) = (\mu_G - d)^2$; this loss function is sometimes referred to as *squared error*, since the loss is the square of the difference between the actual length μ_G of the object and the guessed length d. The members F of Ω in this case correspond to all one-dimensional df's G with mean between a and b and with finite variance, and these cannot be represented naturally in terms of a finite number of real parameters. This is therefore a nonparametric problem.

(ii) The structure of D

(a) Point estimation

(b) Interval estimation and region estimation

(c) Testing hypotheses

(d) Ranking and other multiple decision problems

(e) Regression problems

(f) Other problems

A statistical problem is said to be a problem of *point estimation* if D is the collection of possible values of some real or vector-valued property of F (or θ) which depends on F (or θ) in a reasonably smooth way. Usually D will be an interval of real numbers (perhaps the set of all of real numbers), or a rectangle, or similar simple continuum in the plane, etc. The object is to guess the value of this property ϕ_F of the true F. The loss function will usually equal zero when $d = \phi_F$ and will increase as d gets further from ϕ_F in some appropriate sense. Examples 2.1(a), 2.2, and 3.1 are point estimation problems. If in Example 2.2 we wanted to estimate not only μ but also σ, a decision would be a vector $d = (d', d'')$, the first component being the guessed value of μ and the second component being that of σ. D would be the set of all 2-vectors (d', d'') with $d' \geq h$ and $d'' > 0$. The loss function might be something like

$$W((\mu, \sigma^2), (d', d'')) = A|d' - \mu| + B|d'' - \sigma|.$$

This would also be a problem of point estimation. The statistical procedures in case (a) are often called *estimators* or *point estimators.* The term *point estimation* refers to the nature of the decisions d as points of a continuum, in contrast to case (b).

In Example 2.1(d) the decisions were *intervals* of possible values of θ. Similarly, in general, if there is a real-valued property ϕ_F of F (as in case (a)) which we want to guess at not by guessing a single value but rather by guessing an interval of possible values of ϕ_F, we call this a problem of *interval estimation.* If ϕ_F is vector-valued, e.g., if it consists of the pairs (μ, σ) discussed in the previous paragraph, and if each possible decision is a rectangle of the form $\mathbf{d} = \{(u, v) : d_1' \leq u \leq d_1'', d_2' \leq v \leq d_2''\}$, which means "my guess is that the true (μ, σ) lies in the rectangle $\mathbf{d}$" or, in other words, "my guess is that $d_1' \leq \mu \leq d_1''$ and $d_2' \leq \sigma \leq d_2''$," we might still refer to this as a problem of interval estimation, rectangles being thought of as 2-dimensional intervals. However, suppose we allow decisions of the form "my guess is that $(\mu - 2)^2 + 5(\sigma - 3)^2 < 1$." The set of points $\mathbf{d} = \{(u, v) : (u - 2)^2 + 5(v - 3)^2 < 1\}$ is not a rectangle, but rather the interior of an ellipse. When the decision can be such a region of possible values of ϕ_F, the region no longer necessarily being an interval, we refer to the problem as one of *estimation by regions*, interval estimation being a special case of this. The statistical procedures are often called *interval* or *region estimators.* In some cases, which we will discuss later, they are called *confidence intervals* or *confidence regions.*

If there are ony two possible decisions, as in Examples 2.1(c), 2.3, and 2.4, the problem is often called one of *testing hypotheses*, the statistical procedures being called *tests.* Additional nomenclature commonly employed in such problems will be introduced later.

If there are a finite number of possible decisions, but more than two, the problem is often called a *multiple decision problem.* Example 2.1(b) is of this form. Another common problem of this type is the following: Three types of grain are grown on test plots, and the statistician must decide which variety has or which varieties have the largest mean yield of grain per plant (the observed yield of a plant is the mean yield of the variety plus a chance departure whose expectation is zero). Or perhaps the decision is a statement as to the complete ordering of varieties, of the form "variety B is best, C is next, A is worst." Both of these grain problems are often referred to as *ranking problems.* You should carefully list the possible decisions in either case, for your own benefit.

It is worthwhile to mention at this point a shortcoming of almost all theoretical work in statistics prior to 1939, and of most current textbooks and cookbooks on statistics: that is, that problems other than those of type (a), (b), and (c) are almost never treated. Typically, instead of treating a multiple decision problem as such and selecting an appropriate procedure accordingly, the classical statistics books try to work the problem into one of type (c), which it is not, in order to use the cookbook recipes for testing hypotheses. Example

2.1(b), for example, would be treated in many books by first deciding which of the two decisions d_1 and d_2 to make, then deciding which of the two decisions d_1 and d_3 to make, then somehow combining these two results in an "intuitively reasonable" fashion without ever carefully examining the probabilistic meaning (risk) of this combination of procedures, let alone verifying whether such a combined procedure is really a good one. Similarly, in the example of grain mentioned in the previous paragraph, most practical workers would, even today, begin by formulating the problem as one of type (c), either deciding "there is no great difference among the three mean yields" or "there is some appreciable difference." If the second conclusion is reached, the worker then realizes that this is not at all a satisfactory conclusion, since he or she would like to know which variety is best. The statistician therefore tries to decide, using the same data, whether or not variety A is better than B, whether or not B is better than C, etc., and then tries to combine the conclusions of these various tests into one final statement, again not bothering to compute the probabilistic meaning of this final statement or to justify this combination of tests as a good procedure. How much more logical it is to formulate the problem precisely to start with and to deal with it as a multiple decision problem, instead of going through such contortions in order to use a combination of cookbook tests! Nevertheless, it has only been since the 1950s that statisticians, influenced by the work of Wald, have at all constructed and used procedures which were properly constructed in this manner.

Example 3.2 is an example of a *regression problem*. Many problems dealt with in the analysis of variance and more generally in the design of experiments have the common characteristic that certain elements of the environment, capable of having measurable effects on the experimental outcome, can be controlled and varied by the experimenter. The experimental outcome therefore becomes a function of the environmental setting chosen by the experimenter. If the primary interest of the experimenter is to study the variation in outcomes that results from changes in the controlled environment, then the problem is usually called a *regression problem*. In Example 3.2 the experimenter is described as wishing to study the variation in the expected amount of growth of a plant due to variation in the amount of nitrogen present in the soil.

There is in general much overlap between the idea of a regression problem and the idea of design of experiments, for in designing an experiment the aim is to control and utilize the variable aspects of an experimental setting in such a way as to obtain the best information possible. Example 3.2 also illustrates this situation.

Example 3.2. An experimenter wants to determine the effect of nitrogen on the growth of a certain type of plant. She knows that, as long as the proportion p of nitrogen in the soil is restricted to a certain interval of values $a \leq p \leq b$, the expected height of a plant depends linearly on p, this expected height being $\theta_1 p + \theta_2$ if the plant is grown for a year in soil containing a proportion p of nitrogen. θ_1 and θ_2 are unknown. The experimenter is allowed to grow 10 plants in different plots, the ith plot containing a proportion of nitrogen p_i which she can choose in advance to be any value between a and b ($i = 1, 2, \ldots, 10$). The observed yield in the ith plot will then be $\theta_1 p_i + \theta_2$ plus a chance variation whose expectation is zero. It is desired to estimate θ_1 and θ_2. We shall not list in detail what a typical W or typical assumption on the exact form of the distribution of observed yields might be. We need only note that the accuracy attainable by good estimators of θ_1 and θ_2 (that is, the smallness of the risk) will depend on the choice of the p_i's. For example, if all the p_i's are chosen to be very close to each other, it should be intuitively clear that it will usually be difficult to estimate this linear relationship accurately. The possible choices of the vector $(p_1, p_2, \ldots, p_{10})$ are called *experimental designs*, and a problem of choosing $(p_1, \ldots, p_{10})$ so as to obtain accurate estimators (or, more generally, good statistical procedures) is called a *problem of designing an experiment*. Such problems are characterized by the fact that the experimenter has some control over the form of Ω and perhaps S in advance of the experiment.

You can easily think of problems which do not fall into any of the preceding categories. For example, in the grain problem, the decision may be a statement as to which variety is best, together with a numerical guess of how much better it is than the next best variety; this combines features of problems of types ii(a) and ii(d).

We list here a series of topics of practical importance.

(iii) Other topics
- **(a) Sampling methods**
  - **(1) Fixed sample size procedures**
  - **(2) Two-stage procedures**
  - **(3) Sequential procedures**
- **(b) Cost considerations**
- **(c) Experimental design problems**
- **(d) Mathematical ideas of importance**
  - **(1) Sufficiency**
  - **(2) Randomization**
  - **(3) Asymptotic theory**
- **(e) Prediction problems**

A discussion of *sampling methods* includes a discussion of the way experimental replication is to be used in making the desired decision. In Example 1 the coin is to be flipped 10 times, giving 10 independent replications of the experiment "toss the coin; observe the outcome." The statistical procedures discussed in Example 2.1 are *fixed sample size* procedures characterized by the fact that all replications of the experiment are made before any decisions which involve use of the experimental outcome are made.

The consideration of *sequential procedures* can be viewed as a part of the subject of experimental design, but it is more convenient to view it separately.

Briefly, sequential procedures are ones in which the number of observations to be taken is not determined in advance, but, rather, as experimentation proceeds. For example, in Example 2.3 the middleman might also consider a procedure of the following form: Draw 100 bolts from the barrel; if more than 2 are defective, make decision d_2; if none is defective, make decision d_1; if 1 or 2 out of the 100 are defective, draw 150 additional bolts from the barrel; if these additional bolts are drawn, make decision d_1 or d_2, depending on whether the total number of defectives out of the 250 bolts is ≤ 4 or is ≥ 5. The number of observations is thus no longer fixed, but is now a chance variable. The advantage of using such procedures, as we shall see, is that they often require an *expected* number of observations which is smaller than the number of observations which would be required by a "fixed sample size procedure" in order to achieve the same expected loss as a result of possibly making a wrong decision. Sequential procedures have come into great use since Wald's work on them in the early 1940s. A careful consideration of such procedures should include cost considerations; for example, depending on whether or not the middleman finds it inexpensive to interrupt his sampling of bolts after 100 observations in the preceding two-stage procedure, compared with drawing and inspecting one batch of bolts without interruption, he may or may not prefer the sequential procedure.

The example just discussed is an example of a *two-stage procedure*. The name *sequential procedure* is a generic term applying to a variety of possible procedures wherein the number of observations is a random variable. A different but familiar example of a sequential estimation procedure is the following. Toss a coin repeatedly until a head appears. Count the number N of tosses that were made at the time the first head appeared. Estimate for the value of θ, the probability of a head, $t_N(X_1, X_2, \ldots, X_N) = 1/N$. (Notice the similarity and the difference of this and procedure t_2 in Example 2.1(a).)

The two-stage procedure is characterized as follows. Make a fixed number of replications of the experiment. On the basis of this sample, decide how many additional replications of the experiment to make. It will be seen that this describes the two-stage experiment discussed earlier but does not describe the sequential experiment discussed in the preceding paragraph. In the latter example a decision whether to make more experiments is made after each replication of the experiment. Two-stage methods are best known because of work by C. Stein in the middle 1940s.

It may be true that the various possible experiments cost various amounts of money, in which case the *cost of experimentation* should be added to $E_F W(F, t(X))$ in order to obtain the risk function. The latter will then include expected loss from making a possibly incorrect decision plus cost of experimentation. In such a simple problem as that of Example 2.1(a), it would usually be that we would not be told in advance to toss the coin 10 times but would have to determine the number of tosses in such a way as to balance the increased cost of taking a larger number of observations against the improved accuracy of estimation which can be achieved from more observations.

An interesting aspect of statistical problems is that the effort required to compute and use a "good" procedure may be large, and that the procedure may no longer appear to be so good when we include these *computational costs*. There are many problems for which this is the case, and we will therefore have occasion to consider certain simple procedures which would be inadmissible in terms of expected loss alone but which might be quite good when we consider computational costs associated with finding and using procedures yielding small expected loss.

To the extent that making experiments costs money it is reasonable to consider the choice between possible sequential procedures of sampling to be based in part upon the cost of experimentation. Consideration of the cost of experimentation and the design of good statistical procedures to minimize this cost are usually considered to be aspects of the *design of experiments*. These questions as well as questions of how best to control variation introduced by the environment (as in Example 3.2) are major parts of the subject of design of experiments.

There are some considerations which will apply to all problems and which we will have occasion to treat in this book. For example, we have already mentioned the notion of restricting consideration to admissible procedures and have referred to the need for a criterion to select one from among the many admissible procedures. (Some criteria such as the *minimax* criterion and the *invariance* criterion, which will be defined later, have broad applicability.) We now mention a few other such broad topics.

1. Sufficiency. It will sometimes occur that the data obtained from the experiment can be summarized in a simple way into a more convenient form without "losing any information." For example, in Example 2.1(a) it turns out that, for any procedure t, there is a procedure t' depending only on $\sum_1^{10} X_i$, and such that t' is equivalent to t. Hence, there is no reason to consider procedures such as t_2, and the collection of procedures we must look at is greatly reduced (to functions of the one variable $\sum_1^{10} X_i$ instead of functions of 10 variables X_1, $X_2, \ldots, X_{10}$.) The random variable $\sum_1^{10} X_i$ is said to be a *sufficient statistic* in Example 2.1(a), as well as in 2.1(b)–(d). (The word *statistic* means "function of X.") We shall study this concept in some detail in Chapter 6.

2. Randomization. A surprising fact is that we will sometimes have to consider statistical procedures which are modified from those we have considered previously by allowing us to flip coins or perform other chance experiments (in addition to the experiment whose outcome is observed in the value of X) in order to help us reach a decision. For example, in Example 2.1(c) we might proceed as follows: "If $\left|\sum_1^{10} X_i - 5\right| < 2$, make decision d_1. If $\left|\sum_1^{10} X_i - 5\right| > 2$, make decision d_2. If $\left|\sum_1^{10} X_i - 5\right| = 2$, flip a fair coin (*not* the coin used in the experiment) and make decision d_1 or d_2 according to whether this toss comes out heads or tails." Such procedures are not of great practical significance, but we shall see that their use is sometimes dictated on theoretical grounds. An interesting sidelight is that many experimenters who rebel against using such *randomized procedures* in reaching decisions use them constantly in designing experiments.

3. *Asymptotic theory*. We have mentioned that good procedures are often difficult to compute and may be so complex in form as to make them unwieldy to use. It sometimes happens that, when the sample size n is large, a procedure which is *approximately* optimum according to some criterion is easy to compute and use. The study of such procedures is called the *asymptotic* (as $n \to \infty$) *statistical theory*.

4. *Prediction*. A *prediction problem* is one in which the decision is a guess *not* of some property of F, but rather of some property of a random variable which is observed some time *after* the experiment at hand. For example, having watched the Yankees play baseball all summer, the questions might be "Will the Yankees win the pennant? Will they win the World Series?"

CHAPTER 4

Some Criteria For Choosing a Procedure

Having specified the statistics problem or model associated with an experiment with which we are concerned, we must select a single procedure t to be used with the data from that experiment. Since for most statistical problems there are no uniformly best procedures, some additional criterion (or criteria) is (are) necessary. The purpose of such a criterion is to specify uniquely a statistical procedure t which will be used. Some of the criteria which are often used to select a single procedure t to be used in a particular statistics problem are the following:

1. Bayes criterion
2. Minimax criterion
3. Invariance criterion
4. Unbiasedness criterion
5. Sufficiency criterion
6. Robustness criterion
7. Maximum likelihood and likelihood ratio methods
8. Method of moments
9. Criteria of asymptotic efficiency

In the development of mathematical statistics since 1900, the criteria listed here have been used by workers in the subject as prescriptions for obtaining procedures by methods considered intuitively appealing. For example, there is today a group of workers who consider the use of the Bayes criterion the only reasonable way to determine a statistical procedure. In this book we will continue to maintain a contrary position, that no one of the criteria listed is the right one to use. Instead our viewpoint is that each of the criteria is a mathematical principle which can be used to specify a certain subclass of possible statistical procedures. Which criterion or criteria to use in a particular statistical problem requires further examination of the characteristics of the procedures determined by a criterion.

In most statistical problems criteria 1, 7, and 8 give rise to a uniquely determined statistical procedure, but this is not always the case. Criteria 2, 3, 4, 5, 6, and 9 in general do not give rise to uniquely determined statistical procedures. Instead one obtains a subclass of procedures from which the final choice of a statistical procedure is to be made. In the case of the invariance criterion, the unbiasedness criterion, and the maximum likelihood criterion, there are important examples in which no statistical procedure determined by use of the criterion can be admissible.

In this chapter we will give descriptions of some of these criteria and their use. Some of the criteria will be discussed again in later sections after new ideas have been introduced. See in particular Chapters 6 and 7.

4.1. The Bayes Criterion

The *Bayes criterion* is intuitively one of the most appealing, but its use requires an assumption and a precise piece of information which are additional to the specification of S, Ω, D, and W. The assumption is that the unknown true F (or θ) which we encounter in the experiment at hand is itself a random variable. The piece of information is the precise specification of the probability law (called the *a priori probability law*) of F (or θ).

It is to be emphasized that there is a serious question as to whether F can be considered as a random variable rather than merely as an unknown quantity. In Example 2.2, is it reasonable to assume that nature or some deity "threw dice" in order to decide at what distance μ the star should be? Even if we did accept such a philosophy, when can we possibly obtain knowledge as to what chance law was used in picking μ? On the other hand, in Example 2.3 it might be that the proportion θ of defectives in any barrel can be thought of as a random variable whose makeup depends on the state of mind of the laborer and the state of the machine tool in the particular period when the supplier manufactured the bolts in this barrel. This in itself would be of no use if we did not know the form of the a priori law of θ. However, it might be that the conditions in the supplier's factory have been relatively stable over a long period of time during which the middleman's customers have informed him of the exact number of defectives they found in each barrel. This information might then be used by the middleman to obtain close approximations to the a priori law which governs each future barrel (it being assumed that the θ's that go with different barrels are identically distributed). For example, if in the previous 10,000 barrels it had been found that 4,000 had 1 percent defective and 6,000 had 3 percent defective, the manufacturer would think of the θ of each future barrel as being a random variable which takes on the values .01 and .03 with probabilities .4 and .6, respectively. This is probably not the exact law of θ, but is close enough for practical purposes. Keep in mind that, once the barrel is before you, it has a definite (unknown) proportion of defectives θ; the random character of θ can be thought of as describing the possible makeups of a barrel before it is actually manufactured.

Suppose, then, that we know the a priori probability law ξ in a given problem. Suppose first that ξ is discrete, so that there will be a countable collection of possible values of F, say $F_1, F_2, \ldots$, such that the probability (according to ξ) that $F = F_i$ is

$$p_\xi(F_i),$$

where $p_\xi(F_i) \geq 0$ and

$$\sum_i p_\xi(F_i) = 1.$$

We now think of the statistical problem as consisting of two stages of chance mechanisms: first, the actual F to be assigned to the experiment at hand is chosen (by nature, etc.) according to the probability law ξ; second, once F is chosen, the random variable X, governed by the chosen F, is observed by us. If the random choice of F results in $F = F_i$, and if we use procedure t, the expected loss we incur is $r_t(F_i)$. Since the a priori probability that $F = F_i$ is $p_\xi(F_i)$, the a priori expected loss (hereafter called *a priori risk*) is

$$R_t(\xi) = \sum_i r_t(F_i)p_\xi(F_i).$$

We have exhibited ξ in $R_t(\xi)$ to indicate the dependence of the result on the knowledge of ξ, but since ξ is assumed known, we will simply obtain a real number $R_t(\xi)$ for each procedure t.

The Bayes criterion is to choose t *to minimize* $R_t(\xi)$. Such a choice of t is called a *Bayes procedure relative to* ξ: that is, t^* is Bayes relative to ξ if

$$R_{t^*}(\xi) = \min_t R_t(\xi)$$

or in other words

$$R_{t^*}(\xi) \leq R_t(\xi) \quad \text{for all } t.$$

The Bayes risk $R_{t^*}(\xi)$ of the Bayes procedure t^* relative to ξ is therefore an "average" or expected value. If we think of conducting many experiments, and if F is a random variable, then by the law of large numbers the average risk incurred over a long series of identical and independent experiments should be roughly $R_{t^*}(\xi)$ in value. Use of the Bayes procedure t^* therefore minimizes the average risk over a long series of repeated experiments. An intuitive justification for using a Bayes procedure rests on this observation.

We shall see shortly that the determination of a Bayes procedure is often a simple computation (much simpler, for example, than the computation of a *minimax* procedure). Although this is attractive in itself, it unfortunately presents the statistician with the temptation to assume ξ to be of some known form without much justification merely in order to be able to get an explicit answer quickly (and perhaps to avoid taking a course in mathematical statistics). This can lead to catastrophic consequences, and there have unfortunately been many sinners. Intuitively, if the statistician guesses ξ^* to be the distribution function of F while the actual distribution is ξ, then the Bayes procedure t^* relative to ξ^* will not minimize the expected risk. In a long series of repeated experiments too high a price will be paid.

The following example shows still another reason why the arbitrary choice of an a priori ξ may not make much sense. If F can be parametrized by a finite interval of values θ, say $a \leq \theta \leq b$, and if ξ is not discrete but has a density function

$$f_\xi(\theta),$$

then, by analogy with the preceding, we have

$$R_t(\xi) = \int_a^b r_t(\theta) f_\xi(\theta)\, d\theta.$$

A common choice of ξ "out of thin air" is to choose ξ to be uniform:

$$f_\xi(\theta) = \begin{cases} 1/(b-a) & \text{if} \quad a \leq \theta \leq b, \\ 0 & \text{otherwise.} \end{cases}$$

A Bayes procedure then minimizes the area under $r_t(\theta)$. The intuitive argument often given for this choice is that the ignorance of the true value of θ is "best expressed" by assuming any interval of possible values of θ to have the same a priori probability as any other interval of values of the same length. The fallacy of this argument is that there is nothing natural about one parametrization over another, and that if Ω is parametrized according to a parameter $\phi = g(\theta)$, uniform distribution of ϕ is not the same assumption as uniform distribution of θ. For example, suppose two risk functions looked like this, where $a = 0$, $b = 1$:

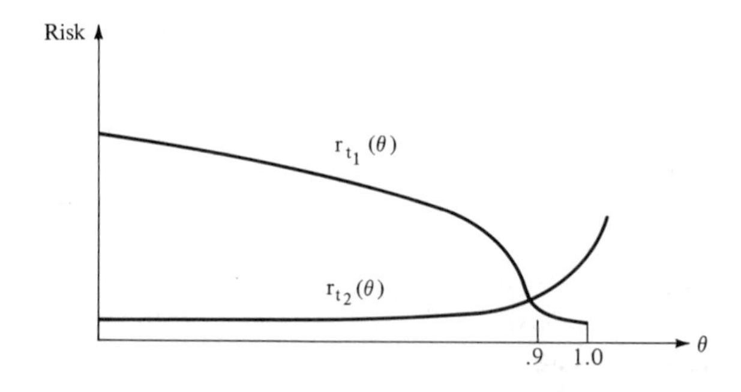

If the underlying probability laws are parametrized instead in terms of ϕ, where

$$\phi = \begin{cases} \theta/9 & \text{if } 0 \le \theta \le .9, \\ 10\theta - 8.9 & \text{if } .9 \le \theta \le 1, \end{cases}$$

then the risk functions of the *same* two procedures for the *same* problem, but graphed in terms of ϕ rather than in terms of θ, look like this:

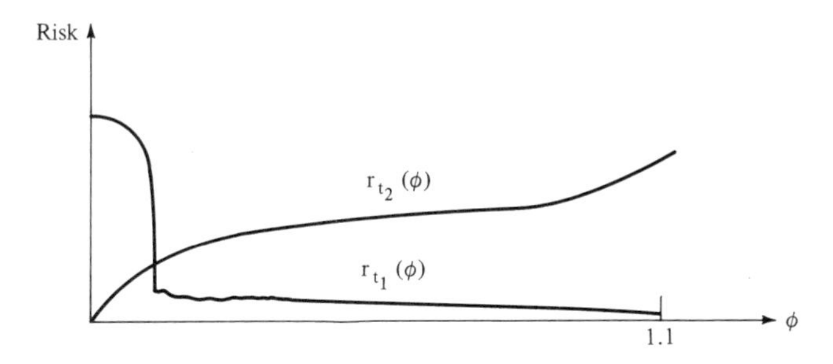

Thus, two naive statisticians, one who likes to parametrize Ω in terms of θ and the other who prefers to parametrize in terms of ϕ, but both of whom automatically assume ξ to be a uniform density (for their respective parameters), will arrive at different conclusions regarding which of t_1 and t_2 is the better procedure. Which is right? *Neither*. The point is that an a priori law is not merely a vague representation of an intuitive feeling of ignorance, but a precise physical datum. Where this datum is not obtainable, or where (more basically) it does not make sense to think of θ as a random variable, one cannot use the Bayes criterion.

Computation of Bayes Procedures

We shall write this out for the case when ξ is discrete on Ω and F is discrete on S. The computation in other cases is similar. We have

$$\begin{aligned} R_t(\xi) &= \sum_F p_\xi(F) r_t(F) \\ &= \sum_F p_\xi(F) \sum_x W(F, t(x)) p_{F;X}(x) \\ &= \sum_x \left[\sum_F W(F, t(x)) p_{F;X}(x) p_\xi(F) \right]. \end{aligned}$$

Consider the expression

$$h_\xi(x, d) = \sum_F W(F, d) p_{F;X}(x) p_\xi(F).$$

For each fixed x, there will be one or more decisions d which will minimize $h_\xi(x, d)$. Let $D_{\xi,x}$ be the set of all such minimizing values d, and let

$$H_\xi(x) = \min_d h_\xi(x, d),$$

so that

$$h_\xi(x,d)\begin{cases} = H_\xi(x) & \text{if } d\in D_{\xi,x},\\ > H_\xi(x) & \text{if } d\notin D_{\xi,x}.\end{cases}$$

In the last line of the expression $R_t(\xi)$, we see that the term in square brackets is just $h_\xi(x, t(x))$. This term, for each x, thus satisfies

$$h_\xi(x,t(x))\begin{cases} = H_\xi(x) & \text{if } t(x)\in D_{\xi,x},\\ > H_\xi(x) & \text{if } t(x)\notin D_{\xi,x}.\end{cases}$$

Thus, on summing over x, we obtain

$$R_t(\xi)\begin{cases} = \sum_x H_\xi(x) & \text{if } t(x)\in D_{\xi,x} \quad \text{for } every\ x,\\ > \sum_x H_\xi(x) & \text{otherwise.}\end{cases}$$

In other words, no procedure can have an a priori risk which is less than $\sum_x H_\xi(x)$, and this lower bound $\sum_x H_\xi(x)$ is attained as the a priori risk of t if and only if $t(x)\in D_{\xi,x}$ for every x. *A procedure* t *is thus Bayes relative to* ξ *if and only if, for every* x, *it assigns a decision* t(x) *which minimizes (over D)* $h_\xi(x, d)$.

If $D_{\xi,x}$ consists of exactly one decision (which may depend on x) for each x, then there will be exactly one Bayes procedure relative to ξ. Otherwise, there can be many Bayes procedures; $R_t(\xi)$ will be the same for all of them, but the risk functions r_t need not be the same.

EXAMPLE 4.1. A coin is tossed twice, with independent and identically distributed tosses. Here $X = (X_1, X_2)$, and $P_\theta\{X_i = 1\} = 1 - P_\theta\{X_i = 0\} = \theta$. (In this book $X_i = 1$ will always mean "heads.") Suppose there are two decisions, d_2 and d_1, meaning, respectively, that "the coin is, or is not, biased toward heads", with

$$W(\theta,d_1)=\begin{cases}0 & \text{if } \theta\le 1/2,\\ w_1 & \text{if } \theta>1/2,\end{cases}$$

$$W(\theta,d_2)=\begin{cases}w_2 & \text{if } \theta\le 1/2,\\ 0 & \text{if } \theta>1/2,\end{cases}$$

where w_1 and w_2 are specified positive constants. Suppose we know a priori that θ is either 1/3 or 3/4, the a priori probability law being

$$p_\xi(1/3) = \gamma,$$

$$p_\xi(3/4) = 1 - \gamma,$$

where $0 < \gamma < 1$, γ being specified. Since

$$\begin{aligned} P_{\theta;X}(x_1,x_2) &= p_{\theta;X_1}(x_1)p_{\theta;X_2}(x_2)\\ &= \theta^{x_1+x_2}(1-\theta)^{2-x_1-x_2}, \end{aligned}$$

we obtain

$$h_\xi(x_1, x_2, d_1) = \sum_\theta W(\theta, d_1) p_{\theta; X}(x_1, x_2) p_\xi(\theta)$$
$$= w_1(1-\gamma)(3/4)^{x_1+x_2}(1/4)^{2-x_1-x_2},$$
$$h_\xi(x_1, x_2, d_2) = w_2\gamma(1/3)^{x_1+x_2}(2/3)^{2-x_1-x_2}.$$

Dividing the first of these expressions by the second, we obtain

$$\frac{9w_1(1-\gamma)}{64w_2\gamma} \cdot 6^{x_1+x_2},$$

and thus $h_\xi(x_1, x_2, d_1)$ is $<$, $=$, or $> h_\xi(x_1, x_2, d_2)$ according to whether this ratio is $<$, $=$, or >1. If for a particular (x_1, x_2) the ratio is <1 then any Bayes procedure must make decision d_1; if for a particular (x_1, x_2) the ratio is >1 then any Bayes procedure must make decision d_2; in the middle case of equality either decision may be taken by a Bayes procedure. To see how Bayes procedures depend on γ, we consider three numerical examples:

(a) If $w_1 = w_2 = 1$ and $\gamma = 1/2$, the unique Bayes procedure is given by

$$t_a(1,1) = d_2,$$
$$t_a(0,1) = t_a(1,0) = t_a(0,0) = d_1.$$

You can check that, for any w_1 and w_2, there are many a priori laws ξ (i.e., many values of γ) relative to which this particular procedure is Bayes. This phenomenon, wherein one procedure is Bayes relative to several different a priori laws, is common when S is discrete.

(b) If $w_1 = w_2 = 1$ and $\gamma = 9/10$, the unique Bayes procedure is given by $t_b(x_1, x_2) = d_1$ for all (x_1, x_2). Intuitively (see following discussion), the a priori probability γ that the true value of θ is $1/3$ is so large that no experimental outcome based on two tosses of the coin can change the a priori feeling that we should (if we conducted *no* experiment) make decision d_1. You should realize that here, as in the other examples, the result depends not only on ξ, but also on the loss function as given by w_1 and w_2.

(c) Suppose $w_1 = w_2 = 1$ and $\gamma = 27/59$. Then, when $(x_1, x_2) = (0, 1)$ or $(x_1, x_2) = (1, 0)$ we have $h_\xi(x_1, x_2, d_1) = h_\xi(x_1, x_2, d_2)$, so that either decision can be made in that case. Hence, any Bayes procedure t_i satisfies

$$t_i(0,0) = d_1 \quad \text{and} \quad t_i(1,1) = d_2,$$

but we can have different Bayes procedures t_i which satisfy the preceding but which are quite different when $(x_1, x_2) = (0, 1)$ or $(1, 0)$; for example,

$$t_1(0,1) = t_1(1,0) = d_1;$$
$$t_2(0,1) = t_2(1,0) = d_2;$$
$$t_3(0,1) = d_1,\ t_3(1,0) = d_2.$$

These three procedures all have the same a priori risk when $\gamma = 27/59$ but have quite different risk functions. Thus, we see that there need not be a *unique* Bayes procedure relative to a given ξ.

We can rewrite our prescription for the form of a Bayes procedure in an intuitively more suggestive way by use of what are called *a posteriori probabilities.* We first describe Bayes' theorem (found in almost any probability book). Suppose we are given (a priori) probabilities $P(C_i)$ (of choosing various urns) and we are given (conditional) probabilities $P(B|C_i)$ of drawing a sample of character B (from the unknown chosen urn if it were urn i). We may then compute a posteriori probabilities

$$P(C_i|B) = P\{\text{Urn is number } i|\text{sample has character } B)$$
$$= \frac{P(C_i)P(B|C_i)}{\sum_j P(C_j)P(B|C_j)}.$$

This formula is called *Bayes' theorem.* In exactly the same way, we now have given a priori probabilities $p_\xi(F)$ of (nature choosing) various possible states of nature, probabilities $p_{F;X}(x)$ of observing a sample value x when the (unknown) "chosen" state of nature is F, and we similarly define the *a posteriori probability that $F = F_0$, given that $X = x$, where ξ is the a priori law,* to be

$$p_\xi(F_0|x) = \frac{p_\xi(F_0)p_{F_0;X}(x)}{\sum_F p_\xi(F)p_{F;X}(x)}.$$

Just as the numbers $p_\xi(F)$ make up a probability law which expresses our a priori knowledge regarding the probabilities of various states of nature, so the numbers $p_\xi(F|x)$ make up a probability law expressing our a posteriori knowledge (changed odds) once we have observed that $X = x$. In fact, thinking of F and X as random variables whose joint probability law is given by $p_\xi(F)p_{F;X}(x)$, we see that $p_\xi(F|x)$ is merely the conditional law of F given that $X = x$.

Suppose we had to make a decision based on no observations, but merely on a given a priori law ξ. The expected loss we would incur if we made decision d would be

$$\bar{h}_\xi(d) = \sum_F W(F,d)p_\xi(F),$$

and we would make some decision d which minimized this quantity. Now, if we observe $X = x$ as previously, we have

$$h_\xi(x,d) = \sum_F W(F,d)p_{F;X}(x)p_\xi(F)$$
$$= \left[\sum_F W(F,d)p_\xi(F|x)\right]\cdot\left[\sum_F p_{F;X}(x)p_\xi(F)\right].$$

The second bracketed factor is positive and does not depend on d, and thus $h_\xi(x, d)$ is minimized by the same values d which minimize

$$\bar{h}_\xi(x, d) = \sum_F W(F, d) p_\xi(F | x).$$

Thus, we have the "*No data principle*": *a Bayes procedure makes a decision in the same way that we would make a decision based on no observations, except that the a priori probabilities* $p_\xi(F)$ *are replaced by the a posteriori probabilities* $p_\xi(F|x)$. This should seem intuitively reasonable (see Problem 4.4). $\bar{h}_\xi(x, d)$ is called the *a posteriori risk* (*expected loss*) from making decision d when $X = x$ and ξ is the a priori law. This quantity, and thus the structure of Bayes procedures, depends on the sample value x only through the values of the a posteriori probabilities. (We shall refer to this fact later, when we discuss sufficient statistics.)

In actual computations, it is usually less effort to work directly with h_ξ rather than with $\bar{h}_\xi$; the latter was introduced mainly as an aid to your understanding.

We now compute several examples.

EXAMPLE 4.2. In Example 2.1(a) if ξ is the uniform density function on $\{0 \leq \theta \leq 1\}$, we have

$$\begin{aligned} h_\xi(x, d) &= \int_0^1 W(\theta, d) f_{\theta; X}(x_1, \ldots, x_{10}) f_\xi(\theta)\, d\theta \\ &= \int_0^1 (\theta - d)^2 \theta^{\sum x_i} (1 - \theta)^{10 - \sum x_i}\, d\theta \\ &= B(\textstyle\sum x_i + 3, 10 + 1 - \sum x_i) - 2dB(\sum x_i + 2, 10 + 1 - \sum x_i) \\ &\quad + d^2 B(\textstyle\sum x_i + 1, 10 + 1 - \sum x_i), \end{aligned}$$

making use of the *beta function* (see Appendix A)

$$B(r, s) = \int_0^1 \theta^{r-1}(1 - \theta)^{s-1}\, d\theta = \frac{\Gamma(r)\Gamma(s)}{\Gamma(r + s)}.$$

Thus, $h_\xi(x, d)$ is a quadratic in d which is minimized by the choice

$$d = \frac{B(\sum x_i + 2, 10 + 1 - \sum x_i)}{B(\sum x_i + 1, 10 + 1 - \sum x_i)} = \frac{\sum x_i + 1}{10 + 2}.$$

(Using the property of the gamma function, $\Gamma(r) = (r - 1)\Gamma(r - 1)$.) This is the unique Bayes procedure with respect to the uniform a priori distribution of θ. Of course, as we have mentioned, it is not to be assumed that θ has such a prior law without real justification.

(It will be proved in Problem 4.5(b) that, for squared error loss, the unique Bayes procedure has a simple description in terms of the posterior law.)

Bayes procedures relative to other prior laws can be computed similarly. A trivial computation occurs if ξ assigns all prior probability to a single point

θ_0 (say). If $0 < \theta_0 < 1$, we obtain (again with $W(\theta, d) = (\theta - d)^2$)

$$h_\xi(x, d) = (\theta_0 - d)^2 \theta_0^{\sum x_i}(1 - \theta_0)^{10 - \sum x_i},$$

and this is minimized uniquely by $d = \theta_0$. (If $\theta_0 = 1/2$, this is the procedure t_4 of Example 2.1(a).) A slightly different result occurs when $\theta_0 = 0$ or 1. For example, if $\theta_0 = 0$, the preceding expression for h is *automatically* 0 if $\sum x_i > 0$, so that any d yields the same minimum value 0 of h if $\sum x_i > 0$; if $\sum x_i = 0$, $h_\xi(x, d) = d^2$ is uniquely minimized by $d = 0$. Thus, when all the prior probability is assigned to the value $\theta = 0$, there are many Bayes procedures, any t' for which $t'(0, 0, \ldots, 0) = 0$ being Bayes *regardless* of the values of $t'(x)$ for other values of x. Some of these are admissible; others are not.

We now give illustrations of detailed Bayes calculations in two more examples. Consult Appendix B on *conditioning*, if necessary.

A 3-hypothesis problem. (See Problem 4.4.) We consider a 3-hypothesis problem but with a loss function more complex than the $0 - 1$ loss function usually employed: Suppose $f_j(x) = je^{-jx}$ (on $S = \{x : x > 0\}$) is the density of X under state of nature $\#j$ for $j = 1, 2, 3$, and that $D = \{d_1, d_2, d_3\}$, but that now the assertion d_3 (my guess is that the state of nature is $\#3$) is more seriously wrong when the true state is $i = 1$ than when it is $i = 2$. This is reasonable since the probability density function $\theta e^{-\theta x}$ is close to $\theta' e^{-\theta' x}$ when θ is close to θ'. The "seriousness" of various losses might be expressed by

$$W\left(1, \begin{matrix} d_1 \\ d_2 \\ d_3 \end{matrix}\right) = \begin{Bmatrix} 0 \\ 1 \\ 3 \end{Bmatrix}; W\left(2, \begin{matrix} d_1 \\ d_2 \\ d_3 \end{matrix}\right) = \begin{Bmatrix} 1 \\ 0 \\ 1 \end{Bmatrix}; W\left(3, \begin{matrix} d_1 \\ d_2 \\ d_3 \end{matrix}\right) = \begin{Bmatrix} 2 \\ 1 \\ 0 \end{Bmatrix}.$$

Thus, it is viewed as 3 times as bad an error, when the true j is 1, to guess it is 3 as to guess it is 2. Then, if $\xi = (\frac{1}{3}, \frac{1}{6}, \frac{1}{2})$, we obtain (note for each j you use the values in the same *row*, of the three Ws):

$$\sum_i W(i, d_j) p_\xi(f_i) f_i(x) = \begin{cases} \text{for } j = 1: 1 \cdot \frac{1}{6} \cdot 2e^{-2x} + 2 \cdot \frac{1}{2} \cdot 3e^{-3x}, \\ \text{for } j = 2: 1 \cdot \frac{1}{3} \cdot e^{-x} + 1 \cdot \frac{1}{2} \cdot 3e^{-3x}, \\ \text{for } j = 3: 3 \cdot \frac{1}{3} \cdot e^{-x} + 1 \cdot \frac{1}{6} \cdot 2e^{-2x}. \end{cases}$$

The comparisons are now messier than for the $0 - 1$ loss function. For example, to find *for which x is d_1 worse than d_2*, we ask: when is $1 \cdot \frac{1}{6} \cdot 2e^{-2x} + 2 \cdot \frac{1}{2} \cdot 3e^{-3x} > 1 \cdot \frac{1}{3} \cdot e^{-x} + 1 \cdot \frac{1}{2} \cdot 3e^{-3x}$?

Subtracting the right side from the left, multiplying through by $6e^x$, and calling $u = e^{-x}$, this becomes:

when is $9u^2 + 2u - 2 > 0$? (4.1)

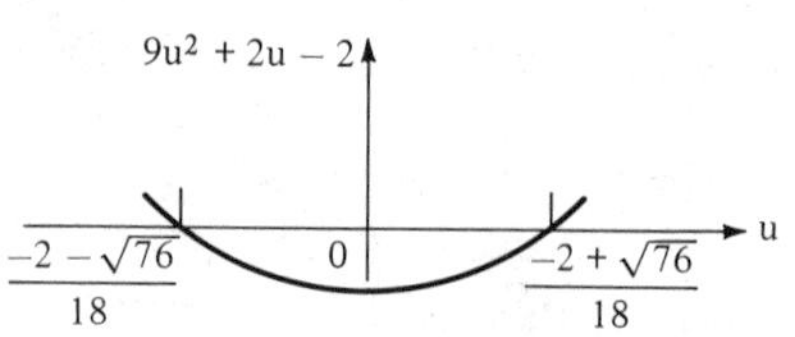

$9u^2 + 2u - 2$

u

$\frac{-2 - \sqrt{76}}{18}$ 0 $\frac{-2 + \sqrt{76}}{18}$

Since we are only concerned with $u > 0$ (because $u = e^{-x} > 0$), the answer to (4.1) is $u > \frac{-2+\sqrt{76}}{18}$; or, finally, since $u = e^{-x} > \frac{-2+\sqrt{76}}{18} \Leftrightarrow -x > \log$ (same) $\Leftrightarrow x < -\log$ (same), d_1 *is worse than* $d_2 \Leftrightarrow x < -\log\frac{-2+\sqrt{76}}{18}$. Thus, the final description, for each j, of the set of x for which each d_j is better than the other two decisions can be a lot of work. If there were different states of nature, (4.1) might become a polynomial inequality of high degree, requiring approximate computation by hand or machine. (If, for example, the three densities were $\theta e^{-\theta x}$ for $\theta = 1, 5, 8$, then (4.1) would be of 7th degree; and if the possible values were incommensurate, e.g., $\theta = 1, \pi, 4$, then (4.1) would be a transcendental function rather than an ordinary polynomial.)

Note that if you were presented with the preceding problem and told that X *had turned out to be some particular value, e.g.,* X = 2, *it would not be necessary to derive the whole Bayes procedure in order to make the Bayes Decision, since you only need* $t_\xi(2)$, *not* $t_\xi(x)$ *for all* x. *Thus, if* X = 2, *you would only have to compare*

$$\tfrac{1}{3}e^{-4} + 3e^{-6}, \qquad \tfrac{1}{3}e^{-2} + \tfrac{3}{2}e^{-6}, \qquad e^{-2} + \tfrac{1}{3}e^{-4},$$

and since the first of these is smallest, the Bayes decision is d_1.

An example illustrating the "data problem–no data problem" principle and calculations like those of Problem 4.5, *but for a different* W: Suppose $\Omega = \{\theta : \theta > 0\}$; that the object is to estimate θ, so that $D = \{d : d > 0\}$; and that $W(\theta, d) = \begin{cases} 1 & \text{if } |\theta - d| > 2 \\ 0 & \text{if } |\theta - d| \le 2 \end{cases}$; that is, the loss is the same for all "bad" decisions (those which misestimate θ by more than 2 units) and is 0 for all "correct" decisions.

Consider the *no data problem* in which there is an a priori density $f_\xi(\theta)$ on Ω. Then we want to choose d to minimize

$$\int_0^\infty W(\theta, d) f_\xi(\theta)\, d\theta = 1 - \int_{d-2}^{d+2} f_\xi(\theta)\, d\theta,$$

or, equivalently, to *maximize*

$$\int_{d-2}^{d+2} f_\xi(\theta)\, d\theta. \quad (4.2)$$

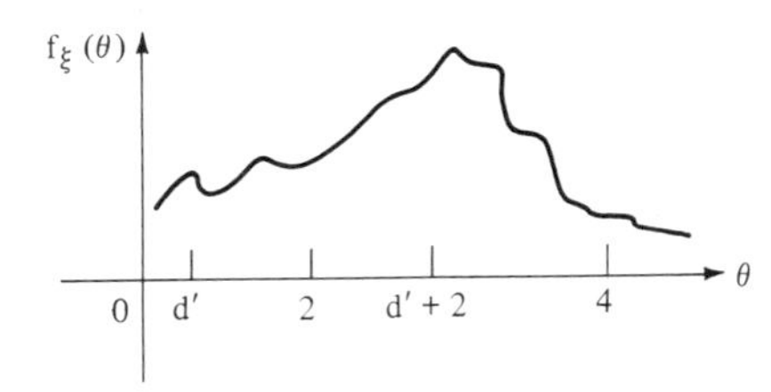

We always will take $d \ge 2$, since the vanishing of $f_\xi(\theta)$ for $\theta < 0$ implies that, if $d = d' < 2$, then (4.2) = area to the left of $d' + 2$, which is always $\le$ the value of (4.2) for $d = 2$, namely, the area to the left of 4. (If f vanishes between $d' + 2$ and 4 it would be that the preceding two areas are equal, but in any event we can choose $d = 2$ as "at least as good as" d' in that case.)

So we want to choose the interval $[d - 2, d + 2]$ of length 4, with $d \geq 2$, to maximize the shaded area at the right.

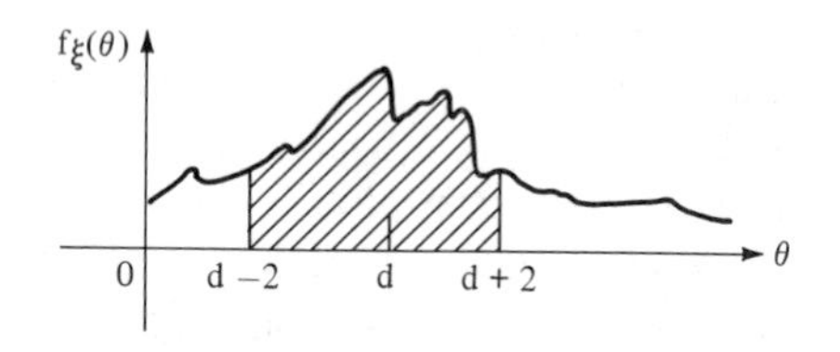

In general this could be a messy problem. Often one looks for elegant solutions if f_ξ is appropriately "nice" in such a problem. We now consider a class of such f_ξ's for which the solution is much simpler than one could generally expect.

Suppose f_ξ is continuous and *unimodal*; that is, there is a *mode* (possibly 0) such that $f_\xi(\theta)$ increases for $\theta < mode$ and decreases for $\theta > mode$. It is an easy calculus exercise to show that (4.2) is maximized by that value $\bar{d}$ for which $f_\xi(\bar{d} - 2) = f_\xi(\bar{d} + 2)$ if $f_\xi(0+) < f_\xi(4)$, and by $\bar{d} = 2$ otherwise. (Note what happens to the area of (4.2) on changing d from the value $d = \bar{d}$ in each of these cases!)

The case $f_\xi(0+) < f_\xi(4)$: The case $f_\xi(0+) \geq f_\xi(4)$:

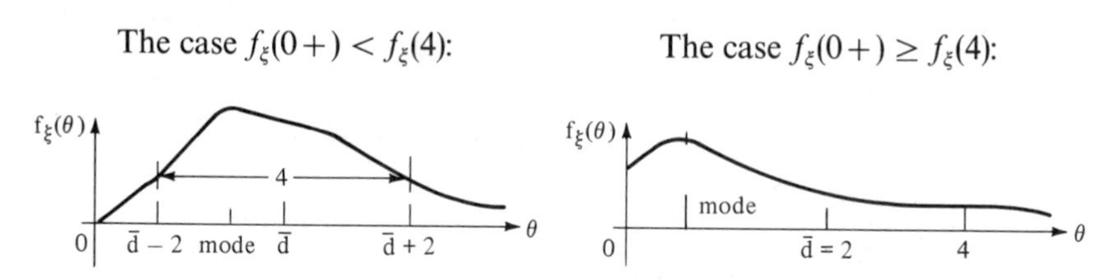

According to the no data principle (discussed previously), in a problem with data, if the a posteriori density of θ (given the data) is unimodal, then the preceding prescription yields a Bayes procedure if you simply replace f_ξ *by the a posteriori density in the description.*

Here is an explicit example of this unimodal type: With the preceding setup of Ω and W suppose $f_\xi(\theta) = \dfrac{c^r \theta^{r-1} e^{-c\theta}}{\Gamma(r)}$ (*gamma density*—see Appendix A), where c and r are positive constants. It is not hard to see that this is unimodal with

$$mode = \begin{cases} (r - 1)/c & \text{if } \; r > 1, \\ 0 & \text{otherwise.} \end{cases} \tag{4.3}$$

Suppose also $f_{\theta;X}(x) = \dfrac{\theta^q x^{q-1} e^{-\theta x}}{\Gamma(q)}$ where q is positive and known; this is the form in our previous example, with $q = 1$. [Actually, if q = some positive integer n, this is the density of the *sum* of n iid random variables each of which has the $q = 1$ (exponential) density. Later we shall see that "sufficiency" in the n-observation problem allows us to consider procedures which depend on the observations only through this sum.]

The posterior density of θ, given that $X = x$, is then

$$f_\xi^*(\theta|x) = f_\xi(\theta)f_{\theta;X}(x)\Big/\int_0^\infty (\text{same})\,d\theta = \frac{\theta^{q+r-1}(x+c)^{q+r}e^{-(x+c)\theta}}{\Gamma(q+r)}.$$

This is also a Γ-density, hence is unimodal.

The Bayes procedure will now be seen to be

$$t_\xi(x) = \begin{cases} 2 & \text{if} \quad q + r \le 1, \\ \text{the solution } d \text{ of } f_\xi^*(d+2|x) = f_\xi^*(d-2|x) & \text{if} \quad q + r > 1. \end{cases} \tag{4.4}$$

For, when $q + r < 1$, we have $f_\xi^*(0+|x) = +\infty$, and when $q + r = 1$ we have $f_\xi^*(\theta|x) = (x + c)e^{-(x+c)\theta}$, which is decreasing in θ, so that $f_\xi^*(0+|x) \ge f_\xi^*(4|x)$ in either of these cases. On the other hand, if $q + r > 1$, we have $f_\xi^*(0+|x) = 0 < f_\xi^*(4|x)$. Thus, the two lines of (4.4) are obtained from our earlier "no-data" conclusion for Γ-prior densities, transformed to use the Γ-posterior density when data are present (assertion in italics).

Finally, the equation of the second line of (4.4) happens to be solvable in simple form in this case. It becomes, for $q + r > 1$,

$$(d+2)^{q+r-1}e^{-(x+c)(d+2)} = (d-2)^{q+r-1}e^{-(x+c)(d-2)},$$

or

$$e^{-4(x+c)} = [(d-2)/(d+2)]^{q+r-1} = \left[1 - \frac{4}{d+2}\right]^{q+r-1},$$

or

$$e^{-4(x+c)/(q+r-1)} = 1 - \frac{4}{d+2};$$

thus, since d is the $t_\xi(x)$ we want,

$$t_\xi(x) = -2 + 4/[1 - e^{-4(x+c)/(q+r-1)}] \quad \text{if} \quad q + r > 1.$$

Note that the form of t_ξ *could have been obtained here directly by minimizing* $\int W(\theta, \bar{d}) f_\xi(\theta) f_{\theta;X}(x)\, d\theta$ *with respect to* $\bar{d}$. *The method we used instead benefits by using the earlier "no data" arithmetic to do this minimizing.* Sometimes it saves time to solve a general no-data problem in this way to get the "form" of the solution (see also Problem 4.4) and then to use the result in specific Bayesian problems with data. Sometimes, as in the 3-hypothesis example just given, we wouldn't really save in calculations.

Remark on risk functions of Bayes procedures. The computation of a Bayes procedure t_ξ relative to a given prior law ξ, illustrated in these last examples, does *not* yield the risk function of t_ξ; having found t_ξ, one must then compute $r_{t_\xi}(F) = E_F W(F, t_\xi(X))$ from scratch. The computation of "the Bayes risk" $R_{t_\xi}(\xi)$ is easier: in the discrete case this is clearly $\sum_x \min_d h_\xi(x,d) = \sum_x h_\xi(x, t_\xi(x))$.

We have seen that the Bayes procedure relative to a given a priori law ξ need not be unique; when there is more than one Bayes procedure relative to ξ, some of them may not be admissible. In part (c) of Example 4.1, the three Bayes procedures discussed are admissible. However, suppose we had a 3-hypothesis problem with 0-or-1 loss function

$$\left(W(F_i, d_j) = \begin{Bmatrix} 0 \text{ if } i = j \\ 1 \text{ if } i \neq j \end{Bmatrix} \quad \text{for } i, j = 1, 2, 3 \right)$$

wherein for one of the points of S, say $X = x_0$, we had $p_{F_1}(x_0) = p_{F_2}(x_0) = 0 < p_{F_3}(x_0)$. If ξ gives 0 a priori probability to F_3, it is easy to see that $h_\xi(x_0, d) = 0$ for all d. Thus, a Bayes procedure t relative to such a ξ can assign *any* decision when $x = x_0$. However, if t is a Bayes procedure for which $t(x_0) = d_1$ or d_2, and if the procedure t' is defined by

$$t'(x) = t(x) \quad \text{if} \quad x \neq x_0$$

$$t'(x) = d_3 \quad \text{if} \quad x = x_0,$$

then $r_{t'}(F_1) = r_t(F_1)$ and $r_{t'}(F_2) = r_t(F_2)$, since $t(X) = t'(X)$ with probability one under F_1 or F_2 (the event $X = x_0$ having probability zero); but $r_t(F_3) - r_{t'}(F_3) = P_{F_3}\{X = x_0\} > 0$. Thus, t' is better than t, and the latter is an inadmissible Bayes procedure.

From a strictly Bayesian point of view (that is, if we really know that the given ξ *is* the a priori law), it is immaterial which of the procedures t_1, t_2, and t_3 we use in part (c) of Example 4.1 and which of t and t' we use in the example of the previous paragraph; for the a priori risk $R_t(\xi)$ is the same minimum value for all Bayes procedures t^* in either case, and only the value of $R_{t^*}(\xi)$ is relevant! From a practical point of view, however, even the most confirmed Bayesian is probably never really certain about the form of ξ in even the most ideal circumstances and therefore would prefer using t' to t in the previous paragraph and would choose among the t_i in part (c) of Example 4.1 according to some additional criterion.

We shall later show that whenever there is a unique Bayes procedure t relative to a given ξ, t is admissible. It is also true that, whenever $p_\xi(F) > 0$ for *all* F in Ω, *every* Bayes procedure relative to that ξ is admissible.

Reasons for Considering Bayes Procedures

The computation of Bayes procedures is of interest even to statisticians who do not believe in the existence of a priori distributions. The reasons for this, as we shall discuss later, are as follows: (i) Under suitable conditions, if we compute the class of all Bayes procedures *relative to all possible* ξ, this class will contain all the admissible procedures and only a few inadmissible procedures; thus, the characterization of the class of all admissible procedures, which seems difficult by a direct approach (wherein we would examine each procedure t^* and investigate its admissibility by comparing r_{t^*} with the r_t's of all other procedures t), can be simplified if we know how to compute Bayes procedures. (ii) It will turn out that one of the methods for obtaining minimax procedures involves the computation of Bayes procedures, as we shall see in Section 4.5.

You should realize, then, when you read articles in which Bayes procedures are computed, that the computations often do not have as their aim the determination of a Bayes procedure which is to be used because the author believes F to have a given a priori law, but rather that the computation of Bayes procedures is often only a *mathematical device* to obtain other results. Such computations could in fact be carried out without ever using the terms *Bayes* and *a priori law* and would create less confusion if carried out in this way. Thus, the first reviewer of Wald's work on this subject, which work was devoted to (i) and (ii) rather than to the Bayesian point of view, erroneously reported that Wald had set the world of statistics back many years by adopting a Bayesian point of view!

Rational Behavior

A large body of work on the axioms of rational behavior has been carried out by L. J. Savage and others. One such set of axioms leads to the conclusion that, in problems wherein Ω and D are both finite sets, any individual who behaves rationally (i.e., consistently with these axioms) is actually acting in a Bayesian fashion relative to some a priori law ξ and some loss (disutility) function W, and that this ξ and W can actually be determined by asking the person a (perhaps infinite) sequence of simple questions of the type "if you make decision d_5, would you rather have it that F_1 is true, or that F_2 and F_3 each have a 50 percent chance of being true?" Many people have taken these results to mean that all statistical problems are solvable on Bayesian grounds, and this misimpression deserves some comment.

First, even in the most ideal problem in which a W and a ξ can be determined on economic or physical grounds, the W and ξ according to which these axioms say that the person acts are not necessarily the correct (physical) ones. In fact, the more careful proponents of this theory refer to the personal (nonphysical) ξ as a "personal" or "intuitive" or "subjective" probability law. *This law is the one such that the individual acts as if it were the actual law; it is not necessarily the actual law according to which he or she should be acting.* Thus, in an experiment conducted by Davidson, Suppes, and Siegel in a simple set of gambling trials wherein the chance device was a fair tetrahedron, people acted consistently in deciding whether or not to bet that a chosen side of the tetrahedron would come up when offered various odds, *but consistently with the idea that the probability of that side's coming up was about .2 rather than with the actual physical value of* .25. These people thus did *not* maximize expected profit or even expected utility with respect to the *actual physical* probabilities which governed the problem. The question then arises, of what good is it to know the "intuitive" ξ according to which the individual acts, if this is not the physical ξ (if one exists) according to which he or she *should* act.

Second, the theory mentioned previously offers no suggestion as to how the intuitive ξ's and W's of two people (or of society) are to be combined in a statistical problem on which the two work together (or with which society is concerned).

Finally, although this theory presents an appealing mathematical result whereby the statistical consumer does not have to come up with W and ξ directly in a complicated problem, but can instead arrive at them methodically by answering a large number of very *simple* questions of the type illustrated, it is nevertheless questionable whether this is of much value in practice. For the questioning procedure would be long; the theory is static, and therefore the procedure would have to be repeated periodically; and it is perhaps just as reasonable to present the consumer with a collection of risk functions of admissible procedures and ask him or her to choose among them, even though this asks one big question instead of an infinite sequence of little questions.

We remark that many Bayesians do not pretend to go through the preceding determination of the precise ξ and W implied by the rational behavior axioms. But, pragmatically, they assert that a useful way of choosing among procedures is the Bayes approach with a ξ that (for simplicity, assuming W given) at least roughly summarizes the statistician's feelings about the chances of encountering various states of nature. These statisticians unfortunately often do not bother to look at the operating characteristic of the procedure obtained in this way. It can easily turn out that the Bayes procedure t_ξ has a high risk in some subset B of Ω where ξ assigns little probability, and that another procedure t' has much smaller risk than t_ξ in that region and only slightly larger risk than t_ξ elsewhere in Ω. This will be discussed again in Section 4.2 and is illustrated in several of the problems.

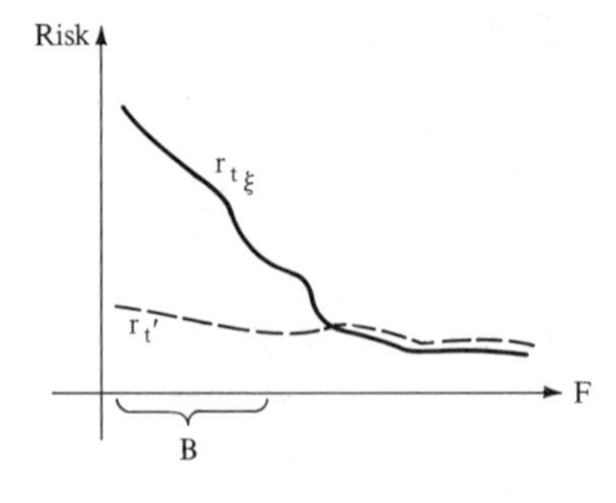

If one is sure of ξ and only interested in minimizing $R_t(\xi)$, one may comfortably use t_ξ. Where ξ is only an approximation to guide the choice of procedure, a closer look at risk functions is needed. This is also true for other criteria for choosing a procedure: statistics is complex enough that no simple recipe will yield a satisfactory r_t in every setting.

The phenomenon illustrated previously, in which an admissible t (here t_ξ) is greatly improved upon (in risk) for some F by a t' which never performs much worse, is called *subadmissibility*.

4.2. Minimax Criterion

A procedure t^* is said to be *minimax* if

$$\max_F r_{t^*}(F) = \min_t \left[\max_F r_t(F) \right];$$

or in other words, if

$$\max_F r_{t^*}(F) \leq \max_F r_t(F) \quad \text{for every } t.$$

The minimax criterion is to choose, for actual use in a statistical problem, an admissible procedure which is minimax. There can easily exist inadmissible procedures which are minimax, and it may be that there are several admissible minimax procedures. The risk functions of several minimax procedures might look like this (risk functions of nonminimax procedures are not drawn in this picture):

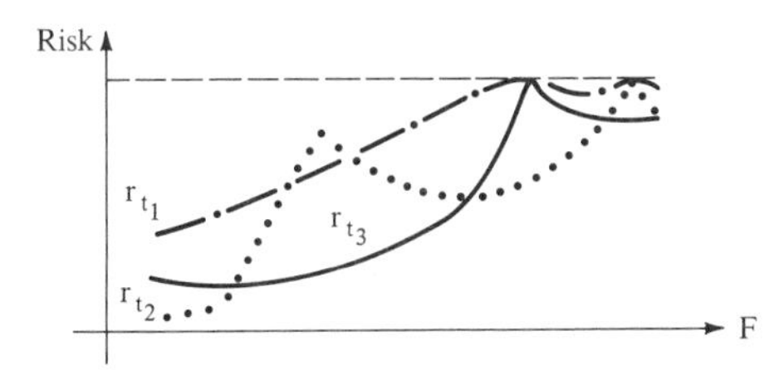

Here t_1 is minimax but inadmissible. Where there is more than one admissible minimax procedure (e.g., t_2 and t_3), we would need some other criterion *in addition* to the minimax criterion to choose among the admissible minimax procedures.

The minimax criterion has often been described as that of the conservative person who wants to avoid the most catastrophic losses (or of the pessimist who is sure that the worst possibilities will occur) and who therefore chooses that procedure for which the worst (over all F) expected loss is as small as possible.

For any criterion, it is possible to construct examples wherein that criterion is inapplicable (as is usually the case for the Bayes criterion) or leads to a result which seems foolish on closer inspection. Thus a statistical problem with $\Omega = \{F_1, F_2, \ldots, F_{100}\}$ may have a unique admissible minimax procedure t^*

with constant risk $r_{t^*}(F_i) = 1$, but there may be another procedure t' for which

$$r_{t'}(F_1) = 1.001, \qquad r_{t'}(F_i) = .002 \quad \text{if} \quad i > 1.$$

This is an instance of the subadmissibility phenomenon mentioned before. The risk function of t' is only very slightly greater than that of t^* at F_1 and is very much smaller at all other values F_i. In such circumstances, most people would rather use t' than t^*, being willing possibly to incur the slight increase in risk under F_1 in return for the possibility of such a great reduction in risk under all other F_i. In Problems 4.19–4.21, this phenomenon will be illustrated. In those examples, there is an interval (a_n, b_n) such that the constant risk of the minimax procedure t_n^* for sample size n is only slightly less than that of some other procedure t_n' when the parameter value lies in (a_n, b_n), and such that $r_{t_n'}$ is less (and often much less) than $r_{t_n^*}$ outside that interval; moreover, $b_n - a_n \to 0$ as $n \to \infty$. In such cases the sequence $\{t_n'\}$ is said to be a *subminimax* sequence.

Since a parametrization of Ω in terms of some other parameter ϕ (see discussion in Section 4.1) could transform the interval (a_n, b_n), for a given n, into a large set of ϕ-values, and its complement into a small set, there is no natural criterion (as was "better than") for choosing t_n' over t_n^* here. Such a choice depends either on an aspect of human behavior for some individuals, which dictates accepting the possibility of a slight increase in risk under some circumstances in return for the possibility of an appreciable decrease in risk under other circumstances, or else upon an at least vague feeling that there is a prior law that is relevant and which assigns appreciable probability to values not in (a_n, b_n). We note also that although for each n there is a ϕ (depending on n) that reparametrizes Ω as described, once a ϕ is chosen and fixed, the subminimaxity phenomenon will take place for sufficiently large n, in terms of the parameter ϕ.

One appealing feature of the minimax criterion is that it does not depend on the particular parametrization used to describe Ω, unlike the criterion of "minimizing the area under r_t" (which we criticized) or the criterion of unbiasedness, which we shall consider later, for the *maximum* of the risk function of any procedure t does not depend on the way in which the members of Ω are labeled.

4.3. Randomized Statistical Procedures

In many settings with finite S (such as in Problems 4.1 and 4.2), there are only a few possible procedures t of the type we have discussed thus far (*nonrandomized procedures*), and thus only a few corresponding risk functions from which to choose. Thus, we might have only four available risk functions of nonrandomized procedures as shown in the diagram. It is a natural idea to try to obtain some attractive looking *average* of these functions by choosing among the t_i by a chance device. Thus, we may think of spinning a spinner whose

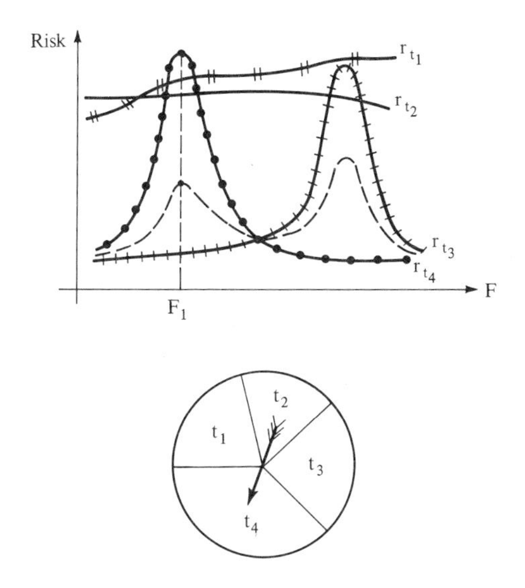

possible outcomes are the names t_i of the nonrandomized procedures, t_i being the spinner's outcome with probability π_i. If we use a spinner with π_i's we view as appropriate and then use the t_i designated by the outcome of a spin, we may view r_{t_i} as the conditional risk given the outcome of the spin; hence from the viewpoint before the spin, the overall expected loss or risk with respect to the chance mechanism consisting of our spin of the spinner and, independent of this, the production of X with law F, is

$$\sum_i \pi_i r_{t_i}(F).$$

For example, choosing t_3 and t_4 with probabilities $\pi_3 = .6$ and $\pi_4 = .4$ results in the broken curve risk function; geometrically, for each value F_1 of F, the point on this curve is the point that divides the vertical segment between the points, $(F_1, r_{t_3}(F_1))$ and $(F_1, r_{t_4}(F_1))$ in the ratio $4:6$ (yielding a point closer to the first of these).

The preceding scheme is called *special or initial randomization.* A more complicated method, called *general randomization,* proceeds as follows: Suppose D is finite, to make the presentation simpler. We observe the value x taken on by X. Then, using a spinner whose outcomes are elements of D with probabilities *depending on* x that we deem appropriate, we use this spinner to choose the decision d that we actually make. Such a *randomized procedure* δ

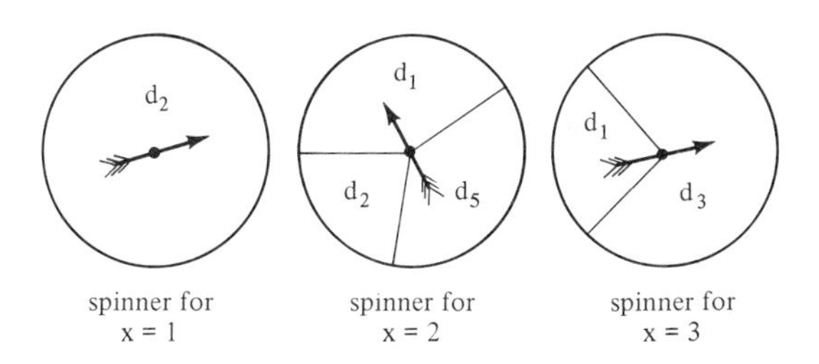

can be viewed as a collection of spinners, one corresponding to each value in S. For the spinner to be used when $X = x$, there is probability $\delta(x, d_i)$ of the outcome d_i, so that $\sum_i \delta(x, d_i) = 1$ for each x. In the illustration, we always make decision d_2 if $X = 1$ ($\delta(1, d_2) = 1$), and choose among d_1, d_2, d_5 when $X = 2$ and between d_1 and d_3 when $X = 3$ (with $\delta(3, d_1) = .28 = 1 - \delta(3, d_3)$).

If, for each x, there is a single element $d(x)$ (depending on x) to which δ assigns probability one, so that $\delta(x, d(x)) = 1$, then the procedure is a "nonrandomized" procedure of the type we have been considering. For, if $X = x$, there is no need to spin any spinner, since with probability one the outcome will be $d(x)$; thus, we might just as well have made the decision $d(x)$ to start with, and the procedure t defined by $t(x) = d(x)$ is of the type we have been considering.

Practical people often object strenuously to the use of genuinely randomized procedures (that is, to ones which are not actually nonrandomized). Part of the objection is caused by a lack of understanding, in mistakenly asserting that "the spin of a spinner has nothing to do with the outcome of the experiment, and that such spinning is therefore foolish"; this argument is not valid, in that it neglects the fact that the choice of *which* spinner is spun *does* depend on the outcome (in fact, we shall see in our discussion of hypothesis testing that such spinning may be necessary). Another aspect of the objection is that two people observing the same experiment and using the same procedure δ may reach different decisions (because their spinners have different outcomes); thus, to record which procedure δ one uses and what the outcome of the experiment is does not determine which decision is reached. This objection has considerable practical appeal; for this reason, as well as because moderate sample sizes often produce enough nonrandomized procedures that the risk function of any randomized procedure can be closely approximated by that of some nonrandomized procedure, one finds little use of randomized procedures in practice. However, it should be mentioned that many of the statisticians who most object to choosing a decision in a random fashion do not hesitate to choose the experimental design in some randomized way, although one could object to random choice of a design on the analogous grounds that different people faced with the same problem and who choose a design in the same random way might end up with different designs and data! If there is no objection to choosing and recording the random choice of a design, there should be no objection to choosing and recording a random choice of decision. In terms of expected loss, there is no objection to doing either.

As we shall discuss under hypothesis testing, randomization is usually necessary only in the discrete case; its use in the discrete case is essentially to break up lumps of probability so as to yield risk functions which cannot be achieved by nonrandomized procedures (as illustrated in the following example). In constructing a procedure which will satisfy a given optimality criterion (e.g., the minimax criterion), it will usually be that genuine randomization need be performed for only a few values in S.

The loss incurred if decision d is made and F is true is $W(F,d)$. If $X = x$, the decision d is made with probability $\delta(x,d)$, so that

$$\sum_d W(F,d)\delta(x,d)$$

is the expected loss incurred if F is true and if $X = x$. Thus, the overall expected loss, or *risk function*, of δ, is

$$\begin{aligned} r_\delta(F) &= E_F\left\{\sum_d W(F,d)\delta(X,d)\right\} \\ &= \sum_x p_{F;X}(x)\sum_d W(F,d)\delta(x,d) \end{aligned}$$

(in the discrete case), and the *operating characteristic of* δ is

$$\begin{aligned} q_\delta(F,d) &= P_F\{\text{make decision } d \text{ using } \delta\} \\ &= \sum_x p_{F;X}(x)\delta(x,d). \\ &= E_F\{\delta(X,d)\}. \end{aligned}$$

EXAMPLE 4.3. Consider the situation in which $S = \{0,1\}$ and $D = \{d_0, d_1\}$. Then there are exactly four possible nonrandomized statistical procedures t_i, which we hereafter suppose to be defined as in Problems 4.1(b) and 4.2(b). An example of a randomized procedure δ is defined by

$$\begin{aligned} &\delta(0,d_1) = 1, \qquad \delta(0,d_0) = 0; \\ &\delta(1,d_1) = 1/3, \qquad \delta(1,d_0) = 2/3. \end{aligned}$$

Suppose p_0 and p_1 are specified values with $0 < p_1 < p_0 < 1$, and that $P_{F_i}\{X = 1\} = p_i = 1 - P_{F_i}\{X = 0\}$, that d_i is the guess that F_i is true, and that the loss is 0 (if correct) or 1 (if incorrect). If you compute r_{t_i} for each t_i and then compute r_δ you will see that δ is not worse than any t_i; in fact, δ can be proved admissible among all randomized procedures. (The demonstration is similar to ones presented in Problems 4.10 and 4.11.) You will also note that if t_2 and t_3 are rewritten as randomized procedures δ_2 and δ_3 (i.e., $\delta_2(0,d_1) = \delta_2(1,d_1) = 1$ and $\delta_3(0,d_1) = \delta_3(1,d_0) = 1$), then our δ can be rewritten as

$$\delta(x,d) = \tfrac{1}{3}\delta_2(x,d) + \tfrac{2}{3}\delta_3(x,d)$$

for all x and d. This suggests that we could replace our general randomization scheme of spinning at the end of the experiment (on observing the value taken on by x) by the special randomization scheme of using a single spinner with its chance trial performed (if desired) before the experiment and which will with probability 1/3 tell us to use t_2 (which is the same as δ_2) and with probability 2/3 tell us to use t_3. This phenomenon is generally true: for any general randomized procedure δ, there is an initial random choice among *nonrandomized* procedures (that is, a special randomization) which will yield the same operating characteristic as δ. Thus, in particular, in the preceding example you will see that

$$r_\delta(F) = \tfrac{1}{3}r_{t_1}(F) + \tfrac{2}{3}r_{t_2}(F) \quad \text{for all } F.$$

It is even easier to show that any *special randomization* may be replaced by a general randomization with the same risk function: If we choose among procedures t_i with probabilities π_i in a special randomization, then, since under t_i the probability of making decision d when $X = x$ is

$$\delta_i(x, d) = \begin{Bmatrix} 1 & \text{if} & d = t_i(x) \\ 0 & \text{if} & d \neq t_i(x) \end{Bmatrix},$$

we see that the overall (initially randomized) probability of making decision d when $X = x$ is $\sum_i \pi_i \delta_i(x, d)$. This last can be seen to be a general randomization procedure. For every x, it yields the same probability of making each decision d as did the given special randomization scheme, and consequently it yields the same risk function as the latter, *no matter what* Ω *and* W *are*! This *equivalence of the two methods of randomization* was first proved by Wald and Wolfowitz. (See Problem 4.18.)

It is necessary to include randomized procedures in order to obtain such useful results as the fact that every admissible procedure is Bayes or a limit of Bayes rules under certain conditions (next section). The practical necessity of using randomized procedures is slight, as indicated previously.

4.4. Admissibility: The Geometry of Risk Points

Admissibility

As we mentioned in Section 4.1, there are two simple conditions, either of which implies the admissibility of a given statistical procedure t. Both of these involve t's being Bayes relative to some ξ; you are therefore reminded that, exactly as is the case in the use of a priori distributions as a mathematical device for proving minimax results in Section 4.5, so also here we do not have to believe in the relevance of the Bayes criterion in an application in order to know that a t which has been proved admissible by using this device *is* admissible.

(1) *If* t *is the essentially unique Bayes procedure relative to some* ξ *(i.e., if any other Bayes procedure* t′ *relative to* ξ *satisfies* t′(x) = t(x) *except possibly on a set of* x's *which has probability zero for each* F), *then* t *is admissible.*

PROOF: We assume the contrary and get a contradiction. If t is Bayes relative to ξ but not admissible, then there is a t'' which is better than t. Clearly t'' must also be Bayes relative to ξ. But since t'' and t are not equivalent, they cannot be essentially the same; i.e., t is not the essentially unique Bayes procedure relative to ξ. □

(2) *Suppose* $\min_t R_t(\xi) < \infty$. *If either* (i) Ω *is discrete and* $p_\xi(F) > 0$ *for each* F *in* Ω, *or if* (ii) Ω *is a Euclidean set and* $r_t(\theta)$ *is continuous in* θ *for each* t *and* $f_\xi(\theta) > 0$ *for each* θ *in* Ω, *then every Bayes procedure relative to* ξ *is admissible.*

PROOF. If t is Bayes relative to ξ and t is not admissible, so that there is a t' for which t' is better than t, then $R_{t'}(\xi) < R_t(\xi)$, contradicting the Bayes character of t; this is clear in case (i) and follows in case (ii) from the fact that, risk functions being continuous, if $r_t(\theta) < r_{t'}(\theta)$ for some θ, then this inequality is satisfied on a set of θ values of positive a priori probability according to f_ξ (an interval if Ω is 1-dimensional, etc.). □

We have seen in the remark on risk functions of Bayes procedures in Section 4.1 that Bayes procedures which are neither unique nor Bayes relative to a ξ with the properties stated in (2) need not be admissible. On the other hand, the preceding results serve to prove admissibility in many important cases (e.g., the 3-hypothesis example of Section 4.1). The results also apply to randomized procedures; regarding (1), if there is more than one randomized Bayes procedure relative to ξ, it is easy to see that there is also more than one nonrandomized Bayes procedure; hence, (1) applies whether we interpret it to refer to uniqueness among all procedures or merely among nonrandomized ones.

Unfortunately, not all admissible procedures can be proved admissible by using one of these results, as we shall see later. Not even all admissible Bayes procedures can be treated by this approach (this is true of the left-hand endpoint of C in the diagram near the end of this section), and the determination of which procedures are admissible is quite difficult in general.

Complete Classes

A collection C of statistical procedures (including perhaps randomized ones) is said to be *complete* (respectively, *essentially complete*) if, for each procedure δ not in C, there is a δ' in C which is better than (respectively, at least as good as) δ. [Of course, the δ' in this definition depends on δ.] If C is complete (or essentially complete) and no proper subset of C has this property, then C is said to be *minimal complete* (or *minimal essentially complete*).

If we want to make some useful statement regarding which procedures a customer might use in a given problem without adopting a specific criterion (as at the beginning of Section 4.1) to select a single procedure for him or her, then we can present the customer with a complete (or essentially complete) class of procedures (or, more meaningfully, with their risk functions). There is no reason for the customer to use a procedure δ outside this class, since for any such δ there is a δ' in the class which is better (or at least as good). However, it is not much help to give too large a complete class; for example, the class of *all* procedures is complete (there being no procedure outside it). Such large complete classes will contain more procedures than it is necessary to present to the customer, including many inadmissible ones. This is why a *minimal* complete or essentially complete class is what we would really like to know, to present to the customer.

It can be shown that, under conditions usually satisfied in a practical problem, *the unique minimal complete class is the class of all admissible procedures.* Since this last class has been remarked to be difficult to characterize in general, the statistician is often interested in a slightly larger class of procedures which will contain some inadmissible procedures (but not "terribly many") and which is easier to characterize.

This latter class involves the use of Bayes procedures; again, it is one of the uses which does not assume the Bayes criterion to be relevant in the problem but which uses Bayes procedures as a mathematical device. *If* Ω *and* D *are finite, the class of all Bayes procedures is complete.* If Ω is not finite, a complete class can usually be obtained by taking limits of sequences of Bayes procedures in an appropriate sense (a similar device will be mentioned in Section 4.5 in connection with minimax proofs). In many important practical problems, all Bayes procedures have some simple characterization which makes the description of a complete class (which is not far from being minimal) fairly simple; we shall mention several such examples under hypothesis testing. (See also Problems 4.10–4.13, 4.15–4.17 on Bayes procedures.)

The relationship between completeness and essential completeness is not hard to see. Every complete class is by definition essentially complete, but not conversely. If C is complete and we delete from C various procedures, each of which is equivalent to or worse than some procedure remaining in the class, then the resulting class is essentially complete. In particular, if C is minimal complete and if we break up C into a union of disjoint sets in each of which the procedures are all equivalent but in different ones of which the procedures are not equivalent, then any minimal essentially complete class is obtained by selecting one procedure from each of these sets.

The reason why the admissible procedures need not form a complete class, and why there may not even exist any admissible Bayes or admissible procedures, is related to the attainment of extrema and can be understood without reference to statistics. For example, if you must name a real number $d(-\infty < d < \infty)$ and pay the penalty e^d if you name the number d, there is no "admissible" procedure: whatever number d you name, the number $d - 1$ would have been a better choice (and "$-\infty$" is not allowed). A statistical analogue would be the estimation of the mean θ of a normal distribution when the loss function is $e^{(d-\theta)}$: there is no admissible procedure, since one "cannot underestimate enough"; for example, no procedure of the form $t(x) = x + b$ (the "invariant" procedures, discussed later) is admissible, since $b + x - 1$ is better. You can see that most practical problems will not have such a loss function.

The proof of the claim made about completeness of Bayes procedures in the case of finite Ω and D is not too difficult. There is a simple geometric picture of the things we have been discussing. If $\Omega = \{F_1, \ldots, F_k\}$, we can represent the risk function of any procedure as a point $r_\delta^* = (r_\delta(F_1), r_\delta(F_2), \ldots, r_\delta(F_k))$ in k-dimensional space. Let $\mathscr{R}$ be the set of all such points (called *risk points*) corresponding to all procedures δ. It is not hard to see that $\mathscr{R}$ is a *convex set*; i.e., if two points are in $\mathscr{R}$, so is the entire line segment joining them. This is because, if two points in $\mathscr{R}$ are the risk points of δ_0 and δ_1, then the randomized procedure $\delta_\alpha = \alpha\delta_1 + (1 - \alpha)\delta_0$ is easily seen to have a risk point a fraction α of the way from $r_{\delta_0}^*$ to $r_{\delta_1}^*$ on the line segment joining them. It can also be proved that $\mathscr{R}$ is *closed*; that is, that $\mathscr{R}$ contains the limit point of any convergent sequence of points in $\mathscr{R}$. (This is also true if D is infinite but W is sufficiently regular.) A typical $\mathscr{R}$ when $k = 2$ might look like this:

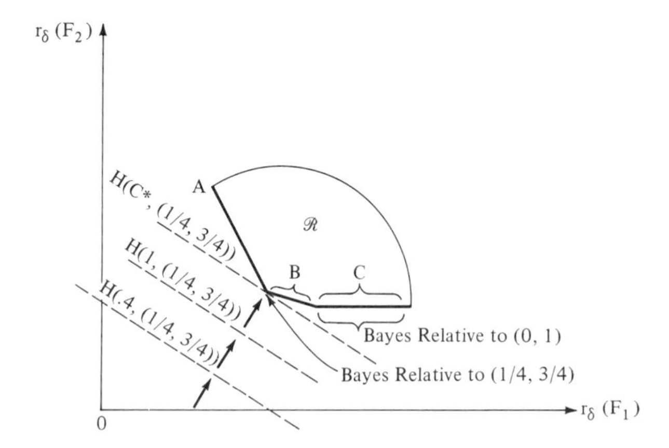

It is useful to keep in mind the two representations of risk functions: the earlier picture of *graphs* of functions r_t on Ω, and the present picture in which r_t is a *point* with coordinates $\{r_t(F), F \in \Omega\}$.

In what follows, we write $\xi_i = p_\xi(F_i)$.

Suppose $\xi = (\xi_1, \xi_2, \ldots, \xi_k)$ is any fixed collection of nonnegative numbers whose sum is one. The collection $H(c, \xi)$ of all points $r = (r_1, \ldots, r_k)$ in k-dimensional space for which $\sum_1^k \xi_i r_i = c$, where c is a specified constant, constitute a hyperplane (line when $k = 2$ or plane when $k = 3$). The direction cosines of a vector perpendicular to such a hyperplane are $\xi_i \Big/ \left(\sum_{j=1}^k \xi_j^2\right)^{1/2}$; these are all nonnegative. The hyperplanes corresponding to different values of c are parallel and can be obtained as c increases by translating a hyperplane of this family in the direction of the vector. When $k = 2$, for example, this translation will either be to the right (if $\xi_1 = 1, \xi_2 = 0$), upward (if $\xi_2 = 1, \xi_1 = 0$), or both (if $\xi_1 > 0, \xi_2 > 0$). Starting with a sufficiently negative c (if $W \geq 0$, $c = 0$ will suffice) and increasing c, there will be a first (smallest) value c^* of c for which $H(c, \xi)$ intersects R nonvacuously. The points in $R \cap H(c^*, \xi)$ are thus risk points of δ's which minimize $\sum_{i=1}^{k} \xi_i r_\delta(F_1)$ among all δ's; that is, they are *Bayes relative to* ξ, and the Bayes risk is c^*. If we now choose a different ξ, we get the Bayes procedures relative to it in the same way, etc.

The risk points of Bayes procedures in the diagram for $k = 2$ have been indicated by a thickened line. The segment marked B contains risk points of admissible procedures which are all Bayes relative to the same ξ satisfying admissibility criterion (2) at the beginning of this section; they are not equivalent, so criterion (1) is not satisfied. The point A in the diagram is meant to illustrate (1) but not (2) being satisfied ($\xi = (1, 0)$). The points C satisfy neither (1) nor (2); all are Bayes, but only the left-hand point of this segment C is admissible; it is better than other points of the segment. For any point which is not Bayes, we can find a point below and/or to the left of it which is Bayes and better. Thus, the Bayes procedures are a complete class, but we would have to eliminate all but the left-hand point of C to obtain a minimal complete class.

Minimax procedures can also be represented in this geometric picture. If we consider the set $B_\lambda = \{(r_1, \ldots, r_k) : \max_i r_i \le \lambda\}$, there will be a smallest value λ^* of λ for which this set intersects R nonvacuously. The picture might look like one of these:

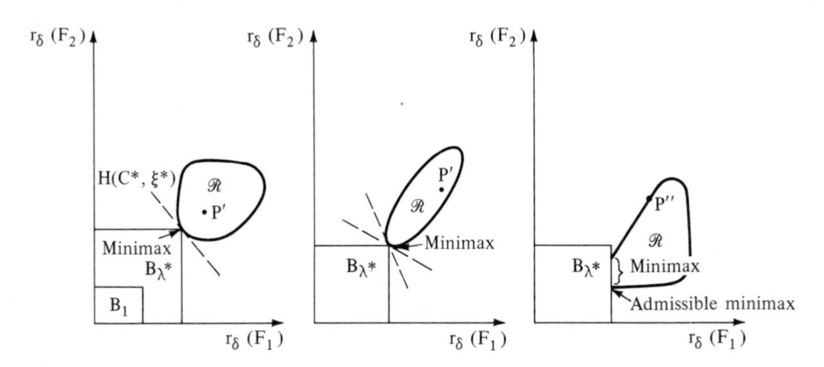

The points of $B_{\lambda^*} \cap R$ are then risk points of procedures for which $\max_i r_\delta(F_i) = \lambda^*$; they are minimax, since no other procedure can have $\max_i r_\delta(F_i) = \lambda' < \lambda^*$ without having $B_{\lambda'} \cap R$ nonvacuous, contradicting the definition of λ^*. Since B_{λ^*} and R are both convex and intersect only on their boundaries, it can be shown that there is a line $L: \xi_1^* r_1 + \xi_2^* r_2 = c$ such that B_{λ^*} lies below and to the left of (or on) L and R lies above and to the right of (or on) L, where $\xi_1^* \ge 0$, $\xi_2^* \ge 0$, $\xi_1^* + \xi_2^* = 1$. Such a ξ^* is then "least favorable" in the sense of the discussion in Section 4.5: a minimax procedure is Bayes relative to it. Our three illustrations show (i) a unique minimax risk point and ξ^*, (ii) several ξ^*'s, (iii) several minimax risk points (all but one are inadmissible). For $k = 3$ one can similarly illustrate the possibility of having several *different* admissible minimax risk points, etc.

It is helpful to understand the geometric interpretation of "better than," admissible, Bayes, minimax, in both representations of risk functions.

4.5. Computation of Minimax Procedures

The determination of a minimax procedure in a statistical problem is usually a much more difficult task than the computation of a Bayes procedure relative to a given ξ. There is no longer anything so simple as the latter computation, wherein $t(x)$ could be determined separately for each fixed x so as to minimize $h_\xi(x, d)$. In fact, a *direct* computation of a minimax procedure would entail the computation of the entire risk function r_t of each procedure t, in order to obtain the quantity

$$r_t^* = \max_F r_t(F),$$

and then the choice of a t for which r_t^* is as small as possible. This would be much too difficult an approach in any statistical problem of interest. There are actually three commonly used *indirect* methods for obtaining minimax procedures, one of which will be given here, the others being described in Section 4.7.

Minimax procedures, as we shall see in examples, often have constant risk functions, although the constancy of $r_t(F)$ is neither necessary nor sufficient for F to be minimax (next paragraph). Work by C. Stein and others since the 1950s has shown that a number of "classical" statistical procedures that arise in multivariate analysis have constant risk, are minimax, but are not admissible. (Others are not even minimax!) It is not known for some of these statistical problems what *is* an admissible minimax statistical procedure, although by general abstract arguments one can show that admissible minimax statistical procedures must exist.

The most applicable method of finding minimax procedures is that described in the next theorem. This makes use of the fact that minimax procedures often have constant risk functions. However, you should understand that it is not always the case that there is a minimax procedure with constant risk, nor is it true that every procedure with constant risk is admissible. Thus, in the first and second diagrams in the figure on page 56, any risk point P' on the 45° line above and to the right of the minimax risk point *also* has constant risk, but is not minimax; in the third diagram there is *no* minimax procedure with constant risk, although there is a rather poor point P'' representing "best among procedures with constant risk."

The Bayes Method of Finding Minimax Procedures

This will sometimes be referred to as "*Method A*". We have mentioned earlier that Bayes procedures have uses as mathematical devices in settings in which we do not really have an a priori law or do not know what it is. One such use is to prove certain procedures are minimax. We shall show that if a procedure t^* which is Bayes relative to some ξ has a particular additional property, then t^* is minimax. If a procedure t^* is proved minimax in this way, it is irrelevant whether or not we believe this ξ to be the a priori law which governs the problem; we know t^* is minimax, and if we adopt the minimax criterion, that is all that matters. The idea of this device is quite simple.

Theorem. *Suppose* t* *is Bayes relative to some* ξ *and that* $r_{t^*}(F)$ *is a constant for* F *in* Ω. *Then* t* *is minimax.*

PROOF. Let ξ denote the a priori law relative to which t^* is Bayes. Then for any procedure t', $R_{t'}(\xi) \geq R_{t^*}(\xi)$. Also, $R_{t^*}(\xi) = \max_F r_{t^*}(F)$, since both sides equal the constant value of $r_{t^*}(F)$. Therefore,

$$R_{t'}(\xi) \geq \max_F r_{t^*}(F).$$

Now $R_{t'}(\xi)$ is an *average* of the values of the risk function $r_{t'}(F)$, so evidently

$$R_{t'}(\xi) \leq \max_F r_{t'}(F),$$

and this combines with the last inequality to show that t^* is minimax. □

A useful generalization of the theorem is this: If t^* is Bayes relative to some ξ and if $R_{t^*}(\xi) = \max_F r_{t^*}(F)$, then t^* is minimax. Note that in the generalization t^* need not have constant risk; the condition

$$R_{t^*}(\xi) = \max_F r_{t^*}(F)$$

is a weaker one, being equivalent to the statement that ξ assigns probability one to the set of F's at which $r_{t^*}(F)$ attains its maximum. The weaker condition suffices since the proof just given uses the constancy of the risk function only to derive that condition.

EXAMPLE 4.4. Suppose X is Bernoulli, $P(X = 1) = \theta$, and suppose $\Omega = \{\theta : 1/3 \leq \theta \leq 2/3\}$. A nonrandomized procedure t for estimating θ is a function of two values $t(0) = a$, $t(1) = b$, and, for squared error loss,

$$r_t(\theta) = (1 - \theta)(a - \theta)^2 + \theta(b - \theta)^2.$$

Suppose ξ puts mass 1/2 at $\theta = 1/3$, mass 1/2 at $\theta = 2/3$. The Bayes risk of t is then

$$\begin{aligned} R_t(\xi) = (\tfrac{1}{2})[(\tfrac{2}{3})(a - (\tfrac{1}{3}))^2 + (\tfrac{1}{3})(b - (\tfrac{1}{3}))^2] \\ + (\tfrac{1}{2})[(\tfrac{1}{3})(a - (\tfrac{2}{3}))^2 + (\tfrac{2}{3})(b - (\tfrac{2}{3}))^2]. \end{aligned}$$

Computing partial derivatives with respect to a and b, setting these equal to zero, and solving gives $a = 4/9$, $b = 5/9$. A direct calculation then shows that $r_t(\theta) = (1/18)^2 + (7/9)(\theta - 1/2)^2$. By the preceding generalization we see that t is minimax.

The calculation of derivatives has shown there could be at most one Bayes procedure for ξ. Therefore t is admissible as well as minimax.

In the preceding example we were able to conclude admissibility. This question is discussed in Section 4.4. Although Bayes procedures are not *always* admissible, we saw that under certain circumstances they *are* admissible.

Application of the preceding results is unfortunately not just a matter of routine. In some applications, we may have no intuitive idea as to which t may be minimax, in which case we have to compute the Bayes procedures relative to various ξ's until we find a ξ relative to which a Bayes procedure has constant risk; this could be a long process of search for the right ξ and t^*. In other applications we may know of a t^* with constant risk and which intuitively seems like a good procedure; in such cases, we still have to hunt for a ξ relative to which the procedure t^* is Bayes, and this may again by a tedious task.

Thus, there is no purely mechanical algorithm for finding a minimax procedure or for proving a procedure to be minimax by using this method. One clue which is sometimes helpful to research workers is that a ξ which does the job, as can be proved under fairly general conditions, is one for which the Bayes risk $\min_t R_t(\xi)$ is a *maximum* over all ξ. In other words, if nature were malevolent and were forced to pick a ξ which would be told to the statistician, it would pick this ξ (known as a *least favorable a priori law*) with which to confront the statistician in order to make the Bayes risk as large as possible. Intuitively, this ξ should thus be spread broadly over Ω rather than concentrated over a small subset of Ω (*unlike* the last example). This qualitative description is sometimes helpful in guessing ξ's, but of course it does not pin down the choice of the right ξ precisely.

One other shortcoming of using this approach is that least favorable ξ's do not exist in many problems, and the method must therefore often be modified in a manner which makes use of a *sequence* $\{\xi_n\}$ of a priori laws, making the computations much worse. [The details of this modified method can be found in Lehmann's *Theory of Point Estimation*] For example, in the "location parameter problem" of estimating the mean of a $\mathcal{N}(\theta, 1)$ law with $\Omega = D = R^1$ (real numbers) and $W(\theta, d) = (\theta - d)^2$, we may intuitively feel, on the grounds mentioned in the previous paragraph, that a good candidate for ξ is the "uniform distribution on the real line Ω." Unfortunately, no such a priori probability law exists, and we must therefore approximate this nonexistent ξ by the uniform law ξ_n from $-n$ to n (or some similar law) in this kind of example. [The modified method then consists of showing that $\lim_{n\to\infty} [\min_t R_t(\xi_n)] = c$, where c is the constant risk of the procedure t^* to be proved minimax.] When Ω is finite, and in certain other cases [as when Ω is compact in an appropriate sense and W is sufficiently regular], at least one least favorable ξ' always exists, and every minimax procedure is Bayes relative to every such ξ'.

Warning: Although minimax procedures in many settings have constant risk functions, this is not always the case (as illustrated by the Bernoulli Example 4.4). Nor is every procedure with constant risk (e.g., every procedure $b + x$ in the normal example just mentioned) minimax. Nor is the least favorable a priori law necessarily a "uniform" distribution (see Example 4.5).

EXAMPLE 4.5. In the binomial problem, Example 4.2 of Section 4.1, suppose we try to find a Bayes procedure with constant risk. Noting the computations done in that example, it is suggestive that we try a density function for θ which is symmetric about $\theta = 1/2$ and which enables the integration in the formula for $h_\xi(x, d)$ to be carried out easily. If we let f_ξ be a symmetric beta-density (Appendix A)

$$f_\xi(\theta) = \theta^{m-1}(1-\theta)^{m-1}/B(m,m) \quad \text{for} \quad 0 \le \theta \le 1,$$

where $m > 0$, then in the second integral in Example 4.2 of Section 4.1 we merely change the exponents of θ and $1-\theta$ from $\sum x_i$ and $10 - \sum x_i$ to $\sum x_i + m - 1$ and $10 - \sum x_i + m - 1$, and the integral can be evaluated as before, yielding

$$t_m(x) = \frac{B(\sum x_i + m + 1, 10 + m - \sum x_i)}{B(\sum x_i + m, 10 + m - \sum x_i)} = \frac{\sum x_i + m}{10 + 2m}$$

as the Bayes procedure. This procedure has risk function

$$\begin{aligned} r_{t_m}(\theta) = E_\theta(t_m - \theta)^2 &= (10 + 2m)^{-2} E_\theta[(\sum x_i - 10\theta) + m(1 - 2\theta)]^2 \\ &= (10 + 2m)^{-2}[10\theta(1-\theta) + m^2(1-2\theta)^2] \\ &= (10 + 2m)^{-2}[(\theta^2 - \theta)(4m^2 - 10) + m^2]. \end{aligned}$$

Thus, if we let $m = \sqrt{10}/2$, we find that t_m does have constant risk and is Bayes relative to some ξ; this procedure $(\sum x_i + \sqrt{10}/2)/(10 + \sqrt{10})$ is thus minimax. (This is precisely the procedure "t_2" of Problem 1 of Chapter 2 when $n = 10$.)

Some additional examples of the use of Method A will be considered later, e.g., under hypothesis testing.

Our next method makes use of a simple result quite similar to that of the previous theorem:

Theorem. *If* t* *is admissible and has constant risk, then* t* *is minimax.*

PROOF. If t^* has constant risk c but is not minimax, then there is a procedure t' for which $\max_F r_{t'}(F) < c$. Since, then, $r_{t'}(F) < r_{t^*}(F)$ for all F, t' is better than t^*. Hence, t^* could not be admissible.

The use of this theorem in place of the previous one when t^* has constant risk replaces the necessity of proving t^* Bayes relative to some ξ by that of proving t^* admissible. Although the proof that a given procedure is admissible is in general not easy (as we have mentioned before), and although a much-used method for proving a procedure admissible is to prove that it is the unique Bayes procedure relative to some ξ (as mentioned in Section 4.1 and proved in Section 4.4), there are a few cases in which admissibility can be proved directly without using the device of Bayes procedures at all. Our second method (B) of minimax proof, to be presented in Section 4.7, is this direct proof of admissibility of certain procedures with constant risk.

As was discussed in Section 4.4 (and also Section 4.1) in some cases not every Bayes procedure is admissible and in some cases not every admissible procedure is Bayes. Thus, Methods A and B are not interchangeable. Method B is particularly useful in certain problems like the location-parameter problem referred to previously, where the minimax t^* is not Bayes relative to any ξ, and where the use of Method A would involve the messier use of a sequence $\{\xi_n\}$ as indicated previously, which would yield the minimax result but not the admissibility. The disadvantage of Method B is that it can be applied successfully for only a few Ω's and W's.

4.6. Unbiased Estimation

The use of a criterion such as unbiasedness or invariance (to be discussed) in a statistical problem will often simplify the selection of a procedure to be used. For, although there will not be a procedure with uniformly smallest risk function among all procedures, when we restrict our consideration to procedures which satisfy a particular criterion there may be a uniformly best procedure among these. For example, there is no uniformly best procedure among those whose risk functions are shown in the first figure of Chapter 2, but if we restrict our consideration to *unbiased* procedures, we remove t_4 (also t_2 and t_3) from consideration, and t_1 is uniformly best among those which remain (in fact, it is uniformly best among all unbiased procedures, not merely better than t_5, as we shall see later).

Two points should be kept in mind regarding the imposition of a criterion of this kind. First, it can happen that, even when we restrict our attention to those procedures satisfying a given criterion, there is no uniformly best one, but instead there are several "crossing" risk functions, just as there were originally. In such a case, an *additional* criterion would be needed to select that procedure which is to be used.

Second, and most important, we must make sure that the criterion does not lead to the selection of bad procedures. It may be that the only procedures satisfying a restriction imposed by a criterion are inadmissible and have risk functions much larger than that of some other procedure which does not satisfy the criterion! We shall see that this can easily happen in even simple problems.

In fact, one can justifiably ask whether, having accepted the viewpoint that our comparison of procedures is to be based entirely on their risk functions, it is reasonable for us now to impose a criterion such as unbiasedness, which is *not* expressible only in terms of the risk function! The answer is that, on the one hand, we are trying to relate the classical development of statistics which is found in most of the literature of the subject (and wherein unbiasedness is overemphasized) to the modern approach; and that, on the other hand, we will see that there is a wide class of statistical problems for which we can prove that certain unbiased procedures have good properties (e.g., are minimax) from our point of view; thus, it will be worthwhile for us to study how such procedures are obtained.

The term *unbiasedness* occurs in both estimation and hypothesis testing (and, in fact, the notion arises for other statistical problems). We discuss it here in the setting of point estimation (see list (ii) in Chapter 3). Thus, we suppose that D is an interval (possibly infinite) of real numbers, or an analogous higher dimensional set, and that a real or vector property $\phi(F)$ is to be estimated.

An estimator t is said to be an *unbiased* estimator of ϕ if $E_F t(X) = \phi(F)$ for all F in Ω. (If ϕ and t are vectors, this equation means that the expectation of each component of t equals the corresponding component of ϕ, for all F.)

The rationale of proponents of the use of unbiased estimators is that the estimator t which we use should have a probability law which changes with F in such a way that this probability law is always closely concentrated about the value $\phi(F)$, and that this aim can be made manifest by having the *mean* of the probability law of t equal to the value $\phi(F)$. The first half of this reasoning is sound; thus, in Example 2.1(a) with $\phi(\theta) = \theta$, an estimator such as t_1 will have a probability function which is closely concentrated about 1/3 if the true θ is 1/3, about 1/2 if the true θ is 1/2, etc. However, the second half of the reasoning is incorrect; for, as is illustrated in many probability texts, the mean of a probability law can be a very poor indication of where the probability mass is located. Thus, we might have an estimator t of a real parameter ϕ for which $P_F\{t(X) = \phi(F) + 100\} = P_F\{t(X) = \phi(F) - 100\} = 1/2$, so that t is unbiased, whereas for some other estimator t' we have $P_F\{|t'(X) - \phi(F)| < 1\} = 1$ for all F without t' being unbiased. In such a case, t' will, *for every possible sample value of* X, be much closer to $\phi(F)$ than will t, no matter what F may be. If every unbiased estimator of ϕ for this problem were as bad as this t, the insistence on the use of an unbiased estimator would seem unjustifiable. In precise terms, for any reasonable W (e.g., if W is an increasing function of $|\phi(F) - d|$), the procedure t will be inadmissible. This recalls the remark made earlier regarding the questionable logic of imposing a criterion such as unbiasedness when we have already decided to base our considerations only on the risk function.

Practical examples wherein the only unbiased estimators are very poor do occur. Although it is true that there are many practical examples in which there is a *reasonable* unbiased estimator, there is no theoretical guarantee that this will be the case in an arbitrary problem; unbiased estimators, where they are good (e.g., minimax) in terms of the risk function, are only known to be so because of further theoretical investigations in particular problems. The theory of unbiased estimation offers no internal insurance that it will lead to good estimators.

Another, even worse, phenomenon can occur; there may exist *no* unbiased estimators at all! For example, in the setting of Example 2.1(a), but for an arbitrary sample size n (not necessarily 10), let us determine which functions $\phi(\theta)$ possess unbiased estimators. If t is an unbiased estimator of ϕ, we have

$$E_\theta t(X_1, \ldots, X_n) = \sum_{x_1, \ldots, x_n} t(x_1, \ldots, x_n)\theta^{\sum x_i}(1-\theta)^{n-\sum x_i}.$$

For each fixed $x_1, \ldots, x_n$, the summand is a polynomial in θ of degree n, and the sum of such polynomials is again a polynomial of degree $\leq n$. Thus, on the basis of n flips of the coin, ϕ cannot possess an unbiased estimator unless ϕ is a polynomial of degree n or less. Suppose, for example, that we wanted to estimate the odds $\phi(\theta) = \theta/(1-\theta)$ of heads against tails. This ϕ is not a polynomial, so the person who insists on using only unbiased estimators would have to refuse to consider such a problem! This intolerant attitude might cost this person their job. Without being more than intuitive, someone who knows of a reasonably good estimator t of θ in this problem might suggest $t/(1-t)$ as a reasonable estimator of $\theta/(1-\theta)$; whether or not the latter is really good, this person would probably keep his or her job, since most bosses will tolerate incompetence but not insubordination.

For another example, suppose in Example 2.1(a) that we *knew* that $1/3 \leq \theta \leq 2/3$, so that $\Omega = \{\theta : 1/3 \leq \theta \leq 2/3\}$, and we wanted to estimate θ, so that $D = \{d : 1/3 \leq d \leq 2/3\}$. Again, one can prove that there are no unbiased estimators, because of the restriction that t is not allowed to take on all values between 0 and 1, but only values between 1/3 and 2/3. (If 1/3 were replaced by b and 2/3 by c throughout this example, where either $b > 0$ or $c < 1$, the same conclusion would hold.)

The reader should recall the earlier discussion for our reason for studying unbiased estimators despite the criticism just given.

For any estimator t whose mean exists for all F, the function

$$b_t(F) = E_F t(X) - \phi(F)$$

is called the *bias* function of t for estimating ϕ. Thus, t is unbiased if and only if $b_t(F) = 0$ for all F.

In almost all of the discussion of unbiasedness in the literature, the criterion of comparison of estimators is their variance. We shall now see what this means from our viewpoint in terms of the risk function. The meaningfulness of the variance as a measure of goodness depends upon two factors: (i) the use of squared error as a loss function and (ii) the restriction to unbiased estimators. To see this, suppose $W(F, d) = (\phi(F) - d)^2$ in a problem for which we are to estimate the parameter ϕ. For any estimator t (not necessarily unbiased) we then have

$$r_t(F) = E_F[t(X) - \phi(F)]^2 = \operatorname{var}_F(t) + [b_t(F)]^2.$$

Hence, if t is unbiased, we obtain $r_t(F) = \operatorname{var}_F(t)$, which explains the meaning of variance as a measure of goodness for unbiased estimators. Clearly, if the loss function were not squared error, we would not obtain this simple relationship, and if t were not unbiased, the term $b_t(F)$ would not vanish, and the risk would not be variance alone. It is easy to give estimators whose variance is zero for all F: merely let $t'(x) = d_0$ for all x, where d_0 is any fixed element of D. Although $\text{var}_F(t')$ is then 0 for all F, this doesn't mean that t' is a very good estimator, since $r_{t'}(F) = [b_{t'}(F)]^2 = [\phi(F) - d_0]^2$, which may be fairly large for most F (note the estimator t_4 in Example 2.1(a) for a specific example).

However, one should not erroneously argue that some unbiased estimator t will have smallest risk "because by making the estimator unbiased we make the term $[b_t(F)]^2$ zero, which is as small as possible." What is wrong with this argument is that it may only be possible to make $b_t(F)$ zero by making $\text{var}_t(F)$ very large, and that some other estimator t^* may have a much smaller variance than t at the expense of only a slight bias, so that $r_{t^*}(F) < r_t(F)$ for all F. Examples of this will be found in Problems 4.25–4.29.

In two of the next three chapters (Chapters 5 and 7) we will study unbiased estimators in more detail.

4.7. The Method of Maximum Likelihood

The method of maximum likelihood is a method for constructing estimators. Because of this inherent limitation J. Neyman and others developed likelihood ratio methods which would apply to statistical problems of testing hypotheses. In the following brief description we will describe only the method of maximum likelihood. The reader is referred to a book like *Testing Statistical Hypotheses* by E. L. Lehmann for a discussion of likelihood ratio methods. A more thorough discussion of maximum likelihood methods is given in Chapter 7.

The method of maximum likelihood assumes Ω can be parametrized by a real or vector parameter θ and that each F in Ω has a density or probability function $p_{\theta;X}(x)$. We will suppose initially that the elements of D are the possible values of the parameter θ.

Suppose then an experiment has been made and the outcome is observed. For each value of θ we examine the number $p_{\theta;X}(x)$. We "may" find a certain subset D_x of D such that if θ_0 is in D_x then

$$p_{\theta_0;X}(x) = \max_{\theta \text{ in } D} p_{\theta;X}(x).$$

We put quotation marks around "may" because in general we will require that the density or probability functions be continuous functions of θ in order to be able to show that the maximum is assumed. To each outcome x we suppose a certain

$$\hat{\theta}(x) \text{ in } D_x$$

is chosen. This value is called a *maximum likelihood estimate* of θ, and an estimator t such that $t(x) = \hat{\theta}(x)$ for all x is called a *maximum likelihood estimator* of θ.

Suppose Ω should be parametrized differently by a parameter $\phi = h(\theta)$. We will show that

$$\hat{\phi}(x) = h(\hat{\theta}(x)).$$

That is, if Ω is reparametrized, then the method of maximum likelihood gives the corresponding estimator of the new parameter.

The function h must have the property that if

$$\phi_1 = h(\theta_1), \qquad \phi_2 = h(\theta_2), \quad \text{and} \quad \phi_1 = \phi_2$$

$$\text{then } \theta_1 = \theta_2.$$

(For if this were not true then df's F_1 and F_2 that were distinguished by the parameter θ would not be distinguished by the parameter ϕ.) Therefore there is a function h^{-1} such that

$$\theta = h^{-1}(\phi).$$

Consequently using the new parameter ϕ the density functions are

$$p_{h^{-1}(\phi);X}(x).$$

If x is observed after an experiment then

$$h^{-1}(\hat{\phi}(x)) = \hat{\theta}(x)$$

or

$$\hat{\phi}(x) = h(\hat{\theta}(x))$$

is a maximizing value of the parameter.

The method of maximum likelihood has been used extensively. Most of the "classical" estimation procedures in multivariate analysis can be obtained by using the method of maximum likelihood. A book like *Introduction to Multivariate Statistical Analysis* by T. W. Anderson is very careful to point out those results that can be obtained by using maximum likelihood or likelihood ratio methods. The extensive use of the method is often justified by various attractive features. We have already noted that the estimator obtained is really an estimator of F in Ω rather than the parameter θ used to describe F. If there is a sufficient statistic the maximum likelihood estimator is necessarily a function of the sufficient statistic. This is discussed in Chapter 7. And it is further argued that $\hat{\theta}(x)$ picks a density function $p_{\theta;X}(x)$ which makes the observed outcome most likely.

From the viewpoint of decision theory, the viewpoint of this book, one fault of the method of maximum likelihood is that in most problems where applicable the method uniquely specifies a statistical procedure as *the* procedure to use without any consideration of the way loss is to be measured. Thus, sometimes the maximum likelihood estimator is good, and sometimes it is very poor, from the viewpoint of decision theory. For instance, in the problem of Example 4.4 the maximum likelihood estimator puts on a poor show: among all estimators $t(x)$ for which $1/3 \le t(0) \le 1/2 \le t(1) \le 2/3$, the ml estimator $\hat{\theta}(x)$ has uniformly largest square error.

Aside from considerations of risk the method of maximum likelihod is restricted in its applications to statistical problems of estimation in which the parametric description of Ω is "nice" enough to ensure that $\hat{\theta}(x)$ is well defined. This limited applicability is not a characteristic of the first six criteria listed at the beginning of this chapter.

4.8. Sample Functionals: The Method of Moments

The method of moments for deriving estimators is considered to be one of the oldest general methods in use. It was introduced by K. Pearson around the turn of the century. Let us suppose $X_1, \ldots, X_n$ are independent and identically distributed; that X_1, the outcome of an experiment, is real-valued; and that certain moments

$$\mu_1' = EX_1, \ldots, \mu_k' = EX_1^k$$

exist. Let $m_1' = \sum_{i=1}^{n} x_i/n, \ldots, m_k' = \sum_{i=1}^{n} x_i^k/n$ be the corresponding sample moments. The *method of moments* uses the following rule. Suppose $g(\mu_1', \ldots, \mu_k')$ is some function of the first k moments of X_1. Use the estimator $g(m_1', \ldots, m_k')$ to estimate the value $g(\mu_1', \ldots, \mu_k')$. Thus, according to the method of moments one should use the sample variance to estimate $\operatorname{var} X_1$. In the example discussed in Section 4.6, the person who decides to use $t/(1-t)$ as an estimator of $\theta/(1-\theta)$ is using the method of moments, if $t = \sum_{i=1}^{n} x_i/n$.

The chief criticism of the method is that, like the method of maximum likelihood, as a rule of action it makes no reference to the way loss is measured. Consequently, one would expect to find examples (analogous to the example in Section 4.7) in which from the viewpoint of risk the method of moments gives a uniformly bad result.

Using asymptotic theory discussed in Chapter 7, one can sometimes introduce a measure of asymptotic efficiency to compare the accuracy of estimators obtained by various methods. Such measures of asymptotic efficiency apply to sequences of estimators rather than to individual estimators, where the nth estimator uses a sample of n observations. We shall see that, in many applications, the sequence of estimators given by the method of moments is not asymptotically efficient, when compared with the sequence of estimators given by the method of maximum likelihood. The latter estimators will at least turn out to be fairly efficient for large sample sizes in many problems, although we shall see in examples that they may be very bad for small sample sizes.

The method of moments is a special case of the following idea: If we have n real observations $X_1, \ldots, X_n$ with common df F, the values $X_1, \ldots, X_n$ taken on by these X_i may be summarized in terms of the *empiric frequency function* $\bar{p}_n$ or *empiric df* $\bar{F}_n$, defined by

$$\bar{p}_n(x; X_1, \ldots, X_n) = \frac{1}{n} \cdot (\text{number of } X_i \text{ equal to } x),$$

$$\bar{F}_n(x; X_1, \ldots, X_n) = \sum_{t \leq x} \bar{p}_n(t; X_1, \ldots, X_n)$$

$$= \frac{1}{n} \cdot (\text{number of } X_i \leq x).$$

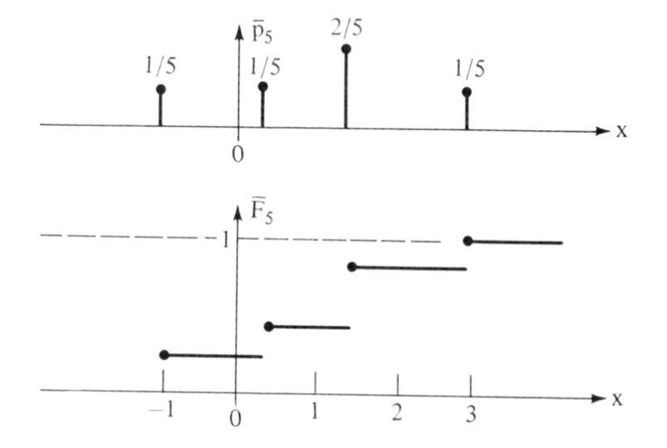

Thus, $\bar{p}_n$ and $\bar{F}_n$ are chance *functions*. Once the X_i are observed, $\bar{p}_n$ and $\bar{F}_n$ take on as "values" functions which are, respectively, an ordinary discrete probability density function (pdf) and the corresponding df. The diagram shows $\bar{F}_n$ and $\bar{p}_n$ for $n = 5$, $X_1 = .3$, $X_2 = X_4 = 1.3$, $X_3 = -1$, $X_5 = 3$. Intuitively, one feels that it is likely that the empiric df $\bar{F}_n$ is close to the true (but unknown) F when n is large, and the *Glivenko-Cantelli theorem* of probability theory asserts this to be correct. Hence, we might use $\bar{F}_n$ to estimate F, and indeed this is often done in the case of nonparametric $\Omega = \{\text{all univariate df's}\}$. If we wish to estimate some *property* of F defined by some functional ψ on Ω, we might use $\psi(\bar{F}_n)$ to estimate $\psi(F)$. For, if ψ is "continuous" in an appropriate sense, then if $\bar{F}_n$ is close to F, $\psi(\bar{F}_n)$ is close to $\psi(F)$. One of the most familiar properties, $\psi(F) =$ first moment of F, would thus be estimated by $\psi(\bar{F}_n) =$ first moment of $\bar{F}_n$ (or $\bar{p}_n$) $= \sum_x x\bar{p}_n(x)$, and since $\bar{p}_n$ assigns mass $\frac{1}{n}$ to x for each X_i taking on the value x, we have $\sum_x x\bar{p}_n(x) = \sum_i \frac{1}{n} X_i = \bar{X}_n$. Similarly, the kth moment of F is estimated by $\sum_i \frac{1}{n} X_i^k$ in this approach. We see that this is exactly the method of moments when $g(\mu_1', \ldots, \mu_k') = \mu_k'$. If $\psi(F) =$ median of F, this approach estimates $\psi(F)$ by $\psi(\bar{F}_n) =$ *sample median*, defined to be any value t

such that $\bar{F}_n(t-) \leq 1/2$ and $\bar{F}_n(t) \geq 1/2$; if n is odd, this value t is the $(n+1)/2^{\text{th}}$ smallest observation, $\frac{n-1}{2}$ observations being larger, and $\frac{n-1}{2}$ being smaller, than this one, if all X_i are different; if n is even, any value between the $\frac{n}{2}^{\text{th}}$ smallest and $\frac{n+2}{2}^{\text{th}}$ smallest can be termed a *sample median*, and the average of the two extreme possibilities is generally used.

The shortcoming of this approach in general, as in the particular case of the sample mean, is that the intuitive motivation ignores the possibility that an estimator of $\psi(F)$ obtained by a different approach may be better.

4.9. Other Criteria

Certain of the criteria listed at the beginning of this chapter require very extensive discussions. These discussions are deferred to later chapters. Thus, Chapter 6 is devoted to a discussion of sufficiency, and Chapter 7 contains discussions of completeness and the application of sufficiency and completeness to problems requiring unbiased estimation, as well as discussions of invariance and asymptotic theory.

Problems

The problems of Chapter 4 are divided into three parts, covering Sections 4.1–4.2, 4.3–4.4, and 4.5–4.8.

I. Problems on Sections 4.1 *and* 4.2. Suggested: As a minimum, work 4.1 or 4.2, 4.4 under the assumption of (c), 4.5(b), one of 4.7, 4.8, 4.9.

4.1. A coin which has probability 1/3 or probability 1/2 of coming up heads (no other values are possible) is flipped once. X is the number of heads obtained on that flip. $D = \{d_0, d_1\}$, where d_i is the decision "my guess is that the coin has probability $1/(2+i)$ of coming up heads." The loss is 1 for an incorrect decision, 0 for a correct decision.
(a) Specify S, Ω, D, W.
(b) There are four possible nonrandomized procedures:

$$t_1(0) = t_1(1) = d_0;$$

$$t_2(0) = t_2(1) = d_1;$$

$$t_3(0) = d_1, \qquad t_3(1) = d_0;$$

$$t_4(0) = d_0, \qquad t_4(1) = d_1.$$

Show that $r_t(F) = P_F\{t \text{ reaches wrong decision}\}$, and use this to find the risk function of each procedure. (There are only two possible values of the argument F.)

(c) Use the results of (b) to determine which of the four procedures among the nonrandomized procedures is (or are)
  (i) admissible;
  (ii) Bayes with respect to the a priori law $p_\xi(1/3) = .1 = 1 - p_\xi(1/2)$;
  (iii) Bayes with respect to the a priori law $p_\xi(1/3) = 3/5 = 1 - p_\xi(1/2)$;
  (iv) minimax.

(d) Is the procedure which was Bayes in (c) (ii) also Bayes relative to any other prior laws? If so, which?

[*Note:* The preceding problem and the next problem illustrate computation of Bayes procedures from risk functions. In settings in which there are many possible t's and F's, it would be hopeless to try to compute every r_t and thus, for a given ξ, every $R_t(\xi)$, for the sake of finding a Bayes procedure. The method of Chapter 4 yields the Bayes procedures relative to a given ξ directly without computing all r_t's. Nevertheless, it is helpful to understand the subject in terms of the risk functions.]

4.2. A coin which has probability 1/2 or probability 1/5 of coming up heads (no other values are possible) is flipped once. X is the number of heads obtained on that flip. $D = \{d_0, d_1\}$, where d_i is the decision "my guess is that the coin has probability $p_i \overset{\text{def}}{=} 1/(3i + 2)$ of coming up heads." The loss is $(j + 1)$ for making decision d_j when it is incorrect (one decision, if incorrect, is twice as serious an error as the other) and is 0 if d_j is correct.

(a) Specify S, Ω, D, W.

(b) There are four possible nonrandomized procedures:

$$t_1(0) = t_1(1) = d_0;$$

$$t_2(0) = t_2(1) = d_1;$$

$$t_3(0) = d_1, \qquad t_3(1) = d_0;$$

$$t_4(0) = d_0, \qquad t_4(1) = d_1.$$

Show that $r_t(p_i) = (2 - i)P_{p_i}\{t \text{ reaches wrong decision}\}$, and use this to find the risk function of each procedure. (There are only two possible values of the argument p_i, so r_t can be thought of as a 2-vector for each t.)

(c) Use the results of (b) to determine *from the risk functions* which of the four procedures among the nonrandomized procedures is (or are)
  (i) admissible;
  (ii) Bayes with respect to the a priori law $p_\xi(1/5) = .9 = 1 - p_\xi(1/2)$;
  (iii) Bayes with respect to the a priori law $p_\xi(1/5) = 5/9 = 1 - p_\xi(1/2)$;
  (iv) minimax;
  (v) minimax but inadmissible.

(d) Is the procedure which was Bayes in (c) (ii) also Bayes relative to any other prior laws? If so, which?

4.3. *Testing between simple hypotheses:* Suppose S is discrete; Ω consists of two possible probability functions of X, say f_0 and f_1; D consists of two elements d_0 and d_1; and

$$W(F_i, d_j) = \begin{cases} w_i & \text{if} \quad i \neq j, \\ 0 & \text{if} \quad i = j, \end{cases}$$

where w_0 and w_1 are given positive numbers. Show that, for any a priori law ξ for which $0 \le p_\xi(f_i) \le 1$, all (nonrandomized) Bayes procedures are of the form

$$(*) \qquad t(x) = \begin{Bmatrix} d_1 \\ d_1 \text{ or } d_0 \\ d_0 \end{Bmatrix} \quad \text{according as } \frac{f_1(x)}{f_0(x)} \begin{Bmatrix} > \\ = \\ < \end{Bmatrix} C,$$

where C is a constant (perhaps 0 or $+\infty$) depending on the w_i's and ξ, but not on x. [*Remark:* In the classical development of hypothesis testing, one specifies a value $\alpha (0 \le \alpha \le 1)$ and then, restricting consideration to tests for which the "size" or "significance level" $P_{f_0}\{t(X) = d_1\}$ is $\le \alpha$, tries to find a test t which *maximizes* the "power" $P_{f_1}\{t(X) = d_1\}$. When $w_0 = w_1 = 1$, this means, in our language, restricting consideration to procedures t with $r_t(f_0) \le \alpha$, and among them seeking any one which *minimizes* $r_t(f_1) = P_{f_1}\{t(X) = d_0\} = 1 -$ power of t. The *Neyman-Pearson* (hereafter N–P) *lemma*, which we shall consider later, asserts essentially that such a test is of the form (*), with C depending on α. This may not seem very surprising, and in connection with Problem 4.17 we will discuss the geometric relationship between the two results. Keep in mind, however, the difference between the statements of goals in the Bayesian and Neyman-Pearson approaches: In the former we state a ξ and minimize $R_t(\xi)$; in the latter we state an α and maximize the power subject to the size of the test being $\le \alpha$. If one varied ξ, one would obtain the Neyman-Pearson tests for various α's, as will be seen in Problem 4.17.]

4.4. [This problem may be simplified by working parts (a) and (b) only under the condition that is stated in part (c).] *A simple* k-*decision problem* (*with* k *urns*). One of k urns is chosen according to known prior probabilities, $p_\xi(i)$ being the probability of choosing urn $i (1 \le i \le k)$. You are not shown which urn was chosen. There are k decisions, d_j meaning "I guess that urn j was chosen." You must make such a guess, and you lose an amount $-m_i < 0$ (that is, you win m_i) if you make a correct guess d_i when urn i was chosen; otherwise you lose 0. (Unless you are playing this game with some fool or relative, you've previously paid some entrance fee for the privilege of making this guess.)

(a) "*No data problem.*" With the preceding setup, show that you maximize your expected gain by using the following rule:

(*) Choose any decision d_{i_0} for which i_0 maximizes $p_\xi(i) m_i$.

For definiteness in comparing this rule with (b), think of taking i_0 as the smallest value which satisfies (*), if more than one choice is optimum.

(b) *Problem with data.* Now suppose that, once the urn is chosen according to p_ξ, a ball is drawn from it, and a discrete characteristic X of the ball is told to you. The probabilities $p_i(x)$, that a ball chosen from urn i yields $X = x$, are known to you. Let $p_\xi(i|x)$ denote the posterior probability that urn i was chosen, given that $X = x$. Show that a Bayes procedure is obtained by the method of Section 4.1.

$$(**) \qquad t(x) = d_{i_0(x)} \quad \text{where} \quad i_0(x) = \text{smallest } i \text{ maximizing } p_\xi(i|x) m_i.$$

[A different $i_0(x)$ that maximizes $p_\xi(i|x) m_i$ may be chosen instead, if more than one choice maximizes, but you should realize that the commonly employed phrase "choose *any* $d_{i_0(x)}$ where $i_0(x)$ maximizes" does not describe a specific procedure, but rather a *class* of procedures.]

Important: The preceding (a) and (b) illustrate a general phenomenon in problems of decision theory stated in Section 4.1. Suppose one first solves the simple "no data" problem in terms of a partition of all possible prior laws ξ into subsets $\{C_d : d \in D\}$ such that, if the given prior law ξ_0 is in C_{d_0}, decision d_0 is optimum. Then a Bayes procedure when data $X = x$ is observed can be obtained by computing the posterior law $\xi^{(x)}$ (say) *and using the prescription of the previous sentence* (same C_d's!) *with* ξ_0 *replaced by* $\xi^{(x)}$. This "updating of ξ_0 in terms of the observation X" is *intuitively* a reasonable thing to do; we have shown that it is also a *correct* procedure of optimization.

(c) If all m_i are equal, to what simple verbal descriptions do the rules of (*) and (**) reduce in terms of "most probable urn"?

(d) If all of the losses w_{ij} (from making decision j when urn i was the chosen one), $i \neq j$, were not equal to the same value 0 as assumed previously, would the same simple description of (*) and (**) have been obtained? [In any event, the *principle* stated at the end of (b) is still valid.]

4.5. Suppose Ω can be labeled according to the values of a real parameter θ, where $a \leq \theta \leq b$. The decisions are $D = \{d : a \leq d \leq b\}$, representing guesses as to the true value of θ. The loss function is $|\theta - d|^r$, where r is a given positive value. The prior density on Ω is f_ξ. Assume all f_θ's or p_θ's are positive throughout S.

(a) Using the principle stated at the end of Problem 4.4(b), show that, if $f_\xi(\theta|x)$ is the posterior density of θ given that $X = x$, then a Bayes procedure is obtained by choosing $t(x) = d'$ to minimize $\int_a^b |\theta - d'|^r f_\xi(\theta|x)\,d\theta$. [If it helps, work only for discrete S. In any event, do *not* try to find a formula for the minimizing d' in this part (a).]

(b) In particular, for "squared error loss" ($r = 2$), show from (a) that the unique Bayes procedure is $t(x) =$ mean of posterior law of θ.

(c) In part (b), suppose the loss is squared *relative* error, $\left(1 - \dfrac{d}{\theta}\right)^2$, instead of squared error. Let $E\{\ |x\}$ denote (conditional) expectation with respect to the posterior law, given that $X = x$. Assume $a \geq 0$, and $a > 0$ in the discrete case. Show that the unique Bayes estimator is $E\{\theta^{-1}|x\}/E\{\theta^{-2}|x\}$, assuming that these expectations are always finite.

(d) For $r = 1$ ("absolute error loss"), show that a Bayes procedure is obtained as any *median* (not necessarily unique!) of the posterior law. Since the crucial result from probability theory used in demonstrating this may be unfamiliar, part of this problem is to prove it:

If g *is a univariate density function with finite first moment,* $\int_{-\infty}^{\infty} |\theta - c|\, g(\theta)\, d\theta$ *is minimized if and only if* c *is a median of* g. [*Hint:* If m is a median and c is not a median, with $c > m$, show (draw it!) that
$\{|\theta - c| - |\theta - m|\} - (c - m)\,\text{sgn}(m - \theta) \geq 0$,

where $\text{sgn}\, u = \begin{cases} 1 & \text{if} \quad u > 0 \\ 0 & \text{if} \quad u = 0 \\ -1 & \text{if} \quad u < 0 \end{cases}$. Moreover, the "$\geq 0$" is "$> 0$" if $m < \theta < c$.

This and $\int_{-\infty}^{\infty} \text{sgn}(m - \theta) g(\theta)\, d\theta = 0$ yield the result.]

4.6. For arbitrary $\Omega = \{F\}$ in the absolutely continuous case where the real random variable X has density f_F if F is true, suppose it is desired to estimate some bounded real functional ψ on Ω (e.g., $\psi(F) = E_F X$, assumed finite for all F), with loss function $w(F)(\psi(F) - d)^2$, where $0 < w(F) \leq B < \infty$. If ξ is any discrete prior law, show that each Bayes estimator relative to ξ is given by

$$(*) \qquad t_\xi^*(x) = \sum_F w(F)\psi(F)f_F(x)p_\xi(F) \Big/ \sum_F w(F)f_F(x)p_\xi(F)$$

for every x (except possibly on a set of probability 0 for every F) for which the denominator of $(*)$ is positive, with an arbitrary definition otherwise (nonunique Bayes procedure). Specialize to the cases $w(F) \equiv 1$ and $w(F) = 1/\psi^2(F)$ treated (for $\psi(\theta) = \theta$) in Problem 4.5(b), (c).

4.7. Show that the procedure $t_{3,n}$ of Problem 2.1 is Bayes relative to the uniform prior density on $[0, 1]$. [Use the conclusion of Problem 4.5(b), if you wish.]

4.8. Show that the procedure $t_{2,n}$ of Problem 2.2 is Bayes relative to the prior density on $(0, \infty)$

$$f_\xi(\lambda) = Ce^{-\lambda(n/2)^{1/2}}\lambda^{(n/2)^{1/2}-1}(1 + \lambda^2),$$

where C is a suitable constant. [You need *not* determine C; it suffices to verify that the given function of λ has finite integral, so that one knows such a C exists. Then modify slightly your proof of 4.5(b), or else use the fact that the formula $(*)$ of 4.6 holds in the present setting with F, $\psi(F)$, $w(F)$, $f_F(x)$, $p_\xi(F)$, $\sum_F$ replaced, respectively, by λ, λ, $(1 + \lambda^2)^{-1}$, $p_\lambda(x)$, $f_\xi(\lambda)$, $\int d\lambda$, which, you should check, are appropriate analogues.]

4.9. Show that the procedures $t_{2,n}$ and $t_{3,n}$ of Problem 2.3 are, respectively, Bayes relative to the prior densities $f_{\xi_1}(\theta) = C_2\phi(\theta - 1)(1 + \theta^2)$ and $f_{\xi_2}(\theta) = C_3\phi(\theta)(1 + \theta^2)$ where ϕ is the standard $\mathcal{N}(0, 1)$ density and the C_i are suitable constants. [You need *not* determine the C_i; it suffices to verify that the given functions of θ have finite integrals, so that one knows such C_i exist. Then modify slightly your proof of Problem 4.5(b), or else use the fact that formula $(*)$ of Problem 4.6 holds in the present setting with F, $\psi(F)$, $w(F)$, $p_\xi(F)$, $\sum_F$ replaced by θ, θ, $(1 + \theta^2)^{-1}$, $f_\xi(\theta)$, $\int d\theta$, which, you should check, are appropriate analogues.]

II. Problems on Sections 4.3 *and* 4.4. Suggested: At least 4.10 or 4.11, and 4.12, 4.15 or 4.13, 4.16; at least read 4.17.

4.10. In Problem 4.1, plot the risk points of the four procedures of part (b) and also of all *randomized* procedures. Answer parts (c) and (d) for the class of randomized procedures. Note any changes from the earlier results, especially in (c) (iv).

4.11. Work Problem 4.10 with Problem 4.1 replaced by Problem 4.2.

4.12. A positive rv X is known to have continuous case density either

$$f_0(x) = \begin{cases} e^{-x} & \text{if } x > 0, \\ 0 & \text{otherwise,} \end{cases}$$

or else

$$f_1(x) = \begin{cases} \frac{1}{2}e^{-(x-1)/2} & \text{if } x > 1, \\ 0 & \text{otherwise,} \end{cases}$$

Here $S = \{x : x > 0\}$; $D = \{d_0, d_1\}$ where d_i means "my guess is that f_i is the true density"; and the loss is one for an incorrect decision and zero for a correct decision. If you do not know it already, convince yourself that $r_t(f) = P_f\{t \text{ is incorrect}\}$ with this W.

[*Note:* In this problem the F_i are continuous, and hence it can be proved (and can be assumed by you) that the set $\mathscr{R}$ of risk points of all randomized procedures coincides with that of all nonrandomized procedures. (Note the difference from Problems 4.1, 4.10 or 4.2, 4.11.) Hence, we will restrict attention to the latter. This problem is devoted to calculating the "lower left-hand boundary" of $\mathscr{R}$, i.e., the Bayes procedures, in a computationally slightly more difficult setting than that of Problem 4.1 or 4.2. For each value γ, $0 \le \gamma \le 1$, you will compute the Bayes procedures relative to the prior law $p_\xi(f_0) = \gamma = 1 - p_\xi(f_1)$.]

(a) If $0 < \gamma < 1$, show that the essentially unique Bayes procedure t_{q_γ} (say) is $t_{q_\gamma}(x) = \begin{Bmatrix} d_1 \\ d_0 \end{Bmatrix}$ if $x \begin{Bmatrix} > \\ \le \end{Bmatrix} q_\gamma$, where $q_\gamma = \max\left(1, 2\log\frac{2\gamma}{1-\gamma} - 1\right)$. Compute $r_{t_q}(f_i)$ for $1 \le q < \infty$. Show that the risk points $(r_{t_q}(f_0), r_{t_q}(f_1))$, $1 \le q < \infty$, can be expressed as the curve $(u, 1 - e^{1/2}u^{1/2})$, $0 < u \le e^{-1}$. [*Lesson from the preceding:* Computation of the *risk function* of a Bayes procedure is an additional calculation and does not fall out of the simpler computation of the Bayes procedure itself.]

(b) When $\gamma = 1$, show that the essentially unique Bayes procedure is $t^*(x) \equiv d_0$. Find its risk point.

(c) When $\gamma = 0$, show that there are many possible Bayes procedures, all of which prescribe $t(x) = d_1$ for $x > 1$, but which can have any form at all when $x \le 1$. Show that the set of risk points of all these Bayes procedures constitute the line segment $\{(u, 0) : e^{-1} \le u \le 1\}$. Essentially only one of these procedures is admissible. Which is it? Is this last procedure also Bayes relative to any other prior laws? (See part (a).)

(d) Use parts (a)–(c) to plot the set of all risk points of all Bayes procedures. Which procedures are admissible? Use (a) to show that the essentially unique minimax procedure is $t_{1.503}$.

4.13. A positive rv X is known to have pdf either

$$f_0(x) = \begin{cases} 2x & \text{if } 0 \le x \le 1, \\ 0 & \text{otherwise,} \end{cases}$$

or else

$$f_1(x) \equiv \begin{cases} 3x^2/8 & \text{if } 0 \le x \le 2, \\ 0 & \text{otherwise.} \end{cases}$$

Here $S = \{x : 0 \le x \le 2\}$, $D = \{d_0, d_1\}$ where d_i means "my guess is that f_i is the true density," and the loss is one for an incorrect decision and zero for a correct decision. If you do not know it already, convince yourself that $r_t(f) = P_f\{t \text{ is incorrect}\}$ with this W. [The *note* just before (a) of Problem 4.12 also applies here.]

(a) If $0 \le \gamma \le 3/19$, show that the essentially unique Bayes procedure t_γ (say) is $t_\gamma(x) = \begin{Bmatrix} d_1 \\ d_0 \end{Bmatrix}$ if $x \begin{Bmatrix} > \\ \le \end{Bmatrix} 16\gamma/3(1-\gamma)$. Compute $r_{t_\gamma}(f_i)$ for $0 \le \gamma \le 3/19$. Putting $\tau = 16\gamma/3(1-\gamma)$, show that the risk points $(r_{t_\gamma}(f_0), r_{t_\gamma}(f_1))$; $0 \le \gamma \le 3/19$ can be expressed parametrically as the curve $(1 - \tau^2, \tau^3/8)$, $0 \le \tau \le 1$. [*Lesson* as in Problem 4.12(a).]

(b) Show that the same procedure $t_{3/19}$ is the essentially unique Bayes procedure relative to every prior law $(\gamma, 1 - \gamma)$ for $3/19 \le \gamma < 1$.

(c) When $\gamma = 1$, show that there are many possible Bayes procedures, all of which prescribe $t(x) = d_0$ for $x \le 1$, but which can have any form at all when $x > 1$. Show that the risk points of all these Bayes procedures constitute the line segment $\{(u, 0) : 1/8 \le u \le 1\}$. Essentially only one of these procedures is admissible. Which is it? Is this last procedure also Bayes relative to any other prior laws? (See part (b).)

(d) Use parts (a)–(c) to plot the set of all risk points of all Bayes procedures. Which procedures are admissible? Use (a) to show that the essentially unique minimax procedure is $t_{.151}$, and find its risk point.

4.14. The analysis in Problems 4.12–4.13 gives in each case the risk points of all procedures of real interest. However, in these 2-state, 2-decision problems it is not hard to compute the full set $\mathscr{R}$, once one knows the Bayes procedures. To see this, define, for each procedure t', a related procedure t'', by

$$t''(x) = \begin{Bmatrix} d_0 \\ d_1 \end{Bmatrix} \quad \text{whenever } t'(x) = \begin{Bmatrix} d_1 \\ d_0 \end{Bmatrix}.$$

Show that $r_{t''} = 1 - r_{t'}$. Thus, a risk point (a, b) is in $\mathscr{R}$ if and only if $(1 - a, 1 - b)$ is in $\mathscr{R}$. From this, draw the "upper right-hand boundary" of $\mathscr{R}$ in either problem.

4.15. *The Neyman-Pearson* approach to 'hypothesis testing" (i.e., to a 2-decision, 2-state setting like that of Problem 4.12 or 4.13) chooses a procedure ("test" in this setting) as follows: One specifies a value $\alpha(0 \le \alpha \le 1)$ and then, restricting consideration to tests for which the "size" or "significance level" $P_{f_0}\{t(X) = d_1\}$ is $\le \alpha$, tries to find a test t which *maximizes* the "power" $P_{f_1}\{t(X) = d_1\}$. This means, in our language, putting the losses $= 0$ or 1 as in Problems 4.12–4.13, restricting consideration to procedures t with $r_t(f_0) \le \alpha$, and among them seeking any one (hereafter called a Neyman-Pearson procedure) which *minimizes* $r_t(f_1) = P_{f_1}\{t(X) = d_0\} = 1 -$ power of t.

(a) In the setting of Problem 4.12 use the picture of 4.12(d) to show that if $\alpha \le e^{-1}$ the Neyman-Pearson (N-P) formulation yields a unique test, and that it is admissible.

(b) If $e^{-1} < \alpha \le 1$, show that there are several NP tests t of different levels, all having $r_t(f_1) = 0$ ("power $= 1$"), but that only one of these is admissible. Thus, show that the specification "find a test of level 1/2 with minimum $r_t(f_1)$" instead of one "of level $\le 1/2$" would have led to an inadmissible procedure. [*Note:* Although the Bayes and N-P approaches are quite different, note that the admissible N-P tests corresponding to *all* α's coincide with the admissible Bayes tests corresponding to *all* ξ's.]

4.16. See the introductory paragraph of Problem 4.15, before (a). Then proceed:

(a) In the setting of Problem 4.13, use the picture of 4.13(d) to show that if $0 \le \alpha \le 1$ the preceding Neyman-Pearson (N-P) formulation yields a unique test, and that it is admissible. Note that the procedure obtained has level exactly α.

(b) To see that a conclusion like that of (a) requires care in its formulation in some problems, suppose that the roles of f_0 and f_1 in the N-P approach are interchanged; thus, subject to $P_{f_1}\{t(X) = d_0\} \leq \alpha$, we want a test that minimizes $P_{f_0}\{t(X) = d_1\}$. Looking at the same diagram of 4.13(d), show that, if $0 \leq \alpha \leq 1/8$, the conclusion of (a) again holds. However, if $1/8 < \alpha \leq 1$, show that there are several N-P tests t of different levels (all $\leq \alpha$) and all having $r_t(f_0) = 0$ ("power = 1"), but that only one of these is admissible. Thus, show that the specification to find a test "of level *exactly* 1/2 with minimum $r_t(f_0)$" instead of one "of level $\leq 1/2$..." would have led to an inadmissible procedure. [*Note:* Although the Bayes and N-P approaches (criteria for choosing the procedure to be used) are quite different, you can see that the class of admissible N-P tests corresponding to *all* α's coincides with the class of admissible Bayes tests corresponding to *all* ξ's.]

4.17. An analysis like that of Problem 4.15 or 4.16 can be carried out for any f_0, f_1 and shows that *any N-P procedure is a Bayes procedure relative to some* ξ. From this and the form of a Bayes procedure, show that

For each α, $0 < \alpha < 1$, there is a procedure which minimizes $r_t(f_1)$ among $\{t : r_t(f_0) \leq \alpha\}$, and each such procedure has the form

$$(*) \qquad t(x) = \begin{Bmatrix} d_1 \\ d_0 \end{Bmatrix} \quad \text{if} \quad f_1(x) \begin{Bmatrix} > \\ < \end{Bmatrix} C_\alpha f_0(x),$$

where C_α is a nonnegative constant.

This is the *Neyman-Pearson lemma.* Note that the behavior of t on $J = \{x : f_1(x) = C_\alpha f_0(x)\}$ is not specified by (*); if that set J has positive probability under f_0, any subset H of that set J for which $P_{f_0}\{H \cup \{x : f_1(x) > C_\alpha f_0(x)\}\} = \alpha$ may be chosen as the subset of J where d_1 is made. In the discrete case, randomization may be needed. More of this in Chapter 8.

4.18. *Equivalence of two methods of randomization* (Section 4.3):

(a) Prove the assertion that the function $\sum_i \pi_i \delta_i(x, d)$ (end of Section 4.3) is a "general randomization" procedure which possesses the properties stated there. [You can also try to prove this for a randomization among nonrandomized procedures $\{t_\alpha, \alpha \in A\}$ where A is no longer finite, and the randomization chooses an α in A according to some probability measure π on A; also, D can be infinite. Full treatment then involves some "measurability" considerations, but you might try it for D discrete and π in the absolutely continuous case on $A = [0, 1]$.]

(b) (More difficult). Now suppose a general randomization procedure δ is given. We want to define A, $\{t_\alpha, \alpha \in A\}$, and π so that special randomization among the t_α according to π on A yields, for every d_j in discrete D, the same probability $\delta(x, d_j)$ as did δ, of making decision d_j when $X = x$. We also assume S is discrete. [The proof involves slightly more measure theory for general S and D.] In the product space $S \times [0, 1)$ of elements (x, α), $0 \leq \alpha < 1$, define $D_j = \left\{(x, \alpha) : \sum_{i=0}^{j-1} \delta(x, d_i) \leq \alpha < \sum_{i=0}^{j} \delta(x, d_i)\right\}$, where the first sum is 0 if

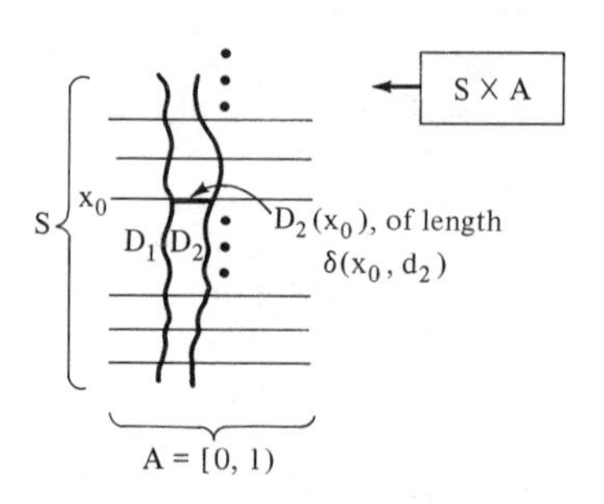

$j = 1$, and $D = \{d_1, d_2, \ldots\}$. We take π to be uniform probability measure ("fair spinner") on $A = [0, 1)$. Note that the horizontal "section" $D_j(x_0) = \{(x, \alpha) : (x, \alpha) \in D_j, x = x_0\}$ in the diagram has length $\delta(x_0, d_j)$, so that the given δ can be implemented by partitioning A into the sets $D_1(x_0), D_2(x_0), \ldots$ when $X = x_0$, and using a fair spinner (uniform measure) on the interval $\{x_0\} \times [0, 1) = \{x_0\} \times \bigcup_j D_j(x_0)$ to choose a decision. Now define $t_\alpha(x) = d_j \Leftrightarrow (x, \alpha) \in D_j$. Verify that the aim of the second sentence of this part (b) is attained.

III. Problems on Sections 4.5–4.8. Suggested: at least one of 4.19–4.21 and one of 4.25–4.29.

Note on minimax problems: In such small problems (2 points in Ω) as 4.12(d) and 4.13(d), little looking was needed to find a minimax procedure. In 4.19, 4.20, and 4.21, where Ω is infinite, a nice analytic *class* of prior laws, relative to which the Bayes procedure is easy to compute, is tried, and it is then found that a member of that class yields a Bayes procedure to which the tool of Section 4.5 can be applied to conclude minimaxity. You should keep in mind that such success is not automatic, and that less fortuitous examples than those chosen for these examples could yield a lack of success on several such efforts, and perhaps only a machine computation of an approximately minimax procedure might be feasible. Problems 4.21–4.22 illustrate some minimax phenomena that can occur when Ω is finite and the possible density functions are not all positive on the same set. Problem 4.23(b) shows the type of development one might have to go through to guess a least favorable ξ and obtain a minimax procedure in a setting that is not so "nice" as those of Problems 4.19, 4.20, and 4.21.

4.19. (a) In the setting of Problems 2.1 and 4.7, find a symmetric beta-prior density $f_\xi(\theta) = \text{const.} \times [\theta(1-\theta)]^{\alpha-1}$ for which (by choice of α) the Bayes estimator for sample size n is $t_{2,n}$. Conclude that $t_{2,n}$ is minimax.
(b) Compare $nr_{t_{1,n}}$ and $nr_{t_{2,n}}$ in terms of "subminimaxity."

4.20. (a) In the setting of Problems 2.2 and 4.8, show that β and c can be chosen in the prior density $f_\xi(\lambda) = \text{const.} \times (1 + \lambda^2)\lambda^{\beta-1}e^{-c\lambda}$ so that $t_{2,n}$ is Bayes relative to ξ. Conclude that $t_{2,n}$ is minimax.
(b) Same statement as in 4.19(b).

4.21. (a) In the setting of Problems 2.3 and 4.9, show that κ can be chosen in the prior density $f_\xi(\theta) = \text{const.} \times (1 + \theta^2)\kappa^{-1}\phi(\theta/\kappa)$, where ϕ is the standard normal density, in such a way that $t_{3,n}$ is Bayes relative to ξ. Conclude that $t_{3,n}$ is minimax.
(b) Compare $nr_{t_{1,n}}$ and $nr_{t_{3,n}}$ in terms of "subminimaxity."

4.22. Suppose $\Omega = \{0, 1, 2\}$, $S = \{x : -1 < x < \infty\}$, and $D = \{d_0, d_1, d_2\}$, the problem being the 3-hypothesis problem with 0-or-1 loss function (see Problem 4.4). Here f_0 and f_1 are as in Problem 4.12, and f_2 is the uniform density on the interval $(-1, 0)$.

(a) Find the Bayes procedure relative to the prior law $\xi = (\xi_0, \xi_1, \xi_2)$ where ξ_i is the prior probability that f_i is true. [Since $f_2(x) > 0$ only when $f_1(x) = f_0(x) = 0$, this problem can make use of the results of 4.12(a) in comparing the three posterior probabilities in the present problem.]

(b) If $\bar{\xi}$ is the prior law relative to which the minimax procedure of Problem 4.12(d) is Bayes (it is not necessary to know the exact value of γ), so that $P_{\bar{\xi}}\{f_2\} = 0$, show that there are many Bayes procedures in the present problem which have maximum risk equal to that of the minimax risk point of the simpler problem of 4.12(d) and which are hence minimax here. Of these minimax procedures, show that only one (whose risk under f_2 is 0) is *admissible minimax.*

(c) Considering the 4-hypothesis problem with f_3 (and corresponding d_3) adjoined to $\{f_i\}$ and $\{d_i\}$ to form the Ω and D for the 4-hypothesis problem, where

$$f_3(x) = \begin{Bmatrix} -200x & \text{if} \quad -.1 < x < 0, \\ 0 & \text{otherwise}, \end{Bmatrix},$$

show that there are several *admissible* minimax procedures with differing risk functions.

4.23. Work Problem 4.22, replacing Problem 4.12 by Problem 4.13 and with $S = (-1, 2]$.

4.24. Suppose the setting of Problems 2.1 and 4.19 with $n = 1$ is altered to *assume* $\Omega = \{\theta : \theta \leq 1/2\}$. One might guess that there is a minimax procedure t^* for this problem that does better than t_2, where t_2 was the minimax procedure for the larger parameter space $\{\theta : 0 \leq \theta \leq 1\}$. In this problem we find t^*. In part (a) you are *told* the "least favorable prior law" ξ^* relative to which t^* is Bayes and use the result of Section 4.5 to prove minimaxity. If you are ambitious and wonder how this ξ^* was dreamed up, instead of working (a) go directly to (b).

(a) If the prior law is $P_\xi\{0 = \frac{1}{2}\} = 2 - \sqrt{2} = 1 - P_\xi\{\theta = 0\}$, use 4.5(b) to show that the Bayes procedure for estimating θ with squared error loss and $D = \{d : 0 \leq d \leq \frac{1}{2}\}$ is

$$t^*(x) = \begin{cases} (\sqrt{2} - 1)/2 & \text{if} \quad x = 0, \\ 1/2 & \text{if} \quad x = 1. \end{cases}$$

From this, show that $r_{t^*}(\theta)$ is a quadratic in θ which attains its maximum on Ω at the two values $\theta = 0$ and $\theta = 1/2$. Thus, t^* is a Bayes procedure relative to a ξ^* which assigns probability one to the set were r_{t^*} attains its maximum. Use Section 4.5 to conclude that t^* is minimax. Compare r_{t^*} with r_{t_2} on Ω and on the Ω of Problem 2.1.

(b) Suppose you didn't know the preceding prior law. We shall find a minimax procedure with the help of the following result of decision theory and game theory for this present setting of (compact) Ω and D and (continuous) W: There exists a ("least favorable") prior law ξ^* such that every minimax procedure is Bayes relative to ξ^*; moreover, for any minimax t^*, ξ^* assigns probability one to the subset of Ω where r_{t*} attains its maximum. You can find a minimax procedure t^* by verifying the following steps:

(i) Every estimator can be expressed in terms of two values J, K where $t(0) = J$, $t(1) = K$. The risk function of every procedure is at most quadratic. The maximum of a quadratic or linear function on Ω (looked at because we hope to use Section 4.5) attains its maximum either (A) on the whole interval if the risk is constant, or (B) at one point, or (C) at the two end points of Ω. Of these, (A) is impossible for the risk function of minimax t^* because a risk function which is a polynomial in θ can be constant on $[0, \frac{1}{2}]$ only if it is constant on $[0, 1]$; t^* and t_2 would then have constant risks; one can show that t^* would then have the same risk as t_2: if t^* were worse, it couldn't be Bayes (contradicting the result stated at the start of (b)); if t_2 were worse it couldn't be minimax in Problem 4.19; but it is easy to see that changing $t_2(1)$ from 3/4 to 1/2 improves r_{t_2} uniformly on $[0, \frac{1}{2}]$, contradicting minimaxity of t^*. As for (B), it is eliminated because, if θ_0 were the single point of maximum risk, the use of Section 4.5 would entail $\xi^*(\theta_0) = 1$ and hence $t^* \equiv \theta_0$ and $r_{t^*}(\theta) = (\theta - \theta_0)^2$ with maximum *not* at θ_0. So (C) must hold for a minimax procedure and we hunt for a prior law which will let us use Section 4.5 for such an $r_{t^*}: P_\xi\{\theta = \frac{1}{2}\} = \alpha \text{ (say)} = 1 - P_\xi\{\theta = 0\}$.

(ii) We must determine α and t^*. If $t^*(0) = J^*$ and $t^*(1) = K^*$, show that 4.5(b) yields $J^* = \alpha/(4 - 2\alpha)$ and $K^* = \frac{1}{2}$. Computing r_{t^*} from this, and solving the equation $r_{t^*}(0) = r_{t^*}(\frac{1}{2})$, yields $\alpha = 2 - \sqrt{2}$ and thus the solution stated in (a). You can check that the coefficient of θ^2 in $r_{t^*}(\theta)$ is positive, so that case (i) (γ) does hold and the tool of Section 4.5 applies. [The technique used to find a minimax procedure here cannot be expected to work in more complicated problems, such as even the Bernoulli case with $n = 1$, since the possible forms of r_{t^*} will not be so simple. Nevertheless, this problem illustrates the type of thinking that goes into finding a minimax procedure in such less simple settings.]

Note on problems 4.25–4.28: The families Ω in these problems are all *scale parameter* families, in which θ is a measure of both the location and dispersion of the density $f_\theta(x) = \theta^{-1} f_1(x/\theta)$. This will be discussed further in Chapter 7. In the present problems it is helpful to note that, if X_1 has density f_θ in such a family, then for $c > 0$ the rv $Z = cX_1$ has density $f_{c\theta}(z)$, as can be seen by the rule for finding the density of such a function Z. In particular, putting $c = \theta^{-1}$ yields $P_\theta\{X_1 < a\} = P_1\{X < a/\theta\} = \int_0^{a/\theta} f_1(u)\,du$ and $E_\theta X_1^\alpha = \theta^\alpha E_1 X_1^\alpha$; that is, calculations can be performed for the "standard" value $\theta = 1$ and used to reduce computations for other values of θ.

4.25. Suppose $X = (X_1, X_2)$ where the X_i's are iid, each with absolutely continuous case density function

$$f_\theta(x) = \begin{cases} 3x^2/\theta^3 & \text{if } 0 < x < \theta, \\ 0 & \text{otherwise.} \end{cases}$$

Here $\Omega = \{\theta : \theta > 0\}$, $D = \{d : d > 0\}$, $W(\theta, d) = (\theta - d)^2$.

(a) Show that each of the following two estimators is an unbiased estimator of θ:

$$t_1(x_1, x_2) = \tfrac{2}{3}(x_1 + x_2);$$

$$t_2(x_1, x_2) = \tfrac{7}{6}\max(x_1, x_2).$$

[Show that $Y = \max(X_1, X_2)$ has density $6y^5/\theta^6$, $0 < y < \theta$; you need *not* compute the density of $X_1 + X_2$ to compute r_{t_2}, later.]

(b) Find the risk function of each of the procedures t_1, t_2, and show that the former is 60 percent larger than the latter. Since it will be shown in Chapter 5 that t_1 is the "best among linear unbiased estimators" of θ, what do you conclude about the general reasonableness of imposing this criterion?

(c) Show that among all procedures of the form $t'_c(x_1, x_2) = c\max(x_1, x_2)$, the procedure $t'_{8/7}$ is uniformly best, and that it is *not* an unbiased estimator of θ. The risk function of $t'_{8/7}$ is thus uniformly better than that of the unbiased estimator $t_2 = t'_{7/6}$. Since it will be seen later that t_2 is the "best among unbiased (not necessarily linear) estimators" of θ, what do you conclude about this criterion? [An aside: There is an analogous possible improvement $c'(x_1 + x_2)$ over t_1 here, but it will also be inferior to $t'_{8/7}$.]

4.26. Work Problem 4.25 with the following modifications: $f_\theta(x) = px^{p-1}/\theta^p$ for $0 < x < \theta$ (and $= 0$ otherwise), where p is a known value >0; in (a), $t_1 = \dfrac{p+1}{2p}(X_1 + X_2)$, and $t_2 = \dfrac{2p+1}{2p}\max(X_1, X_2)$; in (b), $r_{t_1}/r_{t_2} = (2p+2)/(p+2)$; in (c), the best choice of c is $(2p+2)/(2p+1)$.

4.27. Work Problem 4.25 with the following modifications: D and W are as in Problem 4.25 and, with $\Omega = \{\theta : \theta > 0\}$ again, X_1 and X_2 have density

$$f_\theta(x) = \begin{cases} 3\theta^3 x^{-4} & \text{if } x \geq \theta, \\ 0 & \text{otherwise.} \end{cases}$$

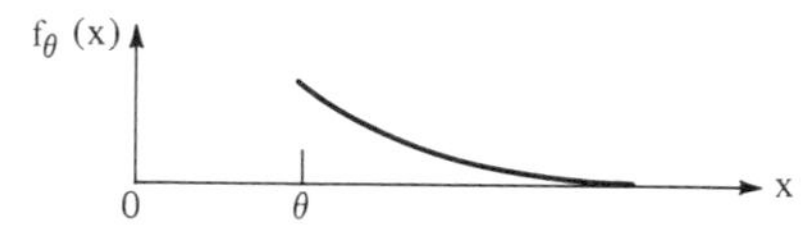

In (a), $t_1 = \tfrac{1}{3}(X_1 + X_2)$ and $t_2 = \tfrac{5}{6}\min(X_1, X_2)$; $Z = \min(X_1, X_2)$ should be shown by you to have density $6\theta^6 z^{-7}$ for $z \geq \theta$ (and 0 otherwise). In (b), $r_{t_1}/r_{t_2} = 4$. In (c), among estimators $t'_c = c\min(X_1, X_2)$, show $c = 4/5$ yields the best risk.

4.28. Work Problem 4.25 with the following modifications: D and W are as in Problem 4.25 and, with $\Omega = \{\theta : \theta > 0\}$ again, X_1 and X_2 have absolutely continuous case density

$$f_\theta(x) = \begin{cases} p\theta^p x^{-p-1} & \text{if } x \geq \theta, \\ 0 & \text{otherwise,} \end{cases}$$

where p is a known value >2. In (a), $t_1 = \dfrac{p-1}{2p}(X_1 + X_2)$ and $t_2 = \dfrac{2p-1}{2p}\min(X_1, X_2)$. In (b), $r_{t_1}/r_{t_2} = (2p-2)/(p-2)$. In (c), among estimators $t'_c = c\min(X_1, X_2)$, show $c = (2p-2)/(2p-1)$ is best. *Add:* (d) If $1 < p \leq 2$, show that unbiased estimator t_1 has infinite variance, whereas t_2 has finite variance and $\dfrac{2p-2}{2p-1}\min(X_1, X_2)$ is even better. If $p = 1$, what estimator would you use?

4.29. Suppose $X_1, X_2, \ldots, X_{50}$ are iid discrete rv's, each with two possible values if $\theta > 0$:

$$P_\theta\{X_i = \theta\} = .02,$$

$$P_\theta\{X_i = 0\} = .98.$$

If $\theta = 0$, $P_\theta\{X_i = 0\} = 1$. Here $\Omega = \{\theta : \theta \geq 0\}$, $D = \{d : d \geq 0\}$, and $W(\theta, d) = (\theta - d)^2$. Thus, the problem is one of estimating θ. In the next chapter we shall often restrict consideration in such problems to linear functions of the X_i's which are unbiased estimators of θ, and will show that the uniformly best procedure of this type must be of the form $b\bar{X}_{50}$ where b is a positive constant and $\bar{X}_{50} = \frac{1}{50}\sum_1^{50} X_i$.

(a) Show that $E_\theta X_i = .02\theta$ and $\text{var}_\theta(X_i) = .0196\theta^2$. Consequently, show that the "best linear unbiased estimator" just described (call it $\bar{t}$) has $b = 50$ and

$$r_{\bar{t}}(\theta) = .98\theta^2.$$

(b) (This part of the problem shows that, even if one keeps the restriction to unbiased estimators, the restriction to use only linear functions of the X_i's in (a) may have been costly.) Define $U = \max_{1 \leq i \leq 50} X_i$. First show that $P_\theta\{U = 0\} = P_\theta\{X_i = X_2 = \cdots = X_{50} = 0\} = .98^{50} = .364$, so that $P_\theta\{U = \theta\} = .636$. Show that the estimator $t' = (1 - .98^{50})^{-1}U$ is an unbiased estimator (*not* linear in the X_i's) of θ for which $r_{t'}(\theta) = .57\theta^2$. Thus, $r_{\bar{t}}$ is over 70 percent larger than $r_{t'}$. [The theory treated in Chapter 7 will show that t' is best unbiased among *all* estimators, linear or not.]

(c) To see that the restriction to unbiased estimators can also be costly, show that the estimator $t^* = U$ has $r_{t^*}(\theta) = .364\theta^2$, so that the $r_{t'}$ of (b) is over 50 percent larger than r_{t^*}. [The theory covered later in Chapter 7 shows that this t^* is what is there called the "best invariant estimator" in the scale parameter problem.]

CHAPTER 5

Linear Unbiased Estimation

[Appendix C is optional reading with Section 5.1. It illustrates various minimization techniques of use here and elsewhere in statistics and in fields like operations research.]

5.1. Linear Unbiased Estimation in Simple Settings

Suppose a real parameter ϕ is to be estimated and that $X = (X_1, \ldots, X_n)$, where the X_i are real. An estimator t is said to be *linear* if it is of the form

$$t(x_1, \ldots, x_n) = a_0 + \sum_{i=1}^{n} a_i x_i,$$

where the a_i are specified constants. Every choice of the a_i's yields a different estimator of this type. Examples are $t(x_1, \ldots, x_n) = \bar{x}_n = \sum_{k=1}^{n} x_k/n$ or $t(x_1, \ldots, x_n) = \pi + x_1 + 2x_2 + \cdots + nx_n$, etc. Examples of estimators which are *not* linear are

$$t'(x_1, \ldots, x_n) = \left[\max_i x_i + \min_i x_i\right] \Big/ 2,$$

$$t''(x_1, \ldots, x_n) = r^{\text{th}} \text{ largest of } x_1, \ldots, x_n,$$

$$t^*(x_1, \ldots, x_n) = n^{-1} \sum_i (x_i - \bar{x}_n)^2,$$

etc. (Actually, t' can be written as a linear estimator if $n = 1$ or 2, as can t'' and t^* if $n = 1$.) $\mathscr{D}_L$ and $\mathscr{D}_{LUE}$ will denote, respectively, the class of all estimators of ϕ which are linear and linear unbiased.

Is there any reason why we should restrict our attention to linear estimators? On the negative side, they will often be unusable since, unless all a_i are 0 for $i > 0$, t may take on values outside D. For example, if we want to estimate the variance of a normal distribution (where t^* may be appropriate) or the mean of a normal distribution when it is known to be positive, a linear estimator could take on (disallowed) negative values if any of $a_1, \ldots, a_n$ were not zero; of course, if we were to estimate the mean of a normal distribution about which nothing was known, so that all real values are allowed as estimates, this difficulty would not arise.

A more serious objection is that, even when linear estimators can be used, there is in general no assurance that any of them will be very good. For example, suppose that $X_1, \ldots, X_n$ are independently distributed with common density

$$f_{\theta;X_1}(x) = \begin{cases} 1 & \text{if } \theta - \frac{1}{2} \leq x \leq \theta + \frac{1}{2}, \\ 0 & \text{otherwise;} \end{cases}$$

that $\Omega = \{\theta : -\infty < \theta < \infty\}$; and that the problem is to estimate θ, with $W(\theta, d) = (\theta - d)^2$. The estimator $\bar{x}_n$ is a reasonable choice *if we restrict our attention to linear estimators:* for example, it will follow from later considerations that, among linear estimators (*not* among *all* estimators) it is minimax and is "best unbiased." Its risk function is easily seen to be $1/12n$ for all θ. However, the estimator t' defined previously (which will turn out to be minimax among *all* estimators) will be seen to have risk function $1/2(n+1)(n+2)$ for all θ. The two estimators coincide when $n = 1$ or 2, but t' is uniformly better than $\bar{x}_n$ for $n \geq 3$, with $r_{t'}(\theta)/r_{\bar{x}_n}(\theta) \to 0$ as an $n \to \infty$.

As another example, suppose $X_1, \ldots, X_n$ are independent with common Cauchy density $1/\pi[1 + (x - \theta)^2]$, where θ is to be estimated. $\bar{x}_n$ is not better than the estimator x_1 (based on a single observation), since in this example both have the same distribution, and no other linear estimator will be much good either. However, the estimator t'' defined previously, with $r = (n+1)/2$ (assuming n odd, for simplicity), which is called the *sample median*, has a probability law which (as we shall see later) becomes concentrated about θ with variance approximately $\pi^2/4n$ when n is large; this estimator is much to be preferred to $\bar{x}_n$, which does not have a finite variance in this example.

Why, then, should we study linear estimators? On the positive side, they are easy to compute. More important, there are two cases wherein we shall later (in the section on minimax procedures) prove that *the best linear unbiased estimator of ϕ is actually minimax among all estimators.* These two cases are ones in which Ω is such that there *is* a uniformly best estimator among all linear unbiased estimators (this will be seen to be true, for example, if $X_1, \ldots, X_n$ are independent and identically distributed, with $E_F X_1 = \phi(F)$, for all F in Ω), and where either (i) the X_i are normally distributed and W is *any* nondecreasing function of $|\phi(F) - d|$, or else (ii) the probability law of the X_i is unspecified but is possibly normal (i.e., the problem is nonparametric), and

W is any strictly convex function of $|\phi(F) - d|$, for example, square error. See Appendix C.1.1.

These two cases are extremely important practical ones and justify our consideration of linear estimators from the modern, not merely from the historical, point of view.

We now treat the simplest cases of linear unbiased estimation. An estimator is linear and unbiased if it satisfies *both* of these restrictions on its form. We shall consider the case where $\phi(F)$ is real and $W(F, d) = [\phi(F) - d]^2$; thus, the risk function of any unbiased estimator t will be $\text{var}_F(t)$.

The simplest possible problem is that in which $X_1, \ldots, X_n$ are uncorrelated random variables with common unknown mean $\phi(F)$, $-\infty < \phi(F) < \infty$, and common finite variance σ_F^2. There can be any other restrictions on F (in particular, on σ_F^2) in the specification of Ω; these will not affect the result. We have $D = \{d: -\infty < d < \infty\}$. For any linear estimator $t(x_1, \ldots, x_n) = a_0 + \sum a_i x_i$ the restriction of unbiasedness states that

$$\phi(F) = E_F t(X_1, \ldots, X_n) = a_0 + \phi(F) \sum_1^n a_i \quad \text{for all } F.$$

Since $\phi = a_0 + A\phi$ can be satisfied for all real numbers ϕ only if $a_0 = 0$ and $A = 1$, we see that any *unbiased* linear estimator is of the form

$$t(x_1, \ldots, x_n) = \sum_{i=1}^n a_i x_i \quad \text{with} \quad \sum_{i=1}^n a_i = 1$$

and any t of this form is unbiased. The risk function of such an estimator is

$$r_t(F) = \text{var}_F(t) = \sigma_F^2 \sum_{i=1}^n a_i^2.$$

Thus, the risk function of any such estimator is proportional to σ_F^2, and a *best estimator* (uniformly in F) of this form will be one for which $\sum_{i=1}^n a_i^2$ is smallest. It is easy to show that *the unique choice of the* a_i *which minimizes* $\sum_{i=1}^n \mathrm{a}_i^2$ *subject to the restriction* $\sum_{i=1}^n \mathrm{a}_i = 1$ *is* $\mathrm{a}_1 = \mathrm{a}_2 = \cdots = \mathrm{a}_n = 1/\mathrm{n}$. See Appendix C.2.1. Hence, *under the preceding assumptions*, the sample mean $\bar{x}_n = \sum_{i=1}^n x_i/n$ is the unique best linear unbiased estimator of $\phi(F)$.

This simple result can be proved in any of several ways. Minimization problems of this type arise quite often, and it is instructive to have a feeling for several methods of attacking such problems. Rather than to interrupt the discussion at this point with a diversion into these methods, we shall give several such methods of proof in Appendix C.

We now consider a slightly more complicated problem. Suppose the assumptions are exactly those stated two paragraphs back, except that instead of assuming the X_i's all have the same variance, we suppose they have perhaps different variances, known to within a proportionality constant. Thus, we assume

$$\text{var}_F(X_i) = h_i v_F,$$

where the numbers h_i are *known positive numbers*. (Whether or not v_F is known is immaterial.) As before, we obtain that any unbiased linear estimator is of the form $t(x_1, \ldots, x_n) = \sum_{i=1}^{n} a_i x_i$, with $\sum_{i=1}^{n} a_i = 1$. But now the risk function of such an estimator is

$$\text{var}_F(t) = v_F \sum_{i=1}^{n} h_i a_i^2.$$

Thus, we minimize $\sum_{i=1}^{n} h_i a_i^2$ subject to $\sum_{i=1}^{n} a_i = 1$. The desired result can be obtained in several ways. For example, we shall obtain it by an analogue of the method of Appendix C.2.5. (You may try another method for practice.) From rumor or intuition or consideration of the case $n = 2$, wherein the minimizing a_i's are seen to be inversely proportional to the h_i's, we may conjecture that this is true for general n, that is, that the best choice is $a_i = c/h_i$, where $c = 1/\sum h_i^{-1}$ (so that $\sum a_i = 1$). To try to prove this conjecture, we write (grouping c/h_i with a_i)

$$\begin{aligned}\sum_{i=1}^{n} h_i a_i^2 &= \sum_{i=1}^{n} h_i[(a_i - ch_i^{-1}) + ch_i^{-1}]^2 \\ &= \sum_{i=1}^{n} h_i(a_i - ch_i^{-1})^2 + \sum_{i=1}^{n} c^2 h_i^{-1} + 2\sum_{i=1}^{n} c(a_i - ch_i^{-1}) \\ &= \sum_{i=1}^{n} h_i(a_i - ch_i^{-1})^2 + c,\end{aligned}$$

since $\sum a_i = c \sum h_i^{-1} = 1$. Thus, we have $\sum_{i=1}^{n} h_i a_i^2 \geq c$, with equality if and only if $a_i = ch_i^{-1}$; since the latter is an allowable choice of the a_i's (satisfying $\sum a_i = 1$), it is the unique solution to our problem. Thus, we have proved that, under the *preceding assumptions*, the best linear unbiased estimator of $\phi(F)$ is $\sum_{i=1}^{n} h_i^{-1} X_i \Big/ \sum_{i=1}^{n} h_i^{-1}$. This explains the prescription you will often find stated in books on this subject that "the best estimate is obtained by weighting the observations inversely proportionally to their variances." In some applications Ω may be of such a restricted nature that $\text{var}_F(X_i)$ is the same known number σ_i^2 for all F, in which case we can put $v_F = 1$ and $h_i = \sigma_i^2$; however, the best linear estimator was obtained without assuming $\text{var}_F(X_i)$ known, but rather only assuming the *ratios* $\text{var}_F(X_i)/\text{var}_F(X_j)$ are known (e.g., define $v_F = \text{var}_F(X_1)$ and $h_i = \text{var}_F(X_i)/\text{var}_F(X_1)$). If all h_i are equal, we obtain the previous result for the case of equal variances.

The nature of our assumptions on Ω in the previous problem has been such that a best linear unbiased estimator existed. This need not be the case. For example, suppose X_1 and X_2 are independent with $E_F X_i = \phi(F)$, $\text{var}_F(X_1) = 1$, and $\text{var}_F(X_2) = q_F$, where q_F is unknown $(0 < q_F < \infty)$. If we knew q_F, the best linear unbiased estimator would be obtained from the result of the previous paragraph. Not knowing q_F, there is no best linear unbiased estimator. For, if $t_a(x_1, x_2) = ax_1 + (1-a)x_2$ (any linear unbiased estimator of $\phi(F)$ has this form for some real a), we obtain (for squared error loss)

$$r_{t_a}(F) = \text{var}_F(t_a(X_1, X_2)) = a^2 + (1-a)^2 q_F.$$

If you plot r_{t_a} as a function of q_F for each fixed a with $0 \le a \le 1$, you will see that the risk functions of any two such procedures t_a cross each other, so there is no uniformly best procedure among the procedures t_a. (You can easily verify that if $a < 0$, t_0 is better than t_a, and that if $a > 1$, t_1 is better than t_a; thus, one would at any rate only consider t_a's with $0 \le a \le 1$, under the present assumptions.)

It is possible to solve many other problems by transforming them into the framework of a problem previously solved. For example, suppose the X_i's were uncorrelated and all had the same variance σ_F^2, but that X_i represents a measurement at time i on the value of an economic or physical quantity $\phi(F)i$ which is growing linearly with time (and was zero at time zero), so that X_i has expectation $\phi(F)i$. (Thus, X_i equals the value $\phi(F)i$ plus a chance error of zero mean and variance σ_F^2.) We are to estimate the "rate of growth" $\phi(F)$. We could attack this problem by saying that since a linear unbiased estimator must be of the form $\sum_{i=1}^n a_i X_i$ with $\sum_{i=1}^n i a_i = 1$ (verify this!) and has variance $\sum_{i=1}^n a_i^2 \sigma_F^2$, our problem is now to minimize $\sum_{i=1}^n a_i^2$ subject to the restriction $\sum_{i=1}^n i a_i = 1$. However, instead of solving this new problem, we can note that, if we write $Y_i = X_i/i$, we have

$$E_F Y_i = \phi(F), \qquad \text{var}_F(Y_i) = \sigma_F^2/i^2,$$

and the Y_i's are also uncorrelated. Now, any linear estimator in terms of the X_i's can be expressed in terms of the Y_i's and vice versa, since $\sum_{i=1}^n a_i X_i = \sum_{i=1}^n b_i Y_i$ where $b_i = i a_i$. Thus, we can find the best linear unbiased estimator of $\phi(F)$ in terms of the Y_i's and then transform this back into terms of the X_i's. But the problem in terms of the Y_i's has been solved two paragraphs back. We have only to set $h_i = 1/i^2$, and we obtain $\sum_{i=1}^n i^2 Y_i \Big/ \sum_{i=1}^n i^2$ as the best linear unbiased estimator. Substituting $Y_i = X_i/i$, we obtain $\sum_{i=1}^n i X_i \Big/ \sum_{i=1}^n i^2$ as the best linear unbiased estimator in terms of the X_i's.

In the computation of a best linear unbiased estimator when correlation is present, the results are different. For example, the estimator t_a with $a < 0$ or

$a > 1$ which we need never consider when there is no correlation (see two paragraphs previously) may now turn out to be useful. For $n > 2$, the solution to this problem is more complicated to write down; it will be mentioned briefly in Section 5.4.

5.2. General Linear Models: The Method of Least Squares

Section 5.1 has introduced the idea of linear estimators. In particular the problems of finding linear unbiased estimators and best estimators were discussed. In order to illustrate these ideas we investigated a simple model in which there was one unknown real parameter ϕ to be estimated. In this and later sections of Chapter 5 we will examine more complicated problems involving the ideas of linearity, best, unbiased. Appendix C discusses methods of minimization with reference to examples discussed in this chapter. Section 5.4 uses the methods of linear algebra to give a concise discussion of the general linear model. Certain assertions made in this section are justified later in Section 5.4 using the methods of linear algebra.

The *general linear model* is a natural generalization of the examples of Section 5.1. We suppose throughout this section that $X_1, \ldots, X_n$ are uncorrelated random variables with common variance σ_F^2. (In certain places explicitly mentioned we will violate this assumption in order to include more problems in the theory.) We make the hypothesis that

$$E_F X_i = \sum_{j=1}^{k} b_{ij}\phi_j(F), \qquad 1 \le i \le n, \tag{5.1}$$

where the numbers b_{ij} are known to the experimenter and the $\phi_j(F)$s are unknown real parameters. Usually the number k of unknown parameters will be less than n in practice. If $k > n$ not all the parameters $\phi_1(F), \ldots, \phi_k(F)$ will be estimable. This will be shown in Section 5.4. The name *general linear model* comes from the fact that the expression on the right side of (5.1) is a linear combination of the unknown parameters.

It will be seen that the name *general linear model* refers to the hypothesis made by the experimenter about the first (and second) moments of the random variables. It *does not* refer to the type of decision to be made by the experimenter. Throughout this chapter we shall consider almost exclusively the question of best linear unbiased estimators of unknown parameters in linear models. The theory which results is a nonparametric theory. The underlying "state of nature" F can be any distribution function so long as the hypotheses of the general linear model are satisfied.

One may wish to make tests of hypotheses about the unknown parameters that occur in the general linear model. Unlike in the estimation problem there is no elegant nonparametric theory for the testing of hypotheses about the unknown parameters. When it is further assumed that $X_1, \ldots, X_n$ have a joint normal distribution, then one can develop an elegant theory for the testing of hypotheses. This theory includes the subject called the *analysis of variance* (Chapter 8). Other testing problems, not usually included in a discussion of the analysis of variance, are dealt with in treatments of multivariate analysis.

We will discuss next several examples of settings in which this model arises. It is common to abbreviate "(best) linear unbiased estimator" as B.L.U.E. and "least squares" as LS.

Univariate Curve Fitting

In the univariate context we suppose the experimenter has control over a variable z (real-valued) which influences the experimental outcome. In particular we suppose the experimenter chooses the value z_i for the random variable X_i. Later we will discuss in great detail the model

$$E_F X_i = \phi_1(F) + \phi_2(F) z_i, \qquad 1 \le i \le n, \tag{5.2}$$

where $\phi_1(F)$ and $\phi_2(F)$ are unknown real parameters, $-\infty < \phi_i(F) < \infty$. Instead of merely having a straight line, a quadratic form such as

$$E_F X_i = \phi_1(F) + \phi_2(F) z_i + \phi_3(F) z_i^2$$

or a more complicated form such as

$$E_F X_i = \phi_1(F) + \phi_2(F) \sin z_i + \phi_3(F) \cos z_i + \phi_4(F) z_i$$

may be reasonable models. These models can all be expressed in the form

$$E_F X_i = \sum_{j=1}^{k} \phi_j(F) g_j(z_i), \tag{5.3}$$

where the g_j's are known real functions of a real variable. This is an example of a linear model. The linearity of the model means that $E_F X_i$ is a linear function of the unknown parameters $\phi_j(F)$ (but not necessarily of the known variable z). Writing $b_{ij} = g_j(z_i)$, we have the form (5.1).

[*Warning:* A model such as $E_F X_i = e^{\phi(F) z_i}$ is *not* of this form. A common treatment of this model is to write $Y_i = \log X_i$ and to use the method of least squares to minimize $\Sigma[y_i - t z_i]^2$. This would yield a best linear unbiased estimator of ϕ based on the Y_i's (it is not linear in the X_i's) if $E_F \log X_i = \log E_F X_i$; but this last is not generally true, so it is difficult to describe any good property of the estimator so obtained. Even if we minimized $\Sigma[x_i - e^{-\bar{t} z_i}]^2$ directly, which is a much more difficult computational problem (for which the estimate t just described may serve as a first approximation in an iterative minimization procedure), this "least squares" estimator $\bar{t}$ would not be unbiased, since the model here is not of the form (5.1) where least squares and linear unbiased estimation are related. In fact, a problem such as this is extremely messy computationally, and from a practical point of view the most useful known result is that $\bar{t}$ is a fairly good estimator when n is very large if the X_i's are independent and normal with common variance, this being a special case of a result on "maximum likelihood estimation," to be discussed later.]

We now consider in greater detail the straight line model (5.2). We describe an experiment. Suppose that the experimenter wants to estimate the relationship, *known to be linear*, between two quantities, say between the percentage of a certain chemical in the soil and the expected height of a particular variety of plant when grown in such soil. (The assumption of linearity might only be valid when the percentage varies over a small range of values being considered.) Suppose $z_1, z_2, \ldots, z_n$ are the percentages of the chemical in n pots which are otherwise the same. The z_i's are assumed to be known exactly. Bulbs of the same variety are planted in the n pots and grown under the same conditions for a specified number of days. X_i is the height of the plant in the ith pot at the end of this period. We assume this height is a sum of an intrinsic value (depending on variety of plant, type of soil, amount of the chemical being studied, conditions of growth) and a chance departure from this intrinsic value, due to slight variations in bulbs, water, sunlight, soil heterogeneity, etc., from pot to pot. The chance variation is assumed to have expectation zero, and the chance variations in different pots are independent with equal variances. In particular, the expected value of the heights X_i and X_j of two plants in pots where the amount of the chemical is the same ($z_i = z_j$) will be the same. We also assume that this expectation is in general given by (5.2). The "state of nature" F is now such that for every F in Ω that $E_F X_i$ can be expressed for all i in terms of two unknown parameters ϕ_1 and ϕ_2, just as it could be expressed in terms of a single unknown parameter ϕ in the setup considered in Section 5.1. For every F, we also have $\text{var}_F(X_i) = \sigma_F^2$; this quantity may be unknown, but is the same for all i. The state of nature F satisfies these conditions (on the joint distribution of $X_1, \ldots, X_n$) for every F in Ω.

We now seek best (for squared error) linear unbiased estimators $t_1(x_1, \ldots, x_n) = a_0 + \sum_{i=1}^{n} a_i x_i$ of $\phi_1(F)$ and $t_2(x_1, \ldots, x_n) = b_0 + \sum_{i=1}^{n} b_1 x_i$ of $\phi_2(F)$. For example, for t_2 to be an unbiased estimator of ϕ_2 it must satisfy

$$\phi_2(F) = E_F t_2(X_1, \ldots, X_n) = b_0 + \phi_1(F) \sum_{i=1}^{n} b_i + \phi_2(F) \sum_{i=1}^{n} b_i z_i$$

for all F. Now, $\phi_2 = b_0 + A\phi_1 + B\phi_2$ can be satisfied for all real values ϕ_1 and ϕ_2 if and only if $A = b_0 = 0$ and $B = 1$. Thus, for t_2 to be unbiased, it must be of the form $\sum_{i=1}^{n} b_i x_i$ with $\sum_{i=1}^{n} b_i z_i = 1$ and $\sum_{i=1}^{n} b_i = 0$. Since $\text{var}_F\left(\sum_{i=1}^{n} b_i X_i\right) = \sigma_F^2 \sum_{i=1}^{n} b_i^2$, our problem reduces to the following form: Choose $b_1, \ldots, b_n$ subject to the restrictions

$$\sum_{i=1}^{n} b_i = 0 \quad \text{and} \quad \sum_{i=1}^{n} b_i z_i = 1, \quad \text{to minimize} \sum_{i=1}^{n} b_i^2.$$

This differs from the problem considered in Section 5.1 in that there are now two linear restrictions on the b_i's in place of one. This problem can be solved in many ways; for example, we could use method C.2.1 of Appendix C by solving the two simultaneous equations of restriction for b_{n-1} and b_n in terms of $b_1, \ldots, b_{n-2}$, substituting the solutions into $\sum_{i=1}^{n} b_i^2$, and minimizing the resulting expression (in terms of $b_1, \ldots, b_{n-2}$) without restrictions. We shall proceed here by a shorter method. First we note that, if all z_i's are equal, say $z_1 = \cdots = z_n = c$, we cannot possibly satisfy the two restrictions $\sum_{i=1}^{n} b_i = 0$ and $c \sum_{i=1}^{n} b_i = \sum_{i=1}^{n} b_i z_i = 1$. Thus, no linear unbiased estimator of ϕ_2 would exist in this case. We shall return to this fact later, for the moment merely assuming that not all z_i's are equal.

To solve the problem, let us consider sets of numbers $b_1, \ldots, b_n$ satisfying

$$\sum_{i=1}^{n} b_i = 0 \quad \text{and} \quad \sum_{i=1}^{n} b_i z_i = 1.$$

Let $\bar{z} = n^{-1} \sum_{i=1}^{n} z_i$. If $\sum b_i = 0$, we can subtract $\bar{z} \sum b_i$ from the quantity $\sum z_i b_i$ of the second restriction without changing the value of this quantity. Hence, the two restrictions can be rewritten as

$$\sum_{i=1}^{n} b_i = 0 \tag{5.4}$$

and

$$\sum_{i=1}^{n} (z_i - \bar{z}) b_i = 1. \tag{5.5}$$

Let us now consider a different minimization problem, namely, that of minimization of $\sum_{i=1}^{n} b_i^2$ subject *only to the restriction* (5.5) (that is, without imposing (5.4)). The solution can be obtained in the manner used to solve the example with the h_i's in Section 5.1; the solution is that the minimum of $\sum b_i^2$ subject to $\sum c_i b_i = 1$ is achieved only by the choice $b_i = c_i \Big/ \sum_{j=1}^{n} c_j^2$. In the present setting we set $c_i = z_i - \bar{z}$ and thus obtain that the minimum of $\sum b_i^2$ subject only to the restriction (5.5) is achieved only by the choice $b_i = (z_i - \bar{z})/\sum_j (z_j - \bar{z})^2$.

Now, it is not generally true that a minimum obtained by ignoring a restriction is the same as the minimum obtained subject to the restriction, as shown in the first paragraph of C.2.1 in Appendix C. The latter minimum is clearly at least as great as the former, since the domain of permitted values of the arguments in the latter case is a subset of those allowed in the former. However, if the minimizing values of the arguments for the former (unrestricted) problem *happen also to satisfy the restriction* they clearly yield a solution to the latter (restricted) problem, since the minimum for the latter problem can never be less than for the former. In our problem, we have obtained a minimum of $\sum b_i^2$ subject to (5.5), ignoring the restriction (5.4). However, our solution $b_i = (z_i - \bar{z}) \Big/ \sum_j (z_j - \bar{z})^2$ happens also to satisfy (5.4), since $\sum_i (z_i - \bar{z}) = 0$. Hence, this is also the solution to the original problem of minimizing $\sum b_i^2$ subject to both (5.4) and (5.5).

Thus, the best linear unbiased estimator of ϕ_2 is

$$t_2^*(x_1, \ldots, x_n) = \sum_i (z_i - \bar{z}) x_i \Big/ \sum_i (z_i - \bar{z})^2.$$

In a similar manner one can prove that the best linear unbiased estimator of ϕ_1 is $t_1^*(x_1, \ldots, x_n) = \bar{x}_n - \bar{z} t_2^*(x_1, \ldots, x_n)$. We shall omit the derivation, which involves showing that the only choice of $a_1, \ldots, a_n$ which minimizes $\sum_{i=1}^n a_i^2$ subject to the restrictions $\sum_{i=1}^n a_i = 1$ and $\sum_{i=1}^n a_i z_i = 0$ is the choice $a_i = n^{-1} - \left[\bar{z}(z_i - \bar{z}) \Big/ \sum_j (z_j - \bar{z})^2 \right]$.

It is of interest to know what the risk functions of these estimators are. We have

$$\begin{aligned} \operatorname{var}_F t_2^*(X_1, \ldots, X_n) &= \operatorname{var}_F \left[\sum_i (z_i - \bar{z}) X_i \Big/ \sum_i (z_i - \bar{z})^2 \right] \\ &= \left[\sum_i (z_i - \bar{z})^2 \right]^{-2} \sum_i \operatorname{var}_F[(z_i - \bar{z}) X_i] \\ &= \sigma_F^2 \Big/ \sum_i (z_i - \bar{z})^2. \end{aligned}$$

Also, since $\operatorname{cov}_F(X_i, X_j) = 0$ if $i \neq j$, we have

$$\begin{aligned} \operatorname{cov}_F[\bar{X}_n, t_2^*(X_1, \ldots, X_n)] &= \operatorname{cov}_F \left[n^{-1} \sum_i X_i, \sum_j (z_j - \bar{z}) X_j \Big/ \sum (z_j - \bar{z})^2 \right] \\ &= n^{-1} \left[\sum (z_j - \bar{z})^2 \right]^{-1} \sum_{i,j} \operatorname{cov}_F[X_i, (z_j - \bar{z}) X_j] \\ &= n^{-1} \left[\sum (z_j - \bar{z})^2 \right]^{-1} \sigma_F^2 \sum_j (z_j - \bar{z}) = 0. \end{aligned}$$

Hence,

$$\begin{aligned}\operatorname{var}_F[t_1^*(X_1,\ldots,X_n)] &= \operatorname{var}_F[\bar{X}_n - \bar{z}t_2^*(X_1,\ldots,X_n)]\\ &= \operatorname{var}_F \bar{X}_n + \bar{z}^2 \operatorname{var}_F[t_2^*(X_1,\ldots,X_n)]\\ &= \sigma_F^2\left[n^{-1} + \bar{z}^2 \Big/ \sum (z_j - \bar{z})^2\right].\end{aligned}$$

Since $\operatorname{cov}(\bar{X}_n, t_2^*) = 0$, we also have

$$\begin{aligned}\operatorname{cov}_F[t_1^*, t_2^*] &= \operatorname{cov}_F[\bar{X}_n - \bar{z}t_2^*, t_2^*] = -\bar{z}\operatorname{var}_F(t_2^*)\\ &= -\sigma_F^2\bar{z} \Big/ \sum (z_j - \bar{z})^2.\end{aligned}$$

Returning for a moment to the condition we found it necessary to assume, that not all z_i's are the same, let us examine more closely the meaning of this condition. Suppose all z_i's were equal, say $z_1 = z_2 = \cdots = z_n = 1$. Then our model would read

$$E_F X_i = \phi_1(F) + \phi_2(F) \quad \text{for all } i.$$

Suppose we could not merely observe the X_i's from which we are to make an inference about the true state of nature F but could even know the actual joint distribution function F of the X_i's; no data could ever tell us more than the exact form of F and would generally tell us much less. Even if we knew F and thus $E_F X_1 = \mu(F)$ (say), there is nothing to tell us "what part" of $\mu(F)$ is $\phi_1(F)$ and what part is $\phi_2(F)$. If we knew that $\mu(F) = 5$, there is absolutely nothing which determines whether $\phi_1(F) = 2$ and $\phi_2(F) = 3$ or whether $\phi_1(F) = 4$ and $\phi_2(F) = 1$, etc. If someone tells you *only* that a certain variety of bulb grows 5 inches on the average and askes you "what part ϕ_2 of this growth is due to the chemical in the soil?" there is no way to give a meaningful answer. In this situation, which is described by saying that the parameters ϕ_1 and ϕ_2 are *not identifiable* in this experiment, there is no hope of estimating ϕ_1 and ϕ_2 by unbiased estimators, since these parameters are not even determined by a complete knowledge of F. In this example (where all z_i's equal unity), we would say that the parameters ϕ_1 and ϕ_2 are *not estimable* ("do not possess linear unbiased estimators"), and $\phi_1 + \phi_2$ *is estimable* (with best linear unbiased estimator $\bar{x}_n$). Looking at the situation graphically in our example, a single point $(1, \mu)$ in the (z, x) plane does not determine a unique line $x = \phi_1 + \phi_2 z$. It only tells us the value of the ordinate corresponding to the abscissa value $z = 1$.

In our discussion thus far, we have considered estimation of the two parameters ϕ_1 and ϕ_2. One might also pose the problem as one of estimation of the whole linear relationship; that is, of the whole line $\alpha_F(z) = \phi_1(F) + \phi_2(F)z$. We could make this precise by saying that, for each fixed z, we want the best (for squared error) linear unbiased estimator T_z (say) of the parameter $\lambda_z(F) = \phi_1(F) + \phi_2(F)z$. It can be shown that the best such estimator is

$$T_z(x_1,\ldots,x_n) = t_1^*(x_1,\ldots,x_n) + t_2^*(x_1,\ldots,x_n)z.$$

That is, if we draw the line with intercept $t_1^*(x_1,\ldots,x_n)$ (the best linear unbiased estimator of the intercept $\phi_1(F)$ of the true line) and slope $t_2(x_1,\ldots,x_n)$ (the best linear unbiased estimator of the slope $\phi_2(F)$ of the true line), the resulting *estimator of the true line* will, for each value z, give a best linear unbiased estimator of the true ordinate value $\lambda_z(F)$ corresponding to this abscissa value z (note that z need not be one of the z_i's). Thus, from the point of view of linear unbiased estimation, the problem of estimating the whole line is solved by estimating ϕ_1 and ϕ_2.

The fact that the best linear unbiased estimator of $\lambda_z(F)$ is $t_1^* + t_2^* z$ where t_1^* and t_2^* are the best linear unbiased estimators of ϕ_1 and ϕ_2 is a special case of the fact that, in the linear models we shall treat in this section, the best linear unbiased estimator of a given linear combination of parameters is the same linear combination of the best linear unbiased estimators of these parameters. We shall use this fact repeatedly. It will be proved in Section 5.4.

In Section 5.1 and the preceding part of this section we have considered in detail several examples of best linear unbiased estimation. Let us stop now and try to summarize the basic problems encountered.

The first question is, *when do there exist linear unbiased estimators*? A concise answer to this question may be given in terms of the methods of linear algebra discussed in Section 5.4. Ensuring that a parameter to be estimated can in fact be estimated by a linear unbiased estimator is a question of experimental design.

Given that a particular parameter is estimable, how can we find all linear unbiased estimators? What are the best linear unbiased estimators? How do we find them? How do we solve the minimization problems involved in proving our answer correct? How do we calculate the minimum variance, i.e., the risk? Finally, how should one construct best linear unbiased estimators of linear combinations of the unknown parameters?

We turn now to a question of the method of least squares. In Section 5.3 we give a discussion of orthogonalization. These are two alternative methods of looking at the questions just raised. And in Section 5.4 we will give complete answers to the preceding questions and relate these answers to the methods of least squares and orthogonalization.

The Method of Least Squares

The best linear unbiased estimators t_1^* and t_2^* obtained for the model (5.2) were derived so as to satisfy a precise criterion of optimality (having minimum variance among linear unbiased estimators, or, which is more in the modern spirit, being minimax, as pointed out in Section 5.1). We now turn to an entirely different approach, an intuitively appealing method of "curve fitting" with no reference to anything like a risk function. We look at this approach first in the setting of our simple example of fitting a straight line. Suppose we think of the points (z_i, x_i), $1 \leq i \leq n$, as being drawn in the plane and ask, "Which line most closely fits this data?" We need some measure of closeness to make this precise. For example, we might, for a given line $x = t_1 + t_2 z$, measure the closeness of this line to the observed data by computing the vertical distance $|x_i - (t_1 + t_2 z_i)|$ from the point (z_i, x_i) to the line, summing this over all i, and then choosing as our "line of best fit" a line which minimizes this sum. There are many other possible "measures of closeness" of a fitted line to data, which you can think of. Another simple possibility is to sum the squares of these distances and then choose the line which minimizes the sum

$$Q = \sum_{i=1}^{n} [x_i - (t_1 + t_2 z_i)]^2. \tag{5.6}$$

This method of obtaining a line to fit the data is called the *method of least squares*, and the values $\bar{t}_1$ and $\bar{t}_2$ for the best fit, considered as functions of $x_1, \ldots, x_n$, are called the *least squares estimators* of ϕ_1 and ϕ_2. This approach was invented by Gauss and by Legendre at the beginning of the 19th century. This method leads to fairly simple computations (much simpler than if we had not squared the distances before summing!). It also has a theoretical interest, as we shall see. The graph of the function Q, considered as a function of t_1 and t_2 (for fixed x_i's and z_i's), is an elliptic paraboloid, and its minimum can be found by solving the two equations

$$\frac{\partial Q}{\partial t_1} = 0, \qquad \frac{\partial Q}{\partial t_2} = 0.$$

We seek values $\bar{t}_1$ and $\bar{t}_2$ of t_1 and t_2, which satisfy these equations. These equations can be rewritten, upon differentiating and rearranging terms, as

$$\begin{aligned} n\bar{t}_1 + n\bar{z}\bar{t}_2 &= n\bar{x}_n, \\ n\bar{z}\bar{t}_1 + \left(\sum z_i^2\right)\bar{t}_2 &= \sum x_i z_i. \end{aligned} \tag{5.7}$$

These two equations in the two unknowns $\bar{t}_1$ and $\bar{t}_2$ have a unique solution provided that $\sum (z_i - \bar{z})^2 \neq 0$ (i.e., that not all z_i's are equal). Solving the equations (e.g., by subtracting $\bar{z}$ times the first equation from the second, etc.), we see that *the least squares estimators* $\bar{\mathrm{t}}_1$ *and* $\bar{\mathrm{t}}_2$ *coincide with the best linear unbiased estimators* t_1^* *and* t_2^* *obtained previously.* Thus, the intuitive approach of fitting a "closest" line to the data in the sense of minimizing Q leads to the same choice of estimators of ϕ_1 and ϕ_2 as did the approach of finding best linear unbiased estimators. This phenomenon can be proved to occur also in the more general linear models we shall consider (see Section 5.4). It is theoretically important because there are many problems in which the minimization of the general analogue of Q is much simpler than a direct attempt to minimize an analogue of $\sum a_i^2$ or $\sum b_i^2$ subject to restrictions like (5.4) and (5.5). (If there were k unknown parameters, we would be faced with k such minimizations, each subject to k restrictions.) Actually we have just said that these two minimization problems are equivalent; the important practical consequence is that the method of least squares usually yields an easier derivation of the best linear unbiased estimators than does a more direct approach such as that first used to obtain t_1^* and t_2^*.

Be sure to keep in mind that the equivalence discussed in the previous paragraph is generally valid only under the assumption that the X_i's are uncorrelated with equal variances. Actually, this is already apparent in the simple case of a single parameter, treated in Section 5.1; the "method of least squares" can be thought of there as "fitting a horizontal line" by minimizing $\sum \{x_i - t\}^2$, and the result $\bar{t} = \bar{x}_n$ was seen not generally to be the best linear unbiased estimator, but was so if the X_i's had equal variances and were uncorrelated. Similarly, in the example with $E_F X_i = \phi_1(F) + \phi_2(F) z_i$ which we have been considering, the best linear unbiased estimator depends on the variances and covariances of the X_i's. The general result will be discussed in Section 5.4. As a simple example, recall the slightly more complicated case considered in Section 5.1, where the X_i's were still uncorrelated, but had possibly unequal variances $h_i v_F$ where the h_i's are *known* positive numbers. Then we can write $Y_i = X_i / h_i^{1/2}$ and $E_F Y_i = \phi_1(F) h_i^{-1/2} + \phi_2(F) z_i h_i^{-1/2}$, and the Y_i's are now uncorrelated and have equal variances. We can therefore find the best linear unbiased estimators of $\phi_1(F)$ and $\phi_2(F)$ in terms of the Y_i's either directly or by the equivalent method of least squares and then transform back to the X_i's. In terms of the method of least squares, the estimators $\bar{t}_1$ and $\bar{t}_2$ must be chosen to minimize

$$Q' = \sum_i [y_i - (t_1 h_i^{-1/2} + t_2 (z_i h_i^{-1/2})]^2 = \sum_i h_i^{-1} [x_i - (t_1 + t_2 z_i)]^2.$$

Thus, in terms of the X_i's, this last form differs from the sum Q of (5.6) in the weights h^{-1}, which were all unity in (5.6). Thus, the method of least squares, which is to minimize (5.6), must be replaced by the method of least *weighted* squares, which is to minimize the sum of squares of vertical deviations of observed points from the fitted line, *weighted inversely as the variances of the* X_i's *at those abscissa values.* This same result holds for more complicated models as long as the X_i's are uncorrelated. The result in the case of correlated X_i's, which is more complicated to write down, will be discussed in Section 5.4.

We now briefly mention certain other settings in which the general linear model is a reasonable hypothesis about the expectations $E_F X_1, \ldots, E_F X_n$.

Multivariate Curve Fitting

Suppose we had been interested in the dependence of the expected growth of a plant on two chemicals, say on the amount v of nitrogen and the amount w of potassium in the soil. The model might be that the expected height of the ith bulb, planted in a pot containing amounts v_i and w_i of these two chemicals, is

$$E_F X_i = \phi_1(F) + \phi_2(F) w_i + \phi_3(F) v_i + \phi_4(F) v_i w_i,$$

or something more complicated. This can be written as

$$E_F X_i = \sum_{j=1}^{k} \phi_j(F) g_j(v_i, w_i). \tag{5.8}$$

In fact, thinking of z_i now as being the vector (v_i, w_i), (5.8) becomes (5.3). Similarly, we could have more than two variables upon which $E_F X_i$ depends.

Analysis of Variance

A wide range of problems which are popularly described by the misleading term *analysis of variance* and which will be discussed later in detail are exemplified by the following (*two-way analysis of variance*): In trying to estimate the performance of r different workers and, at the same time, of s different machines, an efficiency expert lets each worker spend one day on each machine. Let $Y_{h,m}$ = amount produced by operator h on machine m $(1 \le h \le r, 1 \le m \le s)$. The assumption is

$$E_F Y_{h,m} = \alpha_h(F) + \beta_m(F); \tag{5.9}$$

that is, the amount produced by a given operator on a given machine is the sum of an effect due to the operator and an effect due to the machine, plus a chance deviation whose expectation is zero. This model (5.9) can be subsumed under the form (5.1) by writing $n = rs$, $k = r + s$, and $\phi_1 = \alpha_1, \ldots, \phi_r = \alpha_r$, $\phi_{r+1} = \beta_1, \ldots, \phi_{r+s} = \beta_s$. The X_i's can be written as any convenient list of the $Y_{h,m}$'s; for example,

$$X_1 = Y_{1,1}, \ldots, X_s = Y_{1,s}, X_{s+1} = Y_{2,1}, \ldots, X_{(2s)} = Y_{2,s}, \ldots, X_n = Y_{r,s}.$$

We would then have

$$E_F X_1 = \phi_1(F) + \phi_{r+1}(F),$$

$$E_F X_2 = \phi_1(F) + \phi_{r+2}(F),$$

etc. That is, $b_{11} = 1, b_{12} = b_{13} = \cdots = b_{1r} = 0; b_{1,r+1} = 1, b_{1,r+2} = \cdots = b_{1,k} = 0$, and so on. Each b_{ij} is either zero or one, and for each i there are just two b_{ij}'s which are one, namely, $b_{i,h}$ and $b_{i,r+m}$ if $X_i = Y_{h,m}$, these "ones" describing which operator and machine produced X_i. There are many more complicated models of this kind which can similarly be seen to fall within the framework (5.1).

To find the best linear unbiased estimators (for squared error loss) directly in the case of the model (5.1), the X_i's being uncorrelated and with common variance, we could proceed in the way we did in the simple linear example at the beginning of Section 5.2. If t_i^* is the best linear unbiased estimator of ϕ_i, this method would, for example, find $t_1^*(x_1, \ldots, x_n) = \sum_{i=1}^{n} a_i x_i$ by minimizing $\sum a_i^2$ subject to the k restrictions

$$\sum_i b_{i1} a_i = 1; \qquad \sum_i b_{i2} a_i = \sum_i b_{i3} a_i = \cdots = \sum_i b_{ik} a_i = 0.$$

This, as we have mentioned, may be a difficult computation if approached directly. What is the meaning of the method of least squares in this case? Under model (5.3), if we fit the curve $x = \sum_j \bar{t}_j g_j(z)$, the sum of squares of vertical deviations of observed points (z_i, x_i) to fitted points $\left(z_i, \sum_j \bar{t}_j g_j(z_i)\right)$ is $\sum_i \left[x_i - \sum_j \bar{t}_j g_j(z_i)\right]^2$. Similarly, under the model (5.8), writing $z_i = (v_i, w_i)$, this is the sum of squares of vertical deviations of observed points (v_i, w_i, x_i) from corresponding points on the fitted *surface* above the (v, w) plane. In the general case (5.1), if ϕ_j is estimated by $\bar{t}_j$, the "fitted" estimate of $E_F X_i$ is $\sum_j b_{ij} \bar{t}_j$, and the sum of squares of deviations of observed from fitted values is

$$Q = \sum_{i=1}^{n} \left[x_i - \sum_{j=1}^{k} b_{ij} \bar{t}_j\right]^2. \tag{5.10}$$

The method of least squares is to choose the $\bar{t}_j$'s to minimize Q. Exactly as (5.7) was derived from (5.6), we now rewrite the k equations

$$\frac{\partial Q}{\partial \bar{t}_j} = 0, \qquad 1 \le j \le k,$$

obtaining the k linear equations in $\bar{t}_1, \ldots, \bar{t}_k$,

$$\begin{aligned} a_{11}\bar{t}_1 + a_{12}\bar{t}_2 + \cdots + a_{1k}\bar{t}_k &= c_1, \\ a_{21}\bar{t}_1 + a_{22}\bar{t}_2 + \cdots + a_{2k}\bar{t}_k &= c_2, \\ a_{k1}\bar{t}_1 + a_{k2}\bar{t}_2 + \cdots + a_{kk}\bar{t}_k &= c_k, \end{aligned} \tag{5.11}$$

where

$$a_{pq} = \sum_{i=1}^{n} b_{ip} b_{iq} \quad \text{and} \quad c_j = \sum_{i=1}^{n} x_i b_{ij}. \tag{5.12}$$

The equations (5.11) have a unique solution $\bar{t}_1, \ldots, \bar{t}_k$ (that is, all parameters $\phi_1, \ldots, \phi_k$ are estimable) if and only if the determinant of the coefficients

$$\begin{vmatrix} a_{11} & a_{12} & \cdots & a_{1k} \\ a_{21} & a_{22} & \cdots & a_{2k} \\ a_{k1} & a_{k2} & \cdots & a_{kk} \end{vmatrix}$$

is not zero. (Do not worry if this is something you haven't ever seen.) For example, in the case of univariate polynomial curve fitting of degree $k - 1$, with $g_j(z) = z^{j-1}$ $(1 \le j \le k)$, the condition that all ϕ_j's be estimable is that there be at least k different values of the z_i's. The equations (5.11) are usually best solved by Gauss elimination; see also the remark at the end of Section 5.4.

The result alluded to earlier now indicates that, if all parameters are estimable, the least squares estimators $\bar{t}_1, \ldots, \bar{t}_k$ coincide with the best linear estimators $t_1^*, \ldots, t_k^*$. Thus, the solving of the so-called normal equations (5.11) will usually be the easiest method of obtaining the best linear unbiased estimators. When not all ϕ_i are estimable, the situation (to be discussed in Section 5.4) is slightly more complicated, but it is still true that the method of least squares yields the best linear unbiased estimators of all estimable parameters (for example, of $\phi_1 + \phi_2$ in the case of our linear example where all z_i's were unity).

Once each t_j^* is obtained as a linear function of the X_j's, the variances and covariances of the t_j^*'s can be computed as they were in the case of fitting a straight line. The general form of the result will be found in Section 5.4.

Additional Remarks

(1) Regarding the problem of fitting a straight line, note that only the X_i's, not the z_i's, were random variables. The general problem of fitting a straight line when *both* variables are subject to error is more complicated and requires other techniques. A typical model for that situation is that we observe (X_i, Z_i) where $E_F X_i = \phi_1(F) + \phi_2(F) E_F Z_i$.

(2) A very special circumstance of the setting described in the previous sentence is that where $E_F\{X_i | Z_i = z_i\} = \phi_1(F) + \phi_2(F) z_i$. In this case, if the conditional distribution of the X_i's for each fixed set of values of the Z_i's is one in which the X_i's are uncorrelated with equal variances, then the estimators t_1^* and t_2^* which we derived have *conditional* expectation $\phi_1(F)$ and $\phi_2(F)$ for every fixed set of values z_i of the Z_i's; hence, summing over the possible values of the Z_i's, we see that t_i^* and t_2^* are unbiased estimators of ϕ_1 and ϕ_2 (considering *both* the X_i's and Z_i's to be random variables). However, the formulas derived previously for $\text{var}_F(t_i^*)$ now give *conditional* variances (for fixed values z_i of the Z_i's) and would have to be multiplied by the probability function of the Z_i's and to be summed in order to give the unconditional variances.

An important example of this is the case in which the vectors $W_i = (X_i, Z_i)$ are independently and identically distributed bivariate normal vectors. (This does *not* say that the components X_i and Z_i of a single W_i are independent.) In this case ϕ_2 is the slope of the regression line of X_i on Z_i, and t_2^* is an unbiased estimator of it. The conditional variance of t_2^* given that $Z_i = z_i$ $(1 \le i \le n)$ was seen to be $\sigma_F^2 / \sum (z_i - \bar{z})^2$, where σ_F^2, the conditional variance of X_i given that $Z_i = z_i$, equals $\text{var}_F(X_1)(1 - \rho_{X_1, Z_1}^2)$. We shall later see that $U = \sum_1^n (Z_i - \bar{z})^2 / 2\,\text{var}_F(Z_1)$ has the gamma density

$$f_U(u) = \begin{cases} u^{(n-3)/2} e^{-u} / \Gamma[(n-1)/2] & \text{if} \quad u > 0, \\ 0 \quad \text{if} \quad u \le 0. \end{cases}$$

Hence, the unconditional variance of t_2^* is

$$\operatorname{var}_F(t_2^*) = E_F\left\{\frac{\sigma_F^2}{2\operatorname{var}_F(Z_1)}U^{-1}\right\}$$
$$= \frac{1}{n-3}\cdot\frac{\operatorname{var}_F(X_1)(1-\rho_{X_1,Z_1}^2)}{\operatorname{var}_F(Z_1)}$$

if $n > 3$ and is infinite if $n \leq 3$. [*Note:* In general, if $v(z)$ and $e(z)$ are the conditional variance and expectation of X_1 given that $Z_1 = z$, then the unconditional variance of X_1 is $Ev(Z) + E\{[e(Z) - Ee(Z)]^2\}$; the second term vanishes in the present case.]

The term *regression function* is also used to describe the function

$$q_F(z) = E\{X_i|Z_i = z\} = \sum_{j=1}^{k} \phi_j(F)g_j(z)$$

in a setting where the Z_i's are considered not to be random variables, as in the earlier part of this section.

(3) An important practical problem is that of deciding "which degree polynomial to fit." This can be approached in the modern spirit by assigning certain losses both to estimating a curve which is far from the true curve and also to using a high degree polynomial as the fitted curve; since a given set of data can always be "better fit" by a high degree polynomial than by one of lower degree, these two sources of loss work against each other and must be properly balanced by a good procedure. Unfortunately, computation of such procedures is messy. The commonly employed "intuitively appealing" practical methods of deciding which degree curve to fit, which will be mentioned again when we discuss hypothesis testing, have never been verified to have any such good property.

(4) *The residual sum of squares.* We know that if $X_1, \ldots, X_n$ are uncorrelated with common variance σ_F^2, then $E_F\sum_1^n (X_i - \bar{X}_n)^2 = (n-1)\sigma_F^2$. The problem of Section 5.1 can be viewed as a least squares problem with $k = 1$ and $g_1(z) = 1$; i.e., the fitting of a "horizontal straight line" to the data (i, x_i) $(1 \leq i \leq n)$. The minimum of $Q = \sum_1^n (x_i - t)^2$ in this case is achieved by putting $t = \bar{x}_n$.

In our general linear model, the minimum $\bar{Q}$ of Q in (5.10) is achieved when the $\bar{t}_j$'s are the best linear unbiased estimators. $\bar{Q}$, the "residual sum of squares," is the sum of squares of distances of observed points from the "curve of best fit." If there are k estimable parameters ϕ_i, one can show that $E_F\bar{Q} = (n-k)\sigma_F^2$, just as we had when $k = 1$. (See Section 5.4 for a proof.)

The question naturally arises, how do we estimate the precision of our estimators t_i^*, if σ_F^2 is unknown? This problem cannot be treated by linear estimation but will be discussed in Section 5.4 and in Chapter 7. Ignoring for the present any question of optimality, we see that $\bar{Q}/(n-k)$

is an unbiased estimator of σ_F^2. Thus, for example, an unbiased estimator of the quantity $\text{var}_F(t_2^*)$ in the problem of fitting a straight line is given by $\bar{Q}/(n-2)\sum(z_i - \bar{z})^2$.

Just as $\sum(x_i - \bar{x}_n)^2$ is often more easily computed in practice as $\sum x_i^2 - n\bar{x}_n^2$ than as a sum of quantities $(x_i - \bar{x}_n)^2$, so $\bar{Q}$ may be more difficult to compute from (5.10) directly than from the formula

$$\bar{Q} = \sum_1^n x_i^2 - \sum_{i=1}^n \left[\sum_j b_{ij}\bar{t}_j\right]^2,$$

which can be obtained by expanding (5.10) and using the fact that the $\bar{t}_j$'s satisfy (5.11). In this expression, $\bar{Q}$ is often described as that part of the "total sum of squares" $\sum x_i^2$ which is not "explained" by the "sum of squares due to the $\bar{t}_j$'s."

The geometric interpretation of $\bar{Q}$ is discussed in Section 5.4.

(5) Another problem is that of *prediction*; for example, in the setting of (5.8), based on responses $X_1, \ldots, X_n$ at points $(v_1, w_1), \ldots, (v_n, w_n)$, we may want to *predict* the value x_{n+1} that will be obtained as the (random) response in a pot where amounts (v_{n+1}, w_{n+1}) of the two chemicals are used. This is discussed in Section 5.5, where it will be seen that this problem can be reduced to that of *estimating* the *nonrandom* quantity $E_F X_{n+1}$.

5.3. Orthogonalization

(This section treats an optional topic.)

In the simple setting of fitting a straight line considered in Section 5.2, we could make the normal equations easier to solve as follows: Rewrite the model as

$$E_F X_i = \phi_1(F) + \phi_2(F)z_i = \psi_1(F) + \psi_2(F)(z_i - \bar{z}),$$

where $\psi_2(F) = \phi_2(F)$ and $\psi_1(F) = \phi_1(F) + \bar{z}\phi_2(F)$. If u_1 and u_2 are the best linear unbiased estimators of ψ_1 and ψ_2, the normal equations for the model expressed in terms of the ψ_i's become

$$nu_1 = \sum x_i,$$
$$\sum(z_i - \bar{z})^2 u_2 = \sum x_i(z_i - \bar{z}),$$

since $\sum 1\cdot(z_i - \bar{z}) = 0$. These are immediately solved to yield $u_1(x_1, \ldots, x_n) = \bar{x}_n$ and $u_2(x_1, \ldots, x_n) = \sum x_i(z_i - \bar{z})/\sum(z_i - \bar{z})^2$. Using the fact, noted in Section 5.2, that the best linear unbiased estimator of $c_1\psi_1 + c_2\psi_2$ is $c_1u_1 + c_2u_2$, we thus obtain that the best linear unbiased estimator of ϕ_1 (i.e., $\psi_1 - \bar{z}\psi_2$) is $u_1 - \bar{z}u_2$, and that of ϕ_2 (i.e., ψ_2) is of course u_2. These coincide with the estimators t_1^* and t_2^* obtained earlier, but the arithmetic here was slightly simpler than before, because of the simpler, "diagonal" form of the normal equations here.

In the general case (5.3) of curve fitting, we can duplicate the previous procedure, as follows: We try to find constants c_{mj}, $1 \le j < m \le k$, such that the functions

$$\begin{aligned}
\bar{g}_1(z) &= g_1(z), \\
\bar{g}_2(z) &= g_2(z) - c_{21}g_1(z), \\
\bar{g}_3(z) &= g_3(z) - c_{31}g_1(z) - c_{32}g_2(z), \\
&\cdots\cdots \\
\bar{g}_m(z) &= g_m(z) - \sum_{j=1}^{m-1} c_{mj}g_j(z), \\
&\cdots\cdots \\
\bar{g}_k(z) &= g_k(z) - \sum_{j=1}^{k-1} c_{kj}g_j(z),
\end{aligned} \tag{5.13}$$

satisfy the conditions

$$\sum_{i=1}^{n} \bar{g}_r(z_i)\bar{g}_s(z_i) = 0 \quad \text{whenever } r \ne s. \tag{5.14}$$

(It will be enough to satisfy (5.14) for $r < s$, since it is symmetric in r and s; in general, in (5.11) and (5.12), a_{pq} is symmetric in p and q.) Suppose for the moment that we can find such constants c_{mj}. Let us define $\psi_1, \psi_2, \ldots, \psi_k$ to be the linear combinations of $\phi_1, \phi_2, \ldots, \phi_k$ which satisfy the equations

$$\begin{aligned}
\phi_1 &= \psi_1 - c_{21}\psi_2 - c_{31}\psi_3 - \cdots - c_{k1}\psi_k, \\
&\cdots\cdots \\
\phi_j &= \psi_j - \sum_{m=j+1}^{k} c_{mj}\psi_m, \\
&\cdots\cdots \\
\phi_{k-1} &= \psi_{k-1} - c_{k,k-1}\psi_k, \\
\phi_k &= \psi_k.
\end{aligned} \tag{5.15}$$

(It is *not* necessary to solve these equations explicitly for the ψ_j's in terms of the ϕ_j's.) Defining $c_{jj} = -1$ and $c_{mj} = 0$ for $j > m$ $\left(\text{so that we can write } \bar{g}_m = -\sum_{j=1}^{k} c_{mj}g_j \text{ and } \phi_j = -\sum_{m-1}^{k} c_{mj}\psi_m\right)$, we then have, from (5.15) and (5.13),

$$\begin{aligned}
\sum_{j=1}^{k} \phi_j g_j &= \sum_{j=1}^{k}\left(-\sum_{m=1}^{k} c_{mj}\psi_m\right)g_j = \sum_{m=1}^{k}\left(-\sum_{j=1}^{k} c_{mj}g_j\right)\psi_m \\
&= \sum_{m=1}^{k} \psi_m \bar{g}_m.
\end{aligned}$$

Hence, instead of being written as $E_F X_i = \sum_{j=1}^{k} \phi_j(F) g_j(z_i)$, our model can be written as $E_F X_i = \sum_{j=1}^{k} \psi_j(F) \bar{g}_j(z_i)$. The normal equations for the problem in terms of the ψ_j's take a very simple form. For $r \neq s$, the coefficients a_{rs} of (5.11) are all zero, since a_{rs} is given by the left side of (5.14). Hence, assuming all ϕ_i's (hence, all ψ_i's) estimable, the least squares estimator u_j of ψ_j is, from (5.11), simply

$$u_j = \sum_{i=1}^{n} \bar{g}_j(z_i) x_i \Big/ \sum_{i=1}^{n} [\bar{g}_j(z_i)]^2. \tag{5.16}$$

Thus, we have

$$\begin{aligned} \operatorname{var}_F(u_j(x)) &= \left[\sum_i [\bar{g}_j(z_i)]^2\right]^{-2} \operatorname{var}_F\left(\sum_i \bar{g}_j(z_i) X_i\right) \\ &= \sigma_F^2 \Big/ \sum_i [\bar{g}_j(z_i)]^2 \end{aligned}$$

and, for $r \neq s$

$$\begin{aligned} \operatorname{cov}_F(u_r, u_s) &= \frac{\operatorname{cov}_F\left[\sum_i \bar{g}_r(z_i) X_i, \sum_i \bar{g}_s(z_i) X_i\right]}{\sum_i [\bar{g}_r(z_i)]^2 \sum_i [\bar{g}_s(z_i)]^2} \\ &= 0, \end{aligned}$$

by (5.14), since $\operatorname{cov}_F(\sum a_i X_i, \sum b_i X_i) = \sigma_F^2 \sum a_i b_i$ if the X_i's are uncorrelated. These computations are much simpler than the general analogues of those in Section 5.2 for the estimators of the ϕ_j's.

Having found the best linear unbiased estimators u_j of the ψ_j's, we can now find the best linear unbiased estimators t_j^* of the ϕ_j's merely by taking the same linear combinations of the u_j's as the ϕ_j's are of the ψ_j's in (5.15):

$$t_j^* = u_j - c_{j+1,j} u_{j+1} - \cdots - c_{k,j} u_k.$$

Thus, since the u_j's were uncorrelated, we have

$$\operatorname{var}_F(t_j^*) = \operatorname{var}_F(u_j) + \sum_{m=j+1}^{k} c_{mj}^2 \operatorname{var}_F(u_m)$$

and, for $r < s$,

$$\begin{aligned} \operatorname{cov}_F(t_r^*, t_s^*) &= \operatorname{cov}_F\left(-\sum_{m=1}^{k} c_{mr} u_m, -\sum_{m=1}^{k} c_{ms} u_m\right) \\ &= \sum_{m=1}^{k} c_{mr} c_{ms} \operatorname{var}_F(u_m) \\ &= -c_{sr} \operatorname{var}_F(u_s) + \sum_{m=s+1}^{k} c_{mr} c_{ms} \operatorname{var}_F(u_m). \end{aligned}$$

It remains to describe how the c_{mj}'s are determined. According to the first two lines of (5.13), equation (5.14) for $r = 1$ and $s = 2$ is satisfied if

$$\sum_i g_1(z_i)(g_2(z_i) - c_{21}g_1(z_i)) = 0;$$

thus,

$$c_{21} = \sum_i g_1(z_i)g_2(z_i) \Big/ \sum_i [g_1(z_i)]^2.$$

In order to satisfy (5.14) for $s = 3$ and $r = 1$, we must have (since $\bar{g}_1 = g_1$)

$$\sum_i g_1(z_i)[g_3(z_i) - c_{31}g_1(z_i) - c_{32}g_2(z_i)] = 0.$$

If we also satisfy

$$\sum_i g_2(z_i)[g_3(z_i) - c_{31}g_1(z_i) - c_{32}g_2(z_i)] = 0,$$

then the second of these last two equations minus c_{21} times the first yields (5.14) for $s = 3$ and $r = 2$, since $\bar{g}_2 = g_2 - c_{21}g_1$. Thus, we will obtain (5.14) for $s = 3$ and both $r = 1$ and 2, if both of these equations are satisfied; that is, if

$$\sum_i [g_1(z_i)]^2 c_{31} + \sum_i g_1(z_i)g_2(z_i)c_{32} = \sum_i g_1(z_i)g_3(z_i),$$

$$\sum_i g_1(z_i)g_2(z_i)c_{31} + \sum_i [g_2(z_i)]^2 c_{32} = \sum_i g_2(z_i)g_3(z_i).$$

This yields two equations in the two unknowns c_{31} and c_{32}, which we could solve. Continuing in this way, for each $m (2 \le m \le k)$ we determine the c_{mj}'s so as to satisfy

$$\sum_i g_r(z_i)\left[g_m(z_i) - \sum_{j=1}^{m-1} c_{mj}g_j(z_i)\right] = 0, \qquad 1 \le r \le m-1,$$

which yields $m - 1$ linear equations in the $m - 1$ unknowns c_{mj} $(1 \le j \le m - 1)$. This is not the simplest way of determining the c_{mj}'s, since the solving of all these sets of linear equations (as m varies) is not much improvement over solving the original equations (5.11). An often better method is to replace (5.13) by

$$\begin{aligned} \bar{g}_1(z) &= g_1(z), \\ \bar{g}_m(z) &= g_m(z) - \sum_{j=1}^{m-1} \bar{c}_{mj}\bar{g}_j(z), \qquad 2 \le m \le k. \end{aligned} \tag{5.17}$$

We choose the $\bar{c}_{sj}$'s to satisfy (5.14) for $1 \le r \le s - 1$, successively for $s = 2$, then $s = 3$, etc. At the stage $s = m$ we already have satisfied (5.14) for $1 \le r < s \le m - 1$, so that (5.17) and (5.14) yield

$$\bar{c}_{mj} = \frac{\sum_i g_m(z_i)\bar{g}_j(z_i)}{\sum_i [\bar{g}_j(z_i)]^2}, \qquad 1 \le j \le m-1,$$

and there are no general sets of linear equations to solve as there were for the c_{mj}'s. The disadvantage of this method is that the $\bar{g}_j$'s will usually be more complex to carry along for computation of the $\bar{c}_{mj}$'s than were the g_j's in terms of which the c_{mj}'s were expressed (for example, if $g_j(z) = z^{j-1}$, $\bar{g}_j$ will be a polynomial of degree $j-1$ of the more complicated form $z^{j-1} - c_{j,j-1}z^{j-2} - \cdots - c_{j,2}z - c_{j,1}$). Moreover, it is still necessary to obtain the c_{mj}'s from the $\bar{c}_{mj}$'s if we want to obtain the t_j^*'s from the u_j's. (The relationship among the c_{mj}'s and $\bar{c}_{mj}$'s is described in detail in Section 5.4. If the functions $\bar{g}_j$ are kept in functional form (as functions of z) at each stage, it may be easy to read off the c_{mj}'s by comparing the $\bar{g}_j$'s so obtained with (5.13); however, if, as is often more convenient, only the *numbers* $\bar{g}_j(z_i)$ are related, it may be harder to obtain the c_{mj}'s.) However, if we do not want the t_j^*'s but only the fitted curve $\sum_j t_j^* g_j(z)$, it is unnecessary to obtain the t_j^*'s (and, hence, the c_{mj}'s), since this fitted curve can also be written as $\sum_j u_j \bar{g}_j(z)$.

Actually, as described in Section 5.4, either of the preceding procedures really amounts to just a particular methodical way of solving the normal equations (5.11). There are widely available computer programs for solving (5.11), which may proceed by any of several methods but which will obviate the necessity of your finding the c_{mj}'s directly yourself. On the other hand, there are many problems for which the preceding routine is helpful, especially if the same $\bar{g}_j$'s (which depend on the z_i's) can then be *tabled* once and for all and used repeatedly, in place of solving (5.11) in terms of the original g_j's each time. (If you encounter these g_j's only once, or have no tables, Gauss elimination will be a quicker way of solving the normal equations than orthogonalization.)

Many standard statistical tables (e.g., those by Fisher and Yates) table the values $\bar{g}_j(z_i)$ for the case where the z_i's are equally spaced and $g_j(z) = z^{j-1}$. The functions $\bar{g}_j$ in this case are called *orthogonal polynomials.* The procedure of obtaining the $\bar{g}_j$'s from the g_j's is in general called *orthogonalization*, and (5.14) is called the condition of *orthogonality* of the $\bar{g}_j$'s. The use of the word *orthogonal* here is explained in detail in Section 5.4. Each g_j is viewed there as a vector of n components $g_j(z_1), \ldots, g_j(z_n)$. If we subtract from g_j its projection onto the $(j-1)$-dimensional linear space P_j determined by (i.e., consisting of all linear combinations of) $g_1, \ldots, g_{j-1}$ (or $\bar{g}_1, \ldots, \bar{g}_{j-1}$), then the resulting vector $\bar{g}_j$ is orthogonal (perpendicular) to P_j, hence to $g_1, \ldots, g_{j-1}$ (or $\bar{g}_1, \ldots, \bar{g}_{j-1}$).

The residual sum of squares $\bar{Q}$ has a particularly simple form in the case in which orthogonal functions are used. The "sum of squares due to best linear unbiased estimators" can now be expressed (because of (5.14)) as

$$\sum_{i=1}^{n} \left[\sum_{j=1}^{k} b_{ij} u_j \right]^2 = \sum_{j=1}^{k} a_{jj} u_j^2,$$

where $a_{jj} = \sum_i [\bar{g}_j(z_i)]^2$. The expression $a_{jj}u_j^2$, which is called the "sum of squares due to (explained by) the estimator of ψ_j," can also be written as

$$\left[\sum_i \bar{g}_j(z_i)x_i\right]^2 \Bigg/ \sum_i [\bar{g}_j(z_i)]^2.$$

Its expectation is

$$a_{jj}\left[\operatorname{var}_F(u_j) + (E_F u_j)^2\right] = \sigma_F^2 + a_{jj}\psi_j^2,$$

from which the expectation of $\bar{Q}$ given in Section 5.2 can easily be derived.

5.4. Analysis of the General Linear Model

We use matrix notation for brevity. (See Appendix A.) A thorough understanding of this subject requires a basic understanding of linear algebra. More detailed discussions of the subject of this brief resumé can be found in many books; Chapter 1 of Scheffé's *The Analysis of Variance* is a good example.

Basic Results

The model (5.1) can be written as

$$E_F X = B\phi(F),$$

where X is the column vector of X_i's, B is the $n \times k$ matrix of b_{ij}'s, and ϕ is the $k \times 1$ column vector of ϕ_j's. We assume for the moment that the X_i's are uncorrelated with common variance σ_F^2; thus, the $n \times n$ covariance matrix of the X_i's is

$$\operatorname{Cov}_F(X) = E_F\{(X - E_F X)(X - E_F X)'\} = \sigma_F^2 I,$$

where prime denotes transpose, I is the identity, and the expectation of a chance matrix is obtained by taking the expectation of each element.

We shall see that if B has rank k (i.e., if the columns of B are linearly independent), then all ϕ_j's, and any linear combination of them, are estimable. To determine in general (that is, if B does not necessarily have rank k) which "linear parametric functions" $c'\phi = \sum_j c_j\phi_j$ are estimable (where c is the column vector of c_j's), we note that any linear unbiased estimator of such a linear parametric function is of the form $a'X$ (with a the column vector of a_i's), so that $E_F a'X = a'B\phi(F)$. Thus, c' must be of the form $a'B$ (a linear combination of the rows of B) if $c'\phi$ is to be estimable, and it is easy to see that any $c'\phi$ of this form *is* estimable. (If B has rank k, c can therefore be anything.) For future reference, we note also that $B'B$ has the same rank (say, h) as B and has rows which are linear combinations of rows of B. It follows that the vector $a'B$ can be rewritten as $r'B'B$, for a suitable r. To summarize, the following three h-dimensional vector spaces of k-vectors are identical: $\{g : g' = r'B'B$ for some r in $R^k\}$, $\{g : g = a'B$ for some a in $R^n\}$, $\{c : c'\phi$ is estimable$\}$.

There is a simple geometric interpretation of best linear unbiased estimation credited to R. C. Bose. In the n-dimensional vector space S^* of all linear homogeneous functions $a'X$ of X, let H denote the linear subspace consisting of all linear combinations of the X_i's of the form $d'B'X$, for all real column vectors d with k components. The dimension h of H is clearly the same as the rank of B. Let J be the $(n-h)$-dimensional subspace of S^* consisting of all linear combinations of the form $q'X$ for which $q'B = 0$. (It may help you to think instead of the n-dimensional vector space S of all possible vectors X, of the h-dimensional subspace $\bar{H}$ of S consisting of all n-vectors of the form Bd, and of $\bar{J}$, the $(n-h)$-dimensional subspace of S consisting of all n-vectors q for which $q'B = 0$.) J is often called the *error space*, and H is called the *estimation space*.

Since, for any n-vector q, $E_F q'X = q'B\phi(F)$, we see that $E_F q'X = 0$ for all F if $q \in \bar{J}$; conversely, since $c'\phi = 0$ for all ϕ only if $c = 0$, we see that, if $E_F q'X = 0$ for all F, then $q'B = 0$, so that $q \in \bar{J}$. Thus, J is the set of all linear functions of X with expectation 0 for all F.

If $q \in \bar{J}$ and $Bd \in \bar{H}$, we have $d'B'q = 0$, so that $q'X$ and $d'B'X$ are uncorrelated (since $\mathrm{Cov}_F(X) = \sigma_F^2 I$, and hence $\mathrm{cov}_F(p'X, q'X) = \sigma_F^2 p'q$). In short, any two vectors of $\bar{H}$ and $\bar{J}$ are orthogonal, so that the corresponding linear functions of H and J are uncorrelated.

Suppose now that $a'X$ is any linear estimator. We can break up a uniquely into its two projections $\bar{a}$ and $\bar{\bar{a}}$ on $\bar{H}$ and $\bar{J}$, where $\bar{a}$ and $\bar{\bar{a}}$ are orthogonal ($\bar{a}'\bar{\bar{a}} = 0$). Since $a = \bar{a} + \bar{\bar{a}}$, we have $E_F a'X = E_F \bar{a}'X$ and $\mathrm{var}_F(a'X) = \mathrm{var}_F(\bar{a}'X) + \mathrm{var}_F(\bar{\bar{a}}'X)$. Hence, the estimator $\bar{a}'X$ has the same expectation as $a'X$, and $\bar{a}'X$ has a variance which is no greater than that of $a'X$ (and a strictly smaller variance unless $\bar{\bar{a}} = 0$). We conclude that, for any linear unbiased estimator which is not in H, there is another linear unbiased estimator (its projection onto H) which is better; that is, *all best linear unbiased estimators are in* H.

Moreover, if two different functions in H had identical expectations for all F, their difference would be in J, which is impossible since only the function zero is in $H \cap J$. Hence, *every function in* H *is the best linear unbiased estimator of a different estimable parametric function* c'ϕ. If $c'\phi$ is estimable, there is a unique $\bar{a}$ in $\bar{H}$ such that $\bar{a}'X$ is the B.L.U.E. of $c'\phi$. (This can be thought of as establishing the natural isomorphism between the h-dimensional space of k-vectors $\{c : c'\phi \text{ is estimable}\}$ and the h-dimensional space of n-vectors $\{a : a'X \text{ is a B.L.U.E.}\}$: if $c' = r'B'B$ and $a' = r'B'$, then $a'X$ is the B.L.U.E. of $c'\phi$.)

Consider now the method of least squares, which tells us to choose the column vector $\bar{t}$ of $\bar{t}_i$'s so as to minimize $Q = (X - B\bar{t})'\,(X - B\bar{t})$. Differentiating Q partially with respect to each $\bar{t}_i$ and setting the result equal to zero, we obtain k linear equations in the $\bar{t}_i$'s, which can be written as

$$B'B\bar{t} = B'X,$$

these being the *normal equations*. (An alternate derivation will be given in the subsection on the geometric interpretation of Q, where also we will see that the minimum Q is an unbiased estimator of $(n-h)\sigma^2$.) This set of equations will have a $(k-h)$-parameter family ($(k-h)$-dimensional linear space) of solutions; if B has rank k, so does $B'B$, and the solution is unique. It is easy to check, from the quadratic nature of Q, that *every* solution $\bar{t}$ of the normal equations minimizes Q.

Suppose $\bar{t}(X)$ is *any* solution of the normal equations, and that $c'\phi$ is an estimable linear parametric function, which we have seen can be written as $r'B'B\phi$. Then $c'\bar{t}(X) = r'B'B\bar{t}(X) = r'B'X$, so that $c'\bar{t}(X) \in H$, and we have seen that $r'B'X$ is the B.L.U.E. of $r'B'B\phi$. We have established the

Gauss-Markoff Theorem: *If* $\bar{t}(X)$ *is any solution of the normal equations and* $c'\phi$ *is any estimable linear parametric function, then* $c'\bar{t}(X)$ *is the unique best linear unbiased estimator of* $c'\phi$.

Thus, we have proved that the method of least squares yields best linear unbiased estimators. The form of our result shows also that, as mentioned in Section 5.2, if t_1^* and t_2^* are best linear unbiased estimators of two linear parametric functions $\bar{c}'\phi$ and $\bar{\bar{c}}'\phi$, then $t_1^* + t_2^*$ is the best linear unbiased estimator of $(\bar{c} + \bar{\bar{c}})'\phi$. Keep in mind the physical interpretation of $B\bar{t}$ as the n-vector of B.L.U.E.'s of components of EX (least-squares fit at the points of observation, in curve-fitting).

If B has rank k, then $B'B$ is nonsingular and every component of ϕ, and hence every linear parametric function, is estimable. The normal equations have a unique solution $\bar{t}(X) = (B'B)^{-1}\, B'X$ in this case. On the other hand, if B does not have rank k, there will be infinitely many solutions $\bar{t}$ to the normal equations. The components of $\bar{t}$ do not necessarily represent best linear unbiased estimators, since some components of ϕ will not even be estimable! On the other hand, if $c'\phi$ is estimable, then, although there are infinitely many different solutions $\bar{t}$ to the normal equations, *the value of* $c'\bar{t}$ *is the same for all of these solutions* $\bar{t}$ and is the unique best linear estimator of $c'\phi$. Similarly, although $\bar{t}$ may not be unique, the "best fit" $B\bar{t}$ *is* unique.

As an example of the phenomenon just described, consider the example of fitting a straight line discussed in Section 5.2, with all $z_i = 2$, so that $E_F X_i = \phi_1(F) + 2\phi_2(F)$ for every i. Here $k = 2$ and B is an $n \times 2$ matrix whose first column is all 1's and whose second column is all 2's. Thus, $h = 1$. The normal equations are

$$n\bar{t}_1 + 2n\bar{t}_2 = \sum x_i,$$

$$2n\bar{t}_1 + 4n\bar{t}_2 = 2\sum x_i.$$

There are infinitely many solutions; namely, for any real number s,

$$\bar{t} = \begin{pmatrix} \bar{t}_1 \\ \bar{t}_2 \end{pmatrix} = \begin{pmatrix} \bar{x}_n - 2s \\ s \end{pmatrix}$$

satisfies the normal equations. Neither $\bar{t}_1$ nor $\bar{t}_2$ represents a best linear estimator, since neither ϕ_1 nor ϕ_2 is estimable. The only estimable linear parametric functions in this case are of the form $c'\phi$ with c' a multiple of $(1, 2)$. In particular, the best linear unbiased estimator of $\phi_1 + 2\phi_2$ is $\bar{t}_1 + 2\bar{t}_2 = \bar{x}_n$, which is uniquely defined even though $\bar{t}$ was not.

If B has rank k, so that we can write $\bar{t}(X) = (B'B)^{-1}B'X$, we have for the covariance matrix of best linear estimators,

$$\begin{aligned}\mathrm{Cov}_F(\bar{t}) &= E_F\{(\bar{t} - E_F\bar{t})(\bar{t} - E_F\bar{t})'\}\\ &= (B'B)^{-1}B'\,\mathrm{Cov}_F(X)B(B'B)^{-1}\\ &= \sigma_F^2(B'B)^{-1}.\end{aligned}$$

If $h < k$, we cannot invert $B'B$, but if $c'\phi = r'B'B\phi$ and $\bar{c}'\phi = \bar{r}'B'B\phi$ are two estimable linear parametric functions, we can still write for the covariance (variance if $r = \bar{r}$) of their best linear unbiased estimators

$$\begin{aligned}\mathrm{cov}_F[r'B'B\bar{t}, \bar{r}'B'B\bar{t}] &= \mathrm{cov}_F[r'B'X, \bar{r}'B'X]\\ &= r'B'\,\mathrm{Cov}_F(X)B\bar{r} = \sigma_F^2 r'B'B\bar{r} = \sigma_F^2 c'\bar{r}.\end{aligned}$$

Change of Assumptions on $\mathrm{Cov}_F(X)$

Suppose our assumption is changed to

$$\mathrm{Cov}_F(X) = v_F S,$$

where S is a known positive definite matrix and v_F is an unknown scalar. It is a well known fact from matrix theory that there is a nonsingular $n \times n$ matrix A such that $ASA' = I$. Write $Y = AX$. Then

$$\mathrm{Cov}_F(Y) = A\,\mathrm{Cov}_F(X)A' = v_F I$$

and

$$E_F Y = AB\phi(E).$$

Hence, our previous results can be used to find best linear estimators if X is replaced by Y and B by AB. The normal equations thus become

$$(AB)'(AB)\bar{t} = (AB)'Y = (AB)'AX.$$

Since $ASA' = I$, we have $S = A^{-1}(A^{-1})'$, and hence $S^{-1} = A'A$. Thus, finally, the preceding normal equations become

$$B'S^{-1}B\bar{t} = B'S^{-1}X,$$

and it is a solution to these rather than to the original normal equations which is used to obtain best linear unbiased estimators. If S is a diagonal matrix, this is the "least weighted squares" method mentioned in Section 5.2.

Orthogonal Polynomials

(This subsection treats an optional topic.)
The orthogonalization procedure described in Section 5.3 amounts to the following, supposing for simplicity that $h = k$: We leave the first column β_1 of B intact and write $\bar{\beta}_1 = \beta_1$. From the second column β_2 we subtract a multiple c_{21} of the first column, so that the result $\bar{\beta}_2 = \beta_2 - c_{21}\beta_1$ is orthogonal to β_1; i.e., $\beta_1'(\beta_2 - c_{21}\beta_1) = 0$, or $c_{21} = \beta_1'\beta_2/\beta_1'\beta_1$. This amounts to subtracting from

$\bar{\beta}_2$ its projection on β_1. In the same way, $\bar{\beta}_3$ is obtained from β_3 by subtracting its projection on the *plane spanned by* β_1 *and* β_2, so that the result $\bar{\beta}_3 = \beta_3 - c_{31}\beta_1 - c_{32}\beta_2$ is orthogonal to that plane. Since β_1 and β_2 are not necessarily orthogonal, the projection on the plane spanned by them is not merely the sum of the projections on β_1 and on β_2; thus, $c_{31}\beta_1$ and $c_{32}\beta_2$ cannot merely be obtained as projections of β_3 on β_1 and β_2, but rather must be chosen so as to satisfy $\bar{\beta}_3'\beta_1 = \bar{\beta}_3'\beta_2 = 0$, which yields two equations ($\beta_1'\beta_1 c_{31} + \beta_1'\beta_2 c_{32} = \beta_1'\beta_3$ and $\beta_2'\beta_1 c_{31} + \beta_2'\beta_2 c_{32} = \beta_2'\beta_3$) in the two unknowns c_{31} and c_{32}, to be solved. If, instead, we wrote $\bar{\beta}_3 = \beta_3 - \bar{c}_{31}\bar{\beta}_1 - \bar{c}_{32}\bar{\beta}_2$, the determination of $\bar{c}_{31}$ and $\bar{c}_{32}$ would be easier, since $\bar{\beta}_1$ and $\bar{\beta}_2$ are orthogonal and hence the projection on the plane spanned by them is the sum of the projections $\bar{c}_{31}\bar{\beta}_1$ on $\bar{\beta}_1$ and $\bar{c}_{32}\bar{\beta}_2$ on $\bar{\beta}_2$; thus, $\bar{c}_{3j} = \beta_3'\bar{\beta}_j/\bar{\beta}_j'\bar{\beta}_j$. We can continue in this way, at the jth stage obtaining $\bar{\beta}_j$ either by solving $j-1$ equations for the c_{ji}'s $(i<j)$ in $\bar{\beta}_j = \beta_j - \sum_{i<j} c_{ji}\beta_i$, or else from the previously obtained $\bar{\beta}_i$'s in the form $\bar{\beta}_j = \beta_j - \sum_{i<j} \bar{c}_{ji}\bar{\beta}_i$ with $\bar{c}_{ji} = \beta_j'\bar{\beta}_j/\bar{\beta}_i'\bar{\beta}_i$. The final result is the same in either case: We obtain a matrix $\bar{B}$ of columns $\bar{\beta}_j$, where $\bar{B} = BC'$ and

$$C = \begin{Vmatrix} 1 & 0 & 0 & 0 & \cdots & 0 \\ -c_{21} & 1 & 0 & 0 & \cdots & 0 \\ -c_{31} & -c_{32} & 1 & 0 & \cdots & 0 \\ \cdots & \cdots & \cdots & \cdots & \cdots & \cdots \\ -c_{k1} & -c_{k2} & -c_{k3} & \cdots & \cdots & 1 \end{Vmatrix} = \bar{C}_k\bar{C}_{k-1}\cdots\bar{C}_3\bar{C}_2,$$

where $\bar{C}_j$ differs from the $j \times j$ identity only in that the first $j-1$ elements of its jth row are $-\bar{c}_{j1}, \ldots, -\bar{c}_{j,j-1}$.

The columns of $\bar{B}$ are orthogonal, so that $\bar{B}'\bar{B}$ is diagonal. Writing $E_F X = B\phi(F) = \bar{B}\psi(F)$ with $\psi = C^{-1'}\phi$, the normal equations for estimating ψ are already in the diagonal form $\bar{B}'\bar{B}u = \bar{B}'X$. If u is the vector of best linear unbiased estimators of ψ satisfying these equations, then $\bar{t} = C'u$ is the vector of best linear unbiased estimators of ϕ. (Note that it is not necessary to compute C^{-1} in order to obtain $\bar{t}$.)

If $h < k$, a procedure like that described (with perhaps also a permutation of the columns) would yield a $\bar{B}$ whose first h columns are orthogonal (and not zero vectors) and whose last $k-h$ columns are zero vectors.

It is important to see how the discussion of Section 5.3 fits into the preceding framework. It should be understood that the original normal equations $B'B\bar{t} = B'X$ would ordinarily be solved most quickly by Gauss elimination, and the reduction to $\bar{B}$ would be worth employing primarily when tables of $\bar{B}$ are available, and $\bar{B}$ does not have to be computed.

Geometric Interpretation of Q: The Residual Sum of Squares

(The main idea of this section can be obtained, with a minimum of algebra, by considering the special case where B consists of zeros except for its first h rows and columns, which form a nonsingular $h \times h$ submatrix. Then $\bar{B}$ below is just B, transformation of the problem being unnecessary.) If $\bar{t}$ satisfies the normal equations, the value $\bar{Q}$ of Q is just the sum of squares of vertical distances of observed points from the best (least squares) fit and is often called the *residual sum of squares*. This has a simple geometric interpretation. For, Bt is the form of an arbitrary vector in $\bar{H}$, so that $Q = (X - Bt)'(X - Bt)$, which is the square of the length of $X - Bt$, is minimized by taking $B\bar{t}$ to be the projection of X onto $\bar{H}$, so that $(X - B\bar{t}) \in \bar{J}$ and thus $B'(X - B\bar{t}) = 0$ (which is an alternate method of obtaining the normal equations, to minimize Q). It is this nature of $B\bar{t}$ as a projection which makes it unique even though $\bar{t}$ is not. (This geometric method of obtaining the normal equations is analogous to the method of Appendix C.2.2.) Thus, from the Pythagorean principle,

$$X'X = (X - B\bar{t})'(X - B\bar{t}) + (B\bar{t})'(B\bar{t}).$$

By the procedure described in the preceding subsection, we can transform the problem to the form $E_F X = \bar{B}\psi(F)$, where $\bar{B}$ has h linearly independent columns followed by $k - h$ columns of 0's. The estimable linear parametric functions are then precisely the linear combinations of $\psi_1, \ldots, \psi_h$. The least squares estimators $u_1, \ldots, u_h$ are now easily seen to be given by $u_i = \bar{\beta}_i' X / \bar{\beta}_i' \bar{\beta}_i$, since the normal equations are in diagonal form. Since the $\bar{\beta}_i$'s are orthogonal, the u_i's are uncorrelated, and

$$\operatorname{var}_F(u_i) = (\bar{\beta}'_i \bar{\beta}_i)^{-2} \bar{\beta}_i' \operatorname{Cov}_F(X) \bar{\beta}_i = \sigma_F^2 / \bar{\beta}_i' \bar{\beta}_i.$$

Hence,

$$\begin{aligned} E_F\{(B\bar{t})' B\bar{t}\} &= E_F\{(\bar{B}u)'\bar{B}u\} \\ &= E_F\{\Sigma_1^h (\bar{\beta}_i'\bar{\beta}_i) u_i^2\} \\ &= \sum_1^h \operatorname{var}_F[(\bar{\beta}_i'\bar{\beta}_i)^{1/2} u_i] + \sum_1^h [E_F(\bar{\beta}_i'\bar{\beta}_i)^{1/2} u_i]^2 \\ &= h\sigma_F^2 + \psi(F)'\bar{B}'\bar{B}\psi(F). \end{aligned}$$

Similarly,

$$\begin{aligned} E_F\{X'X\} &= \sum_1^n \operatorname{var}_F(X_i) + (E_F X)'(E_F X) \\ &= n\sigma_F^2 + \psi(F)'\bar{B}'\bar{B}\psi(F). \end{aligned}$$

The difference between the last two results is

$$E_F(X - B\bar{t})'(X - B\bar{t}) = (n - h)\sigma_F^2.$$

We shall see later that $\bar{Q}/(n - h)$ is a useful estimator of σ_F^2; we have just proved that it is unbiased.

If $h = k$, the "sum of squares due to estimators", $\bar{t}'B'B\bar{t}$, can be written as

$$\bar{t}'\sigma_F^2[\operatorname{Cov}_F(\bar{t})]^{-1}\bar{t}.$$

Similarly, if $h < k$, so that the collection of all *estimable* linear parametric functions is an h-dimensional linear space, we can select any h linearly independent estimable linear parametric functions $\gamma_i = \sum_j d_{ij}\phi_j$, $1 \le i \le h$. If $\bar{v}$ is the vector of best linear estimators v_i of the γ_i's, we can then again write

$$\bar{v}'\sigma_F^2[\text{Cov}_F(\bar{v})]^{-1}\bar{v}$$

for the sum of squares due to estimators. This is a useful computational form in many applications. If the v_i's are orthogonal (uncorrelated) functions of the X_i's, the matrix $\sigma_F^2[\text{Cov}_F(\bar{v})]^{-1}$ reduces to a diagonal matrix, as it did for the u_i's (for which the ith diagonal element was $\bar{\beta}_i'\bar{\beta}_i$) in the preceding derivation.

Prediction

Suppose a random variable X_{n+1}, uncorrelated with $X_1, \ldots, X_n$, is to be observed in the future, and that $E_F X_{n+1} = \psi_F$ is estimable from $X_1, \ldots, X_n$. It is desired to *predict* the value of X_{n+1} based on $X_i, \ldots, X_n$. For example, in the "curve-fitting" setting X_{n+1} may be the response corresponding to the value z_{n+1} of the controllable variable, in future operation of the model at hand.

If T is the linear "predictor" of X_{n+1} to be used (based on $X_1, \ldots, X_n$), we have for the mean squared error of T in predicting X_{n+1},

$$E_F(T - X_{n+1})^2 = E_F(T - \psi_F + \psi_F - X_{n+1})^2 = E_F(T - \psi_F)^2 + \text{var}_F(X_{n+1}),$$

since linear T is uncorrelated with X_{n+1}. We have no control over $\text{var}_F(X_{n+1})$ and thus can only seek T to make $E_F(T - \psi_F)^2$ small. In particular, we see *that the best linear unbiased predictor* ($E_F T = E_F X_{n+1}$) *of* X_{n+1} *is the B.L.U.E. of* $\psi_F(E_F X_{n+1})$.

Corresponding results can be obtained when the X_i's are correlated.

A Detailed Example to Illustrate Linear Estimation Concepts:

Two-Way Analysis of Variance (*ANOVA*): With 3 kinds of corn and 4 kinds of pesticides, suppose there are 2 observations with each of the 12 possible combinations of corn and pesticide. Let Y_{ijr} = height in decimeters of the rth ($r = 1, 2$) plant on variety $i(1, 2, 3)$ with pesticide $j(1, 2, 3, 4)$. The "observed values" of Y_{ijr} are displayed here. (Of course, since Y_{ij1} and Y_{ij2}, for fixed i and j, are identically distributed, it does not really matter which value in the $(i, j)^{\text{th}}$ box is called Y_{ij1} and which is called Y_{ij2}.)

i \ j	1	2	3	4
1	10 12	15 13	11 8	16 17
2	15 18	17 18	15 19	21 22
3	13 15	16 15	15 14	18 20

The model is that the Y_{ijr} are uncorrelated with common variance σ_F^2 (unknown), and that

$$E_F Y_{ijr} = \alpha_i(F) + \beta_j(F), \tag{5.18}$$

where the α_i's and β_j's are unknown. If we let $\phi' = (\alpha_1, \alpha_2, \alpha_3, \beta_1, \beta_2, \beta_3, \beta_4)$ and let the 24×1 vector X consist of the Y_{ijr} listed in any order, then according to (5.18) the row of B corresponding to Y_{ijr} in the usual model $E_F X = B\phi(F)$, has a 1 in the ith place and in the $(j+3)^{\text{th}}$ place (since there are 3 α_i's) and a 0 elsewhere. (For example, $EY_{23r} = (0100010)\phi = \alpha_2 + \beta_3$.)

We use the "dot" notation for averages:

$$Y_{ij\cdot} = \frac{1}{2}(Y_{ij1} + Y_{ij2}); \qquad Y_{i\cdot\cdot} = \frac{1}{8}\sum_{j,r} Y_{ijr} = \frac{1}{4}\sum_j Y_{ij\cdot};$$

$$Y_{\cdot j\cdot} = \frac{1}{3}\sum_i Y_{ij\cdot}; \qquad Y_{\cdots} = \frac{1}{24}\sum_{i,j,r} Y_{ijr}.$$

Before solving the normal equations (or even writing them), let us give an example of the construction of a B.L.U.E. without solving the normal equations, in the case of a linear parametric function for which an intuitive guess of a reasonable linear unbiased estimator is easy to give: Suppose we want to estimate $\alpha_1 - \alpha_2$. A reasonable guess is that $t^* = \text{const.} \times$ [Sum of all obs'ns on variety 1 − Sum of all obs'ns on variety 2] might do the job. We check the two conditions for this to give a B.L.U.E.:

(i) *Unbiasedness:* $E_F \ \text{const.}\left[\sum_{j,r} Y_{1jr} - \sum_{j,r} Y_{2jr}\right] = \text{const.}\sum_{j,r} E_F(Y_{1jr} - Y_{2jr}) =$ by (5.18) $\text{const.}\sum_{j,r}[\alpha_1(F) - \alpha_2(F)] = \text{const.} \times 8[\alpha_1(F) - \alpha_2(F)]$. Hence, $\text{const.} = \frac{1}{8}$ to give an unbiased estimator; that is, $t^* = \frac{1}{8}\sum_{j,r}(Y_{1jr} - Y_{2jr}) = Y_{1\cdot\cdot} - Y_{2\cdot\cdot}$.

(ii) *Bestness*: One must check that $t^* = a'X$ for a of the form $Bq = q_1 b_1 + q_2 b_2 + \cdots + q_7 b_7$ where the q_i's are scalars and b_j is the jth column of B. A quick inspection shows that $(\frac{1}{8}b_1 - \frac{1}{8}b_2)'X = Y_{1\cdot\cdot} - Y_{2\cdot\cdot}$. Thus, t^* is the B.L.U.E. of $\alpha_1 - \alpha_2$.

Of course, the preceding method can fail as a result of guessing an estimator which is not best. For example, one might be tempted to estimate $\alpha_1 + \beta_2$ by $Y_{12\cdot}$. Then (i) works, but (ii) does not; $Y_{12\cdot}$ is an unbiased estimator of $\alpha_1 + \beta_2$, but is not best. (From what follows it will be clear that $Y_{1\cdot\cdot} + Y_{\cdot 2\cdot} - Y_{\cdots}$ is the B.L.U.E. of $\alpha_1 + \beta_2$.)

Now for $B'B$ and $B'X$. It may help you to write out $EX = B\phi$ in a form such as

$$E\begin{pmatrix} Y_{111} \\ Y_{112} \\ Y_{121} \\ Y_{122} \\ \vdots \\ Y_{342} \end{pmatrix} = \begin{pmatrix} 1001000 \\ 1001000 \\ 1000100 \\ 1000100 \\ \vdots \\ 0010001 \end{pmatrix}\begin{pmatrix} \alpha_1 \\ \alpha_2 \\ \alpha_3 \\ \beta_1 \\ \beta_2 \\ \beta_3 \\ \beta_4 \end{pmatrix}$$

giving X in some order and B explicitly. This is helpful in examples of small n. In larger examples like the present one, it is often possible to obtain the normal equations from the description of B following (5.18). Thus, since there are 8 ones and 16 zeros in each of b_1, b_2, b_3, we have $b_i'b_i = 8$ for $i = 1$, 2, 3, and similarly $b_i'b_i = 6$ for $i = 4, 5, 6, 7$. Since no observation corresponds to two different varieties, if any element of b_1 is 1 the corresponding element of b_2 is 0; thus $b_1'b_2 = 0$, and similarly $0 = b_1'b_3 = b_2'b_3$ and $0 = b_4'b_5 = b_4'b_6 = b_4'b_7 = b_5'b_6 = b_5'b_7 = b_6'b_7$. Finally, there are two plants of variety 1 with pesticide 2, so $b_1'b_4 = 2$ and similarly $b_i'b_j = 2$ for $\left\{\begin{matrix} 1 \le i \le 3 \\ 4 \le j \le 7 \end{matrix}\right\}$. Finally, $b_i'X = 8Y_{1..}$ for $1 \le i \le 3$ and $b_{3+j}'X = 6Y_{.j.}$ for $1 \le j \le 4$. The normal equations are thus

$$\begin{pmatrix} 8 & 0 & 0 & 2 & 2 & 2 & 2 \\ 0 & 8 & 0 & 2 & 2 & 2 & 2 \\ 0 & 0 & 8 & 2 & 2 & 2 & 2 \\ 2 & 2 & 2 & 6 & 0 & 0 & 0 \\ 2 & 2 & 2 & 0 & 6 & 0 & 0 \\ 2 & 2 & 2 & 0 & 0 & 6 & 0 \\ 2 & 2 & 2 & 0 & 0 & 0 & 6 \end{pmatrix} \begin{pmatrix} \tilde{t}_1 \\ \tilde{t}_2 \\ \tilde{t}_3 \\ \tilde{\tilde{t}}_1 \\ \tilde{\tilde{t}}_2 \\ \tilde{\tilde{t}}_3 \\ \tilde{\tilde{t}}_4 \end{pmatrix} = \begin{pmatrix} 8Y_{1..} \\ 8Y_{2..} \\ 8Y_{3..} \\ 6Y_{.1.} \\ 6Y_{.2.} \\ 6Y_{.3.} \\ 6Y_{.4.} \end{pmatrix} = \begin{pmatrix} 102 \\ 145 \\ 126 \\ 83 \\ 94 \\ 82 \\ 114 \end{pmatrix}$$

where we have written $\tilde{t}_1, \tilde{t}_2, \tilde{t}_3$ for the first three elements of a solution t_{LS}, and $\tilde{\tilde{t}}_1, \tilde{\tilde{t}}_2, \tilde{\tilde{t}}_3, \tilde{\tilde{t}}_4$ for the next four elements, so as to make the notation reflect the meanings "row effects" and "column effects."

It is not hard to see that $h = 6$. In fact, this could be seen directly from looking at B, as follows: On the one hand, $k - h \ge 1$ (and hence $h \le 6$) because $(b_1 + b_2 + b_3) - (b_4 + b_5 + b_6 + b_7) =$ (vector of all 1's) $-$ (same) $= 0$. On the other hand, it is easy to produce 6 linearly independent estimable parametric functions:

$$\alpha_1 - \alpha_2, \alpha_2 - \alpha_3; \qquad \beta_1 - \beta_2, \beta_2 - \beta_3, \beta_3 - \beta_4; \qquad \alpha_1 + \beta_1. \quad (5.19)$$

(i) That these are *estimable* follows from the fact that each can be written as $\underset{(1\times 7)}{c'}\ \underset{(7\times 1)}{\phi}$ where c' is a linear combination of rows of $B'B$ or of B; or can be written as the expectation of a linear combination of Y_{ijr}'s (e.g., $\beta_1 - \beta_2 =$ (first minus third row of $B)\phi$ *or* $\beta_1 - \beta_2 = \frac{1}{6}$(fourth minus fifth row of $B'B)\phi$ *or* $\beta_1 - \beta_2 = E(Y_{111} - Y_{121})$ *or* $= E(Y_{.1.} - Y_{.2.})$, etc).

(ii) That the six linear parametric functions exhibited in (5.19) are *linearly independent* can be seen by noting that the first set of two is obviously linearly independent, the next set of three is linearly independent, and these five together are linearly independent since the first two depend on different variables from the next three; finally, each of the first five functions of (5.19) has coefficients summing to 0 $\left(\text{is of the form } c'\phi \text{ with } \sum_1^7 c_i = 0\right)$, *as does therefore any linear combination of them*, so that $\alpha_1 + \beta_1$, with the sum of coefficients $= 2$, cannot possibly be represented as a linear combination of the first five. Thus $\alpha_1 + \beta_1$ is a 6th linear parametric function, linearly independent of the first five of (5.19). [*Remark*: One would often see $\left(8\sum_1^3 \alpha_i + 6\sum_1^4 \beta_j\right)\Big/24$, or some constant multiple of this, listed as the 6th member of (5.19), representing $E\sum_{i,j,r} Y_{ijr}/24 =$ "grand mean"; but for the sake of establishing the value of h, *any* estimable linear parametric function not of the form $c'\phi$ with $\sum c_i = 0$ would do.]

There is thus a $(k-h)$-parameter, or 1-parameter, family of solutions to the normal equations. You shouldn't have to be told how to solve such equations. As an example, though, of how to solve them, on staring at the normal equations for a few seconds one is led to trying to get rid of the 2's by considering differences among the first three rows and among the next four rows:

$$(\tilde{t}_1 - \tilde{t}_2) = Y_{1\cdot\cdot} - Y_{2\cdot\cdot}, \qquad (\tilde{t}_2 - \tilde{t}_3) = Y_{2\cdot\cdot} - Y_{3\cdot\cdot};$$

$$(\tilde{\tilde{t}}_1 - \tilde{\tilde{t}}_2) = Y_{\cdot 1\cdot} - Y_{\cdot 2\cdot}, \quad (\tilde{\tilde{t}}_2 - \tilde{\tilde{t}}_3) = Y_{\cdot 2\cdot} - Y_{\cdot 3\cdot}, \quad (\tilde{\tilde{t}}_3 - \tilde{\tilde{t}}_4) = Y_{\cdot 3\cdot} - Y_{\cdot 4\cdot}.$$

The first two of these imply $\tilde{t}_i = Y_{i\cdot\cdot} - L_1$, and the next three imply $\tilde{\tilde{t}}_j = Y_{\cdot j\cdot} - L_2$, where L_1 and L_2 are functions of the $Y_{\cdot j\cdot}$'s and $Y_{i\cdot\cdot}$'s. Working with the *numbers* on the right side of the normal equations, as one might in practice, one might simply *try* (perhaps without realizing it corresponds to guessing with $L_1 = 0$)

$$\tilde{t}_1 = 102/8, \qquad \tilde{t}_2 = 145/8, \qquad \tilde{t}_3 = 126/8,$$

and then find as a consequence, from the normal equations just given,

$$\tilde{\tilde{t}}_1 = \frac{1}{6}[83 - 2(\tilde{t}_1 + \tilde{t}_2 + \tilde{t}_3)] = \frac{83}{6} - \frac{1}{24}(102 + 145 + 126) = \frac{83}{6} - \frac{373}{24},$$

and similarly

$$\tilde{\tilde{t}}_2 = \frac{94}{6} - \frac{373}{24}, \qquad \tilde{\tilde{t}}_3 = \frac{82}{6} - \frac{373}{24}, \qquad \tilde{\tilde{t}}_4 = \frac{114}{6} - \frac{373}{24}.$$

Checking the normal equations, it is easily seen that these *do* give a solution. Thus, without going through the *theoretical* possibilities (1-parameter family of solutions) we have obtained the particular solution corresponding to $L_1 = 0$, $L_2 = Y_{\cdots}$; similarly, if we had started with trial $\tilde{\tilde{t}}_1 = 83/6$, we would have tried $\tilde{\tilde{t}}_2 = 94/6$, $\tilde{\tilde{t}}_3 = 82/6$, $\tilde{\tilde{t}}_4 = 114/6$, and then would have obtained $\tilde{t}_1 = \frac{102}{8} - \frac{373}{24}$, $\tilde{t}_2 = \frac{145}{8} - \frac{373}{24}$, $\tilde{t}_3 = \frac{126}{8} - \frac{373}{24}$. This would correspond to the theoretical solution with $L_1 = Y_{\cdots}$ and $L_2 = 0$.

From the theoretical point of view we can find the 1-parameter family of solutions by substituting $\tilde{t}_i = Y_{i\cdot\cdot} - L_1$ into the fourth normal equation, yielding (since $\tilde{\tilde{t}}_1 = Y_{\cdot 1\cdot} - L_2$) $L_1 + L_2 = Y_{\cdots}$; thus, the general solution is, for arbitrary L_1,

$$\begin{cases} \tilde{t}_i = Y_{i\cdot\cdot} - L_1, & 1 \le i \le 3, \\ \tilde{\tilde{t}}_j = Y_{\cdot j\cdot} - Y_{\cdots} + L_1, & 1 \le j \le 4. \end{cases} \tag{5.20}$$

We only list this to see what is going on in the way of nonunique solutions of the normal equations: from the point of view of developing either general theoretical results about the form of estimators of estimable linear parametric functions in terms of the Y_{ijr}'s, or of finding numerical solutions for the data at hand directly, one need not know (5.20), but rather just *one solution of the normal equations.*

We pick the first solution for what follows: $\tilde{t}_1 = 102/8$, etc., for the direct solution in numbers, and $\tilde{t}_i = Y_{i\cdot\cdot}$, $\tilde{\tilde{t}}_j = Y_{\cdot j\cdot} - Y_{\cdots}$ for the theoretical development. You should check that the other possible solution exhibited would yield the same results.

An often-used description of estimable parametric functions and their estimators in such analysis of variance (ANOVA) problems will now be given. A *contrast* is a linear parametric function $c'\phi$ with $\sum_1^7 c_i = 0$; in particular a contrast with $c_4 = c_5 = c_6 = c_7 = 0$ is called a *variety* (or *row effect*) *contrast* (e.g., $\alpha_1 - \alpha_2$ or $\alpha_1 + \alpha_2 - 2\alpha_3$), and a contrast with $c_1 = c_2 = c_3 = 0$ is called a *pesticide* (or *column effect*) contrast.

The description (5.19) and the preceding normal equation solution can be used (with the replacement of $\alpha_1 + \beta_1$ by $\psi \overset{\text{def}}{=} \frac{1}{4}\sum_1^4 \beta_j + \frac{1}{3}\sum_1^3 \alpha_i$ as indicated in the remark following (ii) to be customary) to describe *all B.L.U.E.'s*, as follows:

$$\begin{aligned}
&\text{Any variety contrast } \sum_1^3 c_i\alpha_i \text{ with } \sum_1^3 c_i = 0 \text{ has B.L.U.E. } \sum_1^3 c_i Y_{i\cdot\cdot};\\
&\text{Any pesticide contrast } \sum_1^4 c_{3+j}\beta_j \text{ with } \sum_1^4 c_{3+j} = 0 \text{ has B.L.U.E. } \sum_1^4 c_{3+j} Y_{\cdot j\cdot}\\
&\left(\text{since } \sum_1^4 c_{3+j}(Y_{\cdot j\cdot} - Y_{\cdots}) = \sum_1^4 c_{3+j}Y_{\cdot j\cdot} - \underbrace{\left(\sum_1^4 c_{3+j}\right)}_{=0} Y_{\cdots}\right);\\
&\psi \text{ has B.L.U.E. } Y_{\cdots};
\end{aligned} \tag{5.21}$$

Any estimable linear parametric function is a linear combination of a variety contrast, a pesticide contrast and ψ, and its B.L.U.E. is the corresponding linear combination of the three B.L.U.E.'s just listed.

If one deals directly with data and wants numbers without going through (5.21), one simply notes, for example, that $\tilde{\tilde{t}}_2 - \tilde{\tilde{t}}_3 = \frac{3}{24} - \left(\frac{-45}{24}\right) = 2.0$. However, to know that this really represents something of interest and is not a nonestimator (as $\tilde{\tilde{t}}_2$ or $\tilde{\tilde{t}}_2 - 2\tilde{\tilde{t}}_3$ is, each depending on which solution to the normal equations one chooses), one would have to have gone through the theory at least to the point of knowing that "all pesticide contrasts are estimable." Thus 2.0 is meaningful, as the B.L.U.E. of $\beta_2 - \beta_3$.

Computation of $Q_{\min}$

Direct Method. First compute Bt_{LS}. It is easiest to exhibit it *not* as a 24-vector, but as an array corresponding to the 3×4 array with 2 entries per cell of the Y_{ij}'s; since the two entries of Bt in each cell are the same, we list each only once:

Direct numerical evaluation of Bt_{LS}

11.04	etc.	

Est. of $\alpha_1 + \beta_1 = \tilde{t}_1 + \tilde{\tilde{t}}_1$
$= 12.75 + (-1.71) = 11.04$

Theoretical development

$Y_{3..} + Y_{.1.} - Y_{...}$	etc.	

B.L.U.E. of $\alpha_i + \beta_j$ is $\tilde{t}_i + \tilde{\tilde{t}}_j$
$= Y_{i..} + Y_{.j.} - Y_{...}$

Thus,

$$Q_{\min} = \sum_{i,j,r} [Y_{ijr} - (\tilde{t}_i + \tilde{\tilde{t}}_j)]^2 = (10 - 11.04)^2 + (12 - 11.04)^2 + \cdots, \quad \text{(24 terms)}$$

where $\tilde{t}_i + \tilde{\tilde{t}}_j =$ B.L.U.E. of $EY_{ijr} = \underbrace{(\text{row of } B \text{ corresponding to } Y_{ijr})}_{1\times 7}\, \underset{7\times 1}{t_{LS}}$

or, in general theoretical terms, since $\tilde{t}_i + \tilde{\tilde{t}}_j = Y_{i..} + Y_{.j.} - Y_{...}$,

$$Q_{\min} = \sum_{i,j,r} [Y_{ijr} - Y_{i..} - Y_{.j.} + Y_{...}]^2.$$

The preceding is not easiest computationally.

Among Other Methods. Recall, either from geometry (Pythagorean principle; $X - Bt_{LS}$ and Bt_{LS} are orthogonal) or from expanding $Q_{\min}$ and using the normal equations, that

$$Q_{\min} = (X - Bt_{LS})'(X - Bt_{LS}) = X'X - t'_{LS}(B'B)t_{LS}$$

$$= \sum_{i,j,r} Y_{i,j,r}^2 - \left[8\sum_i Y_{i..}^2 + 6\sum_j (Y_{.j.} - Y_{...})^2 + 2 \times 2\sum_{i,j} Y_{i..}(Y_{.j.} - Y_{...})\right];$$

the last sum's coefficient comes from the entry 2 in $B'B$ and the appearance twice of $\tilde{t}_i\tilde{\tilde{t}}_j$ (as of uv in $(u + v)^2$).

Since $\sum_j (Y_{.j.} - Y_{...}) = 0$, the last sum over i, j is 0.

After also expanding $6\sum_j (Y_{.j.} - Y_{...})^2$, we get

$$Q_{\min} = \sum_{i,j,r} Y_{ijr}^2 - 8\sum_i Y_{i..}^2 - 6\sum_j Y_{.j.}^2 + 24Y_{...}^2. \tag{5.22}$$

This expression, also derivable from the last expression for $Q_{\min}$ under "*Direct Method*," is usually better for computational purposes because the $Y_{i..}$'s, etc., have been computed anyway in solving the normal equations, and the big sum of 24 terms $\left(\sum_{ijr}\right)$ is easier in terms of the Y_{ijr} (giving $10^2 + 12^2 + \cdots$) instead of the $(Y_{ijr} - \text{B.L.U.E. of } EY_{ijr})^2$ (giving $(10 - 11.04)^2 + (12 - 11.04)^2 + \cdots$). In the present example

$$\sum Y_{ijr}^2 = 6061,$$

$$8\sum_i Y_{i..}^2 = \frac{1}{8}(102^2 + 145^2 + 126^2) = \frac{47{,}305}{8},$$

$$6\sum_j Y_{.j.}^2 = 35{,}445/6,$$

$$24Y_{...}^2 = 24\left(\frac{373}{24}\right)^2 = \frac{139{,}129}{24},$$

from which we get $Q_{\min} = 33.25$. Hence, the standard unbiased estimator of σ_F^2 is $Q_{\min}/(n-h) = \frac{33.25}{18} = 1.84$.

Since $B'B$ is singular, we cannot invert it to obtain variances of B.L.U.E.'s. [The device of "pseudo-inverses" is sometimes used, but will not be discussed.] However, it is easy directly to compute such variances: Corresponding to (5.21), we have

$$\operatorname{var}_F\left(\sum_1^3 c_i Y_{i..}\right) = \sigma_F^2 \sum_1^3 c_i^2/8$$

(since $\operatorname{var}_F Y_{i..} = \frac{\sigma_F^2}{8}$ and the $Y_{i..}$'s are uncorrelated), and

$$\operatorname{var}_F\left(\sum_1^4 c_{3+j} Y_{.j.}\right) = \sigma_F^2 \sum_1^4 c_{3+j}^2/6, \qquad \operatorname{var}_F(Y_{...}) = \sigma_F^2/24.$$

As for covariances, we have similarly, say, for two estimators $\sum_1^3 c_i Y_{i..}$ and $\sum_1^3 \bar{c}_i Y_{i..}$,

$$\operatorname{cov}_F\left(\sum_1^3 c_i Y_{i..}, \sum_1^3 \bar{c}_i Y_{i..}\right) = \sigma_F^2 \sum_1^3 c_i \bar{c}_i/8.$$

One interesting fact is that it is easy to establish that any *row-contrast estimator* is uncorrelated with any *column-contrast estimator*, and either is uncorrelated with $Y_{...}$: for example, since $\operatorname{cov}_F(Y_{i..}, Y_{.j.}) = (\frac{1}{8})(\frac{1}{6})2\sigma_F^2$ for each i, j (because $Y_{i..}$ and $Y_{.j.}$ have exactly the two Y_{ijr}'s in common), we see that

$$\sum_1^3 c_i = 0 \Rightarrow \operatorname{cov}_F\left(\sum_1^3 c_i Y_{i..}, Y_{.j.}\right) = \sum_{i=1}^3 c_i(\operatorname{cov}_F(Y_{i..}, Y_{.j.})) = 0,$$

the last because $\sum c_i = 0$ implies $\sum c_i \sigma_F^2/24 = 0$. Thus, we have shown that $\sum_1^3 c_i Y_{i\cdot\cdot}$ is uncorrelated with $Y_{\cdot j\cdot}$ and hence with $\sum_1^4 \bar{c}_{3+j} Y_{\cdot j \cdot}$.

This leads to a topic to be discussed under *hypothesis testing*: the representation (5.22) of $Q_{\min}$ can be rewritten

$$X'X - \left[8\sum_1^3 (Y_{i\cdot\cdot} - Y_{\cdot\cdot\cdot})^2 + 6\sum_1^4 (Y_{\cdot j\cdot} - Y_{\cdot\cdot\cdot})^2 + 24 Y_{\cdot\cdot\cdot}^2 \right],$$

and the three terms in brackets are, successively, the squared lengths of the projections of Bt_{LS} onto three *mutually orthogonal* subspaces of $\bar{H}$:

$$\text{a space of dimension 2 consisting of } \left\{ \sum_1^3 q_i b_i, \sum_1^3 q_i = 0 \right\};$$

$$\text{a space of dimension 3 consisting of } \left\{ \sum_4^7 q_i b_i, \sum_4^7 q_i = 0 \right\};$$

$$\text{a space of dimension 1 consisting of multiples of } \sum_1^3 b_i = (1, 1, \ldots, 1)'.$$

In this very symmetric model we thus obtain a simple and useful decomposition of the "fit" Bt_{LS} into three orthogonal vectors corresponding to "row effect," "column effect," "grand mean effect." If the number of observations per cell had not been constant (i.e. 2), such a simple decomposition might not have been obtained, and estimators of row and column contrasts might not have been uncorrelated.

Prediction Problem. Two thousand plants are to be planted with variety 1 and pesticide 1. What is the total yield (sum of heights) to be expected?

Answer. The B.L.U.E. of $2{,}000(\alpha_1 + \beta_1)$ is $2{,}000(\tilde{t}_1 + \tilde{\tilde{t}}_1) = 24{,}400$. The *variance* of this prediction is

$$\operatorname{var}_F(2{,}000(\tilde{t}_1 + \tilde{\tilde{t}}_1))$$

$$= 4{,}000{,}000 \operatorname{var}_F(\tilde{t}_1 + \tilde{\tilde{t}}_1) = 4{,}000{,}000 \operatorname{var}_F(Y_{1\cdot\cdot} + (Y_{\cdot 1\cdot} - Y_{\cdot\cdot\cdot}))$$

$$\overset{(*)}{=} 4{,}000{,}000\left[\operatorname{var}_F Y_{1\cdot\cdot} + \operatorname{var}_F\left(\frac{3}{4} Y_{\cdot 1\cdot} - \frac{Y_{\cdot 2\cdot} + Y_{\cdot 3\cdot} + Y_{\cdot 4\cdot}}{4} \right) \right]$$

$$\overset{(**)}{=} 4{,}000{,}000\sigma_F^2\left[\frac{1}{8} + \frac{1}{6}\left(\frac{9}{16} + \frac{1}{16} + \frac{1}{16} + \frac{1}{16} \right) \right] = 10^6 \sigma_F^2,$$

where $(*)$ used $\operatorname{cov}_F(Y_{1\cdot\cdot}, Y_{\cdot 1\cdot} - Y_{\cdot\cdot\cdot}) = 0$, and $(**)$ used $\operatorname{var}_F(Y_{\cdot j\cdot}) = \sigma_F^2/6$. An *unbiased estimator* of var_F(predictor) is thus $10^6 \times 1.84$. The square root of this (not an unbiased estimator of the standard deviation of $2{,}000(\tilde{t}_1 + \tilde{\tilde{t}}_1)$, but often used to estimate it) is 1.36×10^3, to be compared with the prediction 24.4×10^3: Practitioners might write "predicted yield $\pm$ 2(estimated) standard deviation units $= (24.4 \pm 2.7)10^3$."

Other Representations. The *same model* is often written instead as

$$E_F Y_{ijr} = \mu + \alpha_i + \beta_j$$

(8 parameters), or sometimes in this form with the assumptions $\sum_1^3 \alpha_i = 0$ and $\sum_1^4 \beta_j = 0$, which assumptions can be made because the linear parametric functions $c_1 \sum \alpha_i + c_2 \sum \beta_j$, where c_1 and c_2 are constant scalars, are all unidentifiable (nonestimable), and a 6-dimensional space orthogonal to them (in the space of linear parametric functions) contains all *estimable* linear parametric functions.

This does not change any of the contrast estimates, etc., from what we obtained; it is sometimes a useful computational device to adopt such a modified representation of the model.

Orthogonalization

(This subsection treats an optional topic.)
Returning to the setup of (5.18) with three α's and four β's, we seek a 7×7 nonsingular matrix D (the C' of the subsection on orthogonal polynomials) such that $(BD)'(BD)$ is diagonal. We will depart from the development of that subsection in two ways (as is often done in practice), neither of which affects the final result: (i) for convenience, we need not choose D to be triangular; (ii) D need not have determinant 1. There are many possible D's, and we discuss only one of them.

Mainly, we want the nonzero columns of BD to be orthogonal. Since the *last four columns* of B are already mutually orthogonal, leave them alone: the last four columns of D have 1's on the diagonal and 0's elsewhere. (If we had begun with the *first three columns* in this way, more work would remain ahead.)

Since the inner product of any of the first three columns b_1, b_2, b_3 of B with any of the last four columns is 2 (because of the two observations in row i, column j), we see that $\sum_{i=1}^3 d_{ij} b_i$ (with d_{ij} scalars) has inner product $2 \sum_{i=1}^3 d_{ij}$ with any of the last four columns, and this is 0 if $\sum_{i=1}^3 d_{ij} = 0$. (This actually came out in our earlier calculations, but we pretend here that one is orthogonalizing from scratch.) Also, the inner product of $b_1^* = \sum_{i=1}^3 d_{i1} b_i$ with $b_2^* = \sum_{i=1}^3 d_{i2} b_i$ is $8 \sum_1^3 d_{i1} d_{i2}$, since $b_i' b_i = 8$ and $b_i' b_j = 0$ for $1 \le i < j \le 3$. Thus, b_1^* and b_2^* will be orthogonal to each other and to b_4, b_5, b_6, b_7 if the vectors (d_{11}, d_{21}, d_{31}) and (d_{12}, d_{22}, d_{32}) are orthogonal and each has 0 as the sum of its elements. The latter restriction yields $(1, -1, 0)$ as the simplest choice for (d_{11}, d_{21}, d_{31}).

This vector is obviously orthogonal to scalar multiples of $(1, 1, p)$, and $p = -2$ makes the latter also have 0 as sum of elements. Thus, we choose

$$(c_{11}, c_{21}, \ldots, c_{71}) = (1, -1, 0, 0, 0, 0, 0),$$

$$(c_{12}, c_{22}, \ldots, c_{72}) = (1, 1, -2, 0, 0, 0, 0)$$

for the transposes of the first two columns of D. [If we had wanted to make D have determinant 1, these vectors would have been multiplied by $1/\sqrt{2}$ and $1/\sqrt{6}$, respectively, an unnecessary arithmetical complication; see (ii) following (5.19).]

The remaining (third) column of BD will consist of zeros, as mentioned at the end of the subsection on orthogonal polynomials for the case $h < k$. We now find the third column of D. Since the first three columns of B sum to 1 as do the last four columns, we have $B(1, 1, 1, -1, -1, -1, -1)' = 0$, and thus $(1, 1, 1, -1, -1, -1, -1)'$ can be taken as the third column of D. Thus, we have constructed

$$D = \begin{pmatrix} 1 & 1 & 1 & 0 & 0 & 0 & 0 \\ -1 & 1 & 1 & 0 & 0 & 0 & 0 \\ 0 & -2 & 1 & 0 & 0 & 0 & 0 \\ 0 & 0 & -1 & 1 & 0 & 0 & 0 \\ 0 & 0 & -1 & 0 & 1 & 0 & 0 \\ 0 & 0 & -1 & 0 & 0 & 1 & 0 \\ 0 & 0 & -1 & 0 & 0 & 0 & 1 \end{pmatrix}.$$

By construction, $(BD)'(BD)$ is a diagonal matrix, and we need only compute its diagonal elements. The last four diagonal elements are by construction the same as those of $B'B$, namely, 6. The third diagonal element is 0, since the third column of BD has all 0's. The first column of BD contains eight 1's and eight -1's, so its inner product with itself is 16. The second column has sixteen 1's and eight -2's, so its inner product with itself is 48. Similarly multiplying $(BD)'X$, we obtain as the normal equations for estimating $\psi = D^{-1}\phi$, the diagonalized equations

$$\begin{aligned} 16\bar{\bar{t}}_1 &= 8Y_{1\cdot\cdot} - 8Y_{2\cdot\cdot} \\ 48\bar{\bar{t}}_2 &= 8Y_{1\cdot\cdot} + 8Y_{2\cdot\cdot} - 16Y_{3\cdot\cdot} \\ 0 &= 0 \\ \bar{\bar{t}}_4 &= 6Y_{\cdot 1\cdot} \\ \bar{\bar{t}}_5 &= 6Y_{\cdot 2\cdot} \\ \bar{\bar{t}}_6 &= 6Y_{\cdot 3\cdot} \\ \bar{\bar{t}}_7 &= 6Y_{\cdot 4\cdot} \end{aligned}$$

Thus, $\bar{\bar{t}}' = (\bar{\bar{t}}_1, \bar{\bar{t}}_2, 0, \bar{\bar{t}}_4, \bar{\bar{t}}_5, \bar{\bar{t}}_6, \bar{\bar{t}}_7)$ is immediately obtained. Consequently, $D\bar{\bar{t}}$, which is easy to compute, is a solution to the original normal equations for estimating $\phi = D\psi$, and if $c'\phi$ is any estimable linear parametric function we thus obtain $c'D\bar{\bar{t}}$ as its B.L.U.E.. For example, the B.L.U.E. of $\phi_1 - \phi_3 = \alpha_1 - \alpha_3$ is $(1, 0, -1, 0, 0, 0, 0)D\bar{\bar{t}} = \bar{\bar{t}}_1 + 3\bar{\bar{t}}_3 = Y_{1\cdot\cdot} - Y_{3\cdot\cdot}$, as was also obtained earlier. *Note that it was unnecessary to compute* D^{-1}. If one retains the representation of linear parametric functions one wants to estimate in terms of the original ϕ, there is no need to know what $\psi = D^{-1}\phi$ is, except that in more complicated problems it may give additional help in understanding which parametric functions are estimable. Just to complete the theoretical development, we now compute D^{-1}. One should not blindly compute it from the general formula for an inverse. There are several quicker methods. For example, partition D as $\begin{pmatrix} D_1 & \bar{0} \\ D_2 & I_4 \end{pmatrix}$ where D_1 is 3×3, $\bar{0}$ is a 3×4 block of zeros, and I_4 is the 4×4 identity. Correspondingly, write $D^{-1} = \begin{pmatrix} A_1 & A_3 \\ A_2 & A_4 \end{pmatrix}$ (to be determined). Then the definition of inverse is

$$\begin{aligned} \begin{pmatrix} I_3 & \bar{0} \\ \bar{0} & I_4 \end{pmatrix} = I_7 = DD^{-1} &= \begin{pmatrix} D_1 & \bar{0} \\ D_2 & I_4 \end{pmatrix} \begin{pmatrix} A_1 & A_3 \\ A_2 & A_4 \end{pmatrix} \\ &= \begin{pmatrix} D_1 A_1 & D_1 A_3 \\ D_2 A_1 + A_2 & D_2 A_3 + A_4 \end{pmatrix}, \end{aligned} \tag{5.23}$$

by matrix multiplication in terms of the submatrices. Comparing corresponding entries on the left and right sides of (5.23), the upper-right entries yield $A_3 = 0$, since D_1 is nonsingular. Substituting this into the lower-right entry, we obtain $A_4 = I_4$. The upper-left entries yield $A_1 = D_1^{-1}$, and this substituted into the lower-left yields $A_2 = -D_2 D_1^{-1}$. Thus, we need only compute, by standard methods,

$$D_1^{-1} = \frac{1}{6}\begin{pmatrix} 3 & -3 & 0 \\ 1 & 1 & -2 \\ 2 & 2 & 2 \end{pmatrix}$$

and thus

$$D^{-1} = \begin{pmatrix} D_1^{-1} & 0 \\ -D_2 D_1^{-1} & I_4 \end{pmatrix} = \begin{pmatrix} 1/2 & -1/2 & 0 & 0 & 0 & 0 & 0 \\ 1/6 & 1/6 & -1/3 & 0 & 0 & 0 & 0 \\ 1/3 & 1/3 & 1/3 & 0 & 0 & 0 & 0 \\ 1/3 & 1/3 & 1/3 & 1 & 0 & 0 & 0 \\ 1/3 & 1/3 & 1/3 & 0 & 1 & 0 & 0 \\ 1/3 & 1/3 & 1/3 & 0 & 0 & 1 & 0 \\ 1/3 & 1/3 & 1/3 & 0 & 0 & 0 & 1 \end{pmatrix}.$$

(The preceding amounts to solving the equation $\phi = D\psi$ for ψ in terms of ϕ, where ψ is the new vector of parameters as in the subsection on orthogonal polynomials; any method you see of solving these yields $\psi = D^{-1}\phi$, and it is really ψ that we want.)

Thus, we have

$$\psi = D^{-1}\phi = \begin{pmatrix} (\alpha_1 - \alpha_2)/6 \\ (\alpha_1 + \alpha_2 - 2\alpha_3)/6 \\ (\alpha_1 + \alpha_2 + \alpha_3)/3 \\ \beta_1 + (\alpha_1 + \alpha_2 + \alpha_3)/3 \\ \beta_2 + (\alpha_1 + \alpha_2 + \alpha_3)/3 \\ \beta_3 + (\alpha_1 + \alpha_2 + \alpha_3)/3 \\ \beta_4 + (\alpha_1 + \alpha_2 + \alpha_3)/3 \end{pmatrix},$$

where the third entry is inestimable (since BD has third column zero) and all linear combinations of the other six elements of ψ are estimable. The first two elements span the space of row effect contrasts, and the last four elements are easily seen to span the same space as $\{\frac{1}{4}\sum\beta_j + \frac{1}{3}\sum\alpha_j$ and column effect contrasts$\}$. This checks with the results obtained earlier concerning which parametric functions are estimable.

General Remark on Orthogonalization. Keep in mind that the main purpose is computational and ordinarily would be used mainly to compute tables like those of Fisher and Yates. You would probably find it quickest to use Gauss elimination on the original normal equations if you were in the field with (i) no calculator or computer, (ii) no tables of BD, (iii) a need to find B.L.U.E.'s or Bt_{LS} on the spot. For, otherwise, use a calculator along with tables of BD, or use a computer to solve the normal equations in terms of the original B!

Another Computational Remark. Some presentations of linear estimation emphasize the *generalized inverse* (gi) of $B'B$; when $h < k$, this is a matrix that can be used just as $(B'B)^{-1}$ is in obtaining a solution $\bar{t} = (B'B)^{-1}B'X$ when $h = k$. It is not necessary to know about generalized inverses to understand linear estimation, and in fact the simple geometric picture is often obscured in presentations that emphasize gi's. Computationally one uses Gauss elimination as a more efficient approach to solving the normal equations, just as when $h = k$; a main use of the inverse of gi is thus to solve the normal equations in settings that arise repeatedly and in which a simple formula for the gi (or inverse) can be put into the computer memory.

Problems

The problems are divided into three parts. Part I is on Section 5.1, and II and III are on the simpler and more complex applications of Section 5.4, respectively.

I. *Problems on Section 5.1* Suggested: 5.1, 5.2, or 5.8–5.9, 5.3 or 5.14, 5.4 or 5.5, 5.11 or 5.12.

5.1. Suppose $X_1, \ldots, X_n$ are uncorrelated random variables with

$$\begin{aligned} E_F X_i &= c_i \phi_F, \\ \text{var}_F X_i &= q_i v_F, \end{aligned} \qquad (*)$$

where the c_i and q_i are known. The loss function is squared error. Show that the uniformly best linear unbiased estimator of ϕ_F is $[\sum c_i^2/q_i]^{-1}\sum c_i X_i/q_i$ and that it has variance $v_F[\sum c_i^2/q_i]^{-1}$. [*Hint:* First reduce the problem to one in which the random variables have the *same* unknown mean; this case is treated in Section 5.1; this type of reduction of a problem to one of known form is exemplified in a special case at the end of Section 5.1.] Several of the following problems give applications of the preceding result, which can be worked even if you didn't work Problem 5.1.

5.2. This problem illustrates the sensitivity, to the model's assumptions, of "what is a good experiment." In various applications, experience has shown that the c_i and q_i of (*) in the previous problem are related in a simple way. For example, a common biological model is that in which the ith of N genetically identical organisms are subjected to a known amount $b_i > 0$ of a certain treatment (chemical, radiation, etc.). A response X_i (blood pressure, level of some secretion, etc.) is measured. A positive parameter ϕ_F measures the average rate of response of the organism, per unit of treatment:

$$E_F X_i = b_i \phi_F.$$

The variance is often known to be proportional to some known power p of the mean:

$$\text{var}_F(X_i) = c_F \cdot (E_F X_i)^p,$$

where c_F is also unknown. Suppose an experimenter knows the model is valid only in the range $2 \le b_i \le 100$ of the treatment level. He has 500 units of the chemical used in this treatment, so he can choose N and the b_i, subject to $2 \le b_i \le 100$ and $\sum_{i=1}^{N} b_i = 500$. Show that, if the B.L.U.E. of ϕ_F is used,

(a) if $p = 2$, the best choice of the experiment has $N = 250$;
(b) if $p = 0$, the best choice of the experiment has $N = 5$;
(c) if $p = 1$, the choice of N and the b_i's doesn't matter. [Use the conclusion of Problem 5.1, even if you didn't work it.]

5.3. This problem illustrates in a simple setting that as a result of a departure from the model of (*) of Problem 5.1 there can fail to be a uniformly best member of $\mathscr{D}_{LUE}$. In the context of Problem 5.2, suppose the instrument that was to measure the response X_i breaks, and a substitute instrument can only record whether or not $X_i > 20$. Thus, the experimenter can only observe the indicator Y_i of that event (1 or 0 depending on whether or not it occurs). He or she decides only to waste two observations on this inferior instrument, the second at a lower dose than the first. From previous experience with such doses on other organisms, he or she knows that $P_F\{Y_2 = 1\} = \frac{1}{2}P_F\{Y_1 = 1\}$. The experimenter cannot hope to estimate the original ϕ_F but is satisfied to estimate $\theta_F = P_F\{Y_1 = 1\}$.

(a) Show that this setup does *not* fall within the model of Problem 5.1.
(b) Show that every linear unbiased estimator of θ based on Y_1, Y_2 is of the form $t_\alpha = (1-\alpha)Y_1 + 2\alpha Y_2$ for some constant α. [For the sake of obtaining a simple example, we are allowing decision values >1 even though $0 \le \theta \le 1$: if both Y_1 and $Y_2 = 1$, t_α takes on the value $1 + \alpha$.]
(c) Show that $r_{t_\alpha}(\theta) = \theta\{[1 - 2\alpha + 3\alpha^2] - \theta[1 - 2\alpha + 2\alpha^2]\}$. Show that there is no uniformly smallest risk function among the procedures $\{t_\alpha, 0 \le \alpha \le 1\}$

by showing that, for each fixed value θ_0 ($0 \le \theta_0 \le 1$), the t_α which minimizes $r_{t_\alpha}(\theta_0)$ is that with $\alpha = (1 - \theta_0)/(3 - 2\theta_0)$. [In Problem 5.1 the *same* member of the class of linear unbiased estimators minimized $r_t(F_0)$ for each F_0.] Plot a few of these "crossing risk functions" on $\{\theta : 0 \le \theta \le 1\}$.

5.4. A certain model of the national economy assumes that the *labor index of the economy* (LIE) grows linearly with time, being $J + i\phi_F$ in the ith year of the present administration. Economists do not observe the LIE, but rather observe, during the ith year, a random variable P_i with $E_F P_i = J + i\phi_F$. Here J is assumed known for the sake of making a simple problem and has been described in White House circles as "the dismal level inherited from the previous administration," and ϕ_F is the "rate of increase achieved by the new, progressive ball team now that the new coach has made it perfectly clear how he plays the game."
(a) Some administration economists assume the P_i's are uncorrelated, each with the same variance. Suppose (in appropriate units)

$$J = 1.02$$

$$P_1 = 1.08$$

$$P_2 = 1.11$$

$$P_3 = 1.18$$

$$P_4 = 1.14.$$

Reduce the data to the setup of Problem 5.1 and estimate ϕ_F.
(b) Some nonadministration economists do not think the P_i are uncorrelated. They think $P_i = J + (X_1 + \cdots + X_i)$ where the X_i's (growth during year i) are uncorrelated, each with the same mean and variance. Reduce the data of (a) to the setup of Problem 5.1 and estimate ϕ_F under the present model. [You can use the results of Problem 5.1. without having worked it.]
(c) The administration and its opponents both issue predictions of what they estimate the LIE will be "after two terms under this president" ($i = 8$). What are these predictions?

5.5. An environmental organization, the American Nation Team Investigating Great Lakes On Pollution (ANTI-GLOP), decides to study its subject under the assumption that a certain index of water quality changes linearly with time, being of the form $A + t\Phi_F$ during the tth month that they study a given lake. Here A is assumed known for the sake of making a simple problem. The "rate of environmental decay," Φ_F, is not known exactly but is estimated from observations on weight of algae in a standard volume of water, the units being chosen so that if Y_t is the amount of algae in the tth sample, then $E_F Y_t = A + t\Phi_F$.
(a) Some ecologists assume the Y_t's are uncorrelated, each with the same variance. Suppose (in appropriate units)

$$A = 100$$

$$Y_1 = 110$$

$$Y_2 = 118$$

$$Y_3 = 122$$

$$Y_4 = 134$$

$$Y_5 = 143.$$

Reduce the data to the setup of Problem 5.1 and estimate ϕ_F.

(b) Some more sophisticated ecologists do not think that the Y_t's are uncorrelated. They think $Y_t = A + X_t + \frac{1}{2}X_{t-1} + \frac{1}{4}X_{t-2} + \cdots + \frac{1}{2^{t-1}}X_1$, reflecting some (decreasing) dependence on the past; here the X_t are uncorrelated, each with the same variance. Reduce the data of (a) to the setup of Problem 5.1, and estimate ϕ_F, under the present model. [*Hint:* Consider $Y_{t+1} - cY_t$ for appropriate c when $t \geq 1$, together with Y_1.]

(c) Both groups of ecologists are interested in predicting (i.e., in estimating) what the index will be at the end of this year ($t = 12$). What are their two predictions?

5.6. We now assume model (a) ($p = 2$) of Problem 5.2, which is quite common since it occurs in the case where ϕb_i is a "scale parameter;" that is, where the density of X_i is of the form $(b_i\phi)g(x/b_i\phi)$ *with the form of g known*. The value of c_F in Problem 5.2 is known (from the form of g) to be c. If the loss function for estimating ϕ is $(\phi - d)^2$, find the risk function of the estimator kt^* where k is constant and t^* is the estimator concluded to be optimum in Problem 5.1, assuming $p = 2$ in Problem 5.2, when n beetles are used (not necessarily the optimum n). Show that the choice $k = [1 + cn^{-1}]^{-1}$ gives the uniformly best estimator *among estimators of the form* kt*, so that, in particular, the unbiased estimator t^* is inadmissible. [This is analogous to the improvement achieved in Problems 4.25–4.28, where θ is a scale parameter, except that here we are improving on the analog of t_1 rather than t_2 there. The conclusion of inadmissibility of the B.L.U.E. here depends on the scale parameter model and is not an indictment of the B.L.U.E. in other settings.]

5.7. In the setting of Problem 5.6, suppose that $t^*(X_1, \ldots, X_n)$ is *any* unbiased estimator of θ (not necessarily linear, so something would have to be known of the form of g to yield such a t^*, as it did to yield the t_2's of Problems 4.25–4.28). Suppose $\text{var}_\theta t^* = L\theta^2$ where L is a known constant (depending on t^* but not on θ). Show that the uniformly best estimator *among those of the form* Ct* is obtained for $C = (L + 1)^{-1}$. In particular, t^* is inadmissible. How "bad" do you think $r_{t^*}(\theta)/r_{t^*/(L+1)}(\theta)$ could be for various g's?

5.8. An experimenter wishes to determine the expected lifetime θ of a certain aquatic organism when it lives in water with a "standard" amount of pollutant. To this end, he can observe the lifetimes X_i of various organisms of this type, living under various laboratory conditions. It is assumed that the X_i are independent and that, if the ith organism lives in a concentration of ($k_i \times$ standard) pollution, its expected lifetime is θ/k_i. Moreover, the density function of X_i is exponential:

$$f_{X_i;\theta}(x) = \begin{cases} (k_i/\theta)e^{-k_i x/\theta} & \text{if } x > 0, \\ 0 & \text{if } x \leq 0. \end{cases}$$

Use the result of Problem 5.1 to find the best linear unbiased estimator of θ. Show that the variance of this estimator does not depend on the k_i's.

5.9. Another experimenter wants to estimate the expected death rate ϕ per unit of time (proportional to $1/\theta$ of Problem 8.5). Instead of observing varying lifetimes as in Problem 8.5, he prepares n vessels, each containing a large number of organisms, and observes the number X_i of organisms which die in one unit of time in the ith vessel wherein the concentration of pollutant is ($k_i \times$ standard) with $0 \leq k_i \leq 2$. (Even he cannot stand the odor if $k_i > 2$.) It is assumed that the X_i are independent Poisson variables with X_i having expectation $k_i\phi$. It is desired to estimate ϕ with squared error loss. Use the result of Problem 5.1 to prove that the best linear unbiased estimator of ϕ is $\sum X_i/\sum k_i$. What is the variance of this estimator? Design of experiments problem: What choice of $k_1, \ldots, k_n$ would you make in planning this experiment?

5.10. Suppose X_1 and X_2 have common unknown mean ϕ_F and have known variances σ_1^2 and σ_2^2 and known covariance σ_{12}. The object is to estimate ϕ_F, the loss function being squared error.

(a) If $\sigma_1^2 - 2\sigma_{12} + \sigma_2^2 \neq 0$, show that there is a unique best linear unbiased estimator (B.L.U.E.) and find it.

(b) In case (a), can it ever be that the coefficient of X_1 or of X_2 in the best linear estimator is negative? (This assigning of a negative weight to an observation, in computing an average, is contrary to most people's intuitive expectations —think about it!) If so, when? [*Hint:* The sign of $\sigma_1^2 - 2\sigma_{12} + \sigma_2^2$ can be determined from the fact that this quantity is the variance of $X_1 - X_2$.]

(c) If $\sigma_1^2 - 2\sigma_{12} + \sigma_2^2 = 0$, use the fact just noted to show that $\sigma_1^2 = \sigma_2^2 = \sigma_{12}$, and from this conclude that there are formally infinitely many best linear unbiased estimators $a_1X_1 + a_2X_2$ (that is, infinitely many pairs (a_1, a_2) yielding a B.L.U.E.) in this case. What are they? Show that these B.L.U.E.'s are actually all equal, with probability one.

5.11. Two physics students measure the distance a rock falls in one second (starting at rest). The first student makes 5 measurements, averaging 16.1 feet. The second student makes 10 measurements, averaging 15.9 feet. The first student is a better experimenter, and each of her measurements has 1/4 the variance of each of his measurements. What would *you* estimate the distance to be? (state carefully all assumptions, criteria imposed, etc.)

5.12. At a leading university, each of the student newspaper, the Faculty Senate and the university faculty conducts its own poll on preference for the quarter system. They find, respectively, that 43%, 72%, and 60% of all students and faculty favor the quarter system. In past polls of this nature the standard deviation of Senate polls is believed to be about 40% of that of corresponding faculty polls. The student newspaper claims to be the most accurate, with "one-tenth the Senate's error," but in the present instance some unknown proportion of ballots in the newspaper poll had their results tabulated in the wrong category. Use the result of Problem 5.1 to combine these data to obtain your own estimate of the popularity of the quarter system. State all assumptions you make.

5.13. Two astronomers measure the brightness Φ_F of a well-known comet at the same moment. Astronomer i's measurement, X_i, has expectation Φ_F and variance σ_{Fi}^2.

(a) This illustrates that the model of Problem 5.1 and of the remainder of Chapter 5, wherein there exists a *uniformly best* estimator within the class of linear unbiased estimators, really uses the assumed variance structure; without such an assumption, we are still plagued with "crossing risk functions" in the reduced class. To see this, suppose σ_{F1}^2 is known to be 1 but that σ_{F2}^2 can be any positive value. Find the risk function of the general linear unbiased estimator $t_a(X_1, X_2) = (1 - a)X_1 + aX_2$ as a function of σ_{F2}^2 (squared error loss). Show that there is no uniformly best t_a; for example, Problem 5.1 determines which value a_b of a yields smallest risk function when $\sigma_{F2}^2 =$ any specified value $b > 0$, and note that this a_b depends on b.

(b) Suppose you knew absolutely nothing about Astronomer 2 except that he used his favorite antique instrument, claimed to have originated with Galileo. Which t_a would you use, and why?

5.14. *If one uses a procedure in* $\mathscr{D}_L$, here is a justification of unbiasedness (written out in the simplest setting): Suppose $X_1, \ldots, X_n$ are uncorrelated with common mean θ_F and common variance σ_F^2. It is known that $\sigma_F^2 < C < \infty$, where C is known, but θ_F can be any real value.

(a) For squared error loss in estimating θ_F, compute the risk function of a general linear estimator $a_0 + \sum_1^n a_i X_i$. If the criterion of using *only estimators with bounded risk function* is imposed on $\mathscr{D}_L$, show that this results in the restriction to L.U.E.'s (and, in this example, to use of the B.L.U.E.).

(b) Does the same argument work if the model is altered to assume only $\sigma_F^2 = c\theta_F^2 > 0$, with c known? (See Problem 5.6.)

II. *First Problems on Section 5.4.* These problems illustrate the important concepts of linear estimation in the simple but important setting of *one way analysis of variance (ANOVA) with equal numbers of replications.* (Hypothesis testing in ANOVA will be treated in Chapter 8.) This is a common "bread-and-butter" topic. The numerical examples here and in part III of the problems are strongly recommended to aid in understanding the theoretical results.

5.15. Suppose Y_{ij}, $1 \le i \le 3$, $1 \le j \le 5$, are 15 independent random variables with common variance σ_F^2 and $E_F Y_{ij} = \theta_0(F) + \theta_i(F)$. (Thus, the Y_{ij}'s may represent the heights of 15 animals of the same species grown with three different kinds of carcinogenic "nutrients" ($i = 1, 2, 3$). The parameter θ_0 may have been thought of as the "average height in the absence of any added nutrient," and θ_1 is the "average effect of nutrient 1."* For each fixed i, the subscript j merely numbers 5 different animals grown with nutrient i.) This setup is often called "one-way analysis of variance (ANOVA) with equal replications," although we shall postpone the hypothesis testing which is often the statistical problem considered in this setting. Here we treat problems of *estimation*.

(a) Let $X_1, \ldots, X_{15}$ be the Y_{ij}'s written in any order. In the notation of this chapter, letting $\phi' = (\theta_0, \theta_1, \theta_2, \theta_3)$, what is B? Compute $B'B$ and show that it has rank 3 (for example, by showing that the columns of B are linearly *dependent*, but finding a subset of 3 linearly independent columns).

* We will see later that this intuitive meaning does not suggest what quantities can be estimated from the data. If, as in the present experiment, we do not have a "control" observation (no nutrient) whose expectation is θ_0, we cannot estimate θ_0 or θ_1 separately.

(b) Show that none of the functions θ_i is estimable ($i = 0, 1, 2, 3$) (e.g., by showing that the row space of B does not contain that c' for which $c'\theta = \theta_i$). Show that the function $\psi_0 = 15\theta_0 + 5\theta_1 + 5\theta_2 + 5\theta_3$ is estimable and that any function γ of the form $\sum_1^3 c_i\theta_i$ for which $\sum_1^3 c_i = 0$ is also estimable. (Such a γ is called a *contrast of nutrient effects.*) Show that these functions γ (or their coefficient vectors) form a linear space of dimension 2, each member of which is orthogonal to ψ_0 (or its coefficient vector), so that the linear space spanned by (consisting of all linear combinations of) ψ_0 and the γ's has dimension 3 and thus, by part (a), includes *all* estimable linear parametric functions.

(c) Find the B.L.U.E. of ψ_0 and of $\gamma = \sum_1^3 c_i\theta_i$ where $\sum_1^3 c_i = 0$, and find the variance of each. Find an unbiased estimator of σ_F^2. These results should be written in terms of the original Y_{ij}'s, in which form they will be more useful in practice than if left in terms of X_i's. (For, the data appear as a 3×5 array, as in part (d). Working part (d) at the same time may help in understanding part (c).)

(d) Suppose the Y_{ij} (in millimeters) are as follows:

	j = 1	2	3	4	5
i = 1	380	290	325	410	350
2	480	380	435	475	390
3	410	450	375	370	480

(d.1) To illustrate *nonuniqueness* of a least squares solution: One can choose among the 1-space of solutions by choosing a convenient linear combination of t_i's to be 0. Compute *each* of the following solutions (t_0, t_1, t_2, t_3):

(i) Often people solve this model by putting $t_0 = 0$. (Then use the last 3 normal equations.) Then $t_1 = 351$, $t_2 = 432$, $t_3 = 417$.

(ii) Another often used solution is that with $t_1 + t_2 + t_3 = 0$. Then (top normal equation) $t_0 = 400$, and (from the other 3 normal equations) $t_1 = -49$, $t_2 = 32$, $t_3 = 17$.

But both of these give the same estimates of *estimable* linear parametric functions. For example, show that $\theta_1 - \theta_2$ is estimated by $t_1 - t_2 = -81$ whether one uses the solution (i) or (ii). (This is the estimated difference between the potential effect of nutrient 1 and nutrient 2 on a single animal.)

(d.2) Show the corresponding Bt_{LS}, from either (i) or (ii), is

351	351	351	351	351
432	432	432	432	432
417	417	417	417	417

where instead of writing Bt_{LS} as a 15×1 vector we have written the 15 numbers in a 3×5 array corresponding to that of the given Y_{ij}'s. Note the difference between the Y_{ij}'s and the corresponding elements of Bt_{LS}. Each pair of corresponding elements has the same expectation, the element of Bt_{LS} being the B.L.U.E. of the expectation of the corresponding Y_{ij}.

(d.3) Compute $Q_{\min}$ in each of two ways:

(i) $\sum_{i,j}(Y_{ij} - \text{corresponding element of } Bt_{\mathrm{LS}})^2 = (29)^2 + (-61)^2 + \cdots + (63)^2 = 26,430.$

(ii) $\sum_{i,j} Y_{ij}^2 - (Bt_{\mathrm{LS}})'(Bt_{\mathrm{LS}})$ where $\sum Y_{ij}^2 = (380)^2 + \cdots + (480)^2$

and

$(Bt_{\mathrm{LS}})'(Bt_{\mathrm{LS}})$

$$= \begin{cases} (351)^2 + (351)^2 + \cdots + (417)^2 + (417)^2 \\ \text{or } t'_{\mathrm{LS}}(B'B)t_{\mathrm{LS}} = (0, 351, 432, 417)(B'B)(0, 351, 432, 417)' \\ \text{or } t'_{\mathrm{LS}}B'X = (0, 351, 432, 417) \text{ (right side of normal equations).} \end{cases}$$

Note: With fractional entries (usual), instead of the data given here to make your life easier, method (ii) would be computationally simpler than (i).

(d.4) Show that an unbiased estimator of σ_F^2 is $Q_{\min}/(n-h) = 2202.5$. (Note that $\sqrt{2202.5} = 46.9$, which might be used to estimate σ_F, is *not* unbiased: $E\sqrt{U^2} \neq \sqrt{EU^2}$.)

(e) This part is meant to illustrate that one can often find B.L.U.E.'s without going through the solution of the normal equations: If you can *guess* a linear unbiased estimator and show that (i) it is in the estimation space, and (ii) its expectation is what you are trying to estimate, then you have found the B.L.U.E.. In the setting of the present problem, suppose we are not interested in the whole solution of finding B.L.U.E.'s of all estimable parametric functions, but *only* in estimating $\theta_1 - \theta_2$. One might easily guess that one should try an estimator of the form const. $\times \left(\sum_{j=1}^{5} Y_{1j} - \sum_{j=1}^{5} Y_{2j}\right)$, based on the difference between totals from the two nutrients in question. Show quickly that (i) this is indeed a linear combination of $b_1'X$ and $b_2'X$ where b_i is the ith column of B. Show (ii) that "const." may be chosen to give the desired expectation $\theta_1 - \theta_2$. Note that this solution agrees with that of part (c), but you did not have to solve the normal equations.

5.16. (More difficult.) If possible, carry out analogues of 5.15 (a), (b), and (c) for the same model but with $1 \leq j \leq n_i$, $1 \leq i \leq k$, where the n_i's can be different numbers. (That is, different fertilizers can be used different numbers of times; this is called *one-way ANOVA with unequal replications.*) *Hint:* Show that the functions $Y_{i\cdot} = n_i^{-1}\sum_i Y_{ij}$, $1 \leq i \leq k$, are linearly independent and span the estimation space, and thus find what are the functions $E_F \sum_1^k a_i Y_{i\cdot}$.

III. *Second set of problems on Section 5.4.* These problems illustrate the variety of settings that can be treated in the linear estimation setup. *Suggested:* 5.17, 5.18,

5.19–5.20 or 5.23, 5.24–5.25 or 5.26–5.27 or 5.28–5.29, possibly omitting some of the calculations in numerical examples.

5.17. $X_1, X_2, \ldots, X_n$ are independent with common unknown variance σ_F^2 and with $E_F X_i = z_{1i}\phi_1(F) + z_{2i}\phi_2(F)$. The numbers (z_{1i}, z_{2i}), $1 \leq i \leq n$, are *known*.

(a) Show that there exist linear unbiased estimators of *both* ϕ_i's provided that

$$\left(\sum_1 z_{1i}^2\right)\left(\sum_1 z_{2i}^2\right) \neq \left(\sum_1 z_{1i}z_{2i}\right)^2. \qquad (*)$$

[*Optional*: Show that this last condition (*) (under which $h = k$ in the notation of Section 5.4) means that the points (z_{1i}, z_{2i}) in the plane do *not* lie on a line through the origin.]

(b) In the case when B.L.U.E.'s of both ϕ_i's exist, compute $B'B$ and $B'X$ and state how, in terms of these, you could find the B.L.U.E.'s t_1 and t_2, and how you could find their variances. Don't bother to invert $B'B$, etc. [*Optional*: Also find the covariance between them.] Note that the variances of the t_i are not completely known, since σ_F^2 is unknown.

(c) Determine how you would find an unbiased estimator of σ_F^2 in the situation of (b), and how you would find (from the preceding) an *unbiased estimator* of the quantity $\text{var}_F(t_2)$ of (b).

Various applications of the model of Problem 5.17 Problems 5.18, 5.19, 5.22, 5.24, 5.26, and 5.28 illustrate the variety of seemingly different practical problems which have the same mathematical treatment; all fit into the general linear model with $k = 2$. In each case, try if possible to solve the normal equations and give explicit formulas for the t_i of 5.17(b).

5.18. One of the simplest examples of the setup of Problem 5.17 is that in which there are n animals of the same species, $n = n_1 + n_2$, and n_i animals are fed with nutrient i. The expected weight of an animal fed nutrient i is ϕ_i. Show how to choose the z_{ij}'s as 0's and 1's, and then to relabel the X_i's as $Y_{ij}(1 \leq j \leq n_i)$ to make this setting coincide with that of Problem 5.16 with $k = 2$ and ϕ_j here $=$ $\theta_0 + \theta_j$ there. What is the B.L.U.E. of $\phi_1 - \phi_2$?

5.19. The setting can be that of fitting a *surface* which shows the dependence of some phenomenon as a *sum of effects of two different stimuli*. For example, $E_F X_i$ may represent the expected average daily weight of milk from a cow if she is fed z_{1i} units of feed per day and is exposed to z_{2i} hours per day of hard rock music. Suppose in this setting that $n = 12$, and that a different cow is treated with each of the possible combinations of amount of feed 7, 8, 9, or 10, and amount of music 0, 1, or 2. (The cow may be able to stand more than 2 hours of this music, but

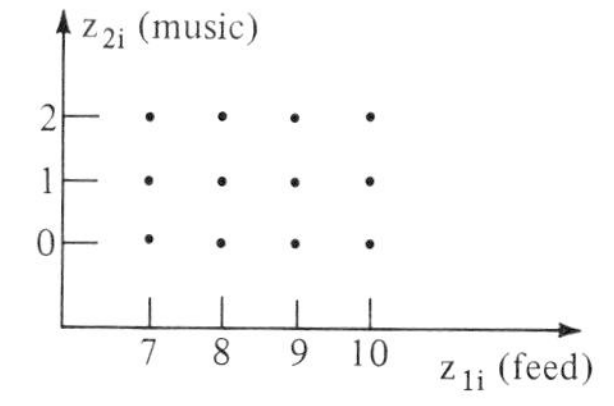

the experimenter cannot.) You can think of the cows as sitting, chewing, and listening on the 12 points (r, s), $7 \le r \le 10$, $0 \le s \le 2$, in the feed-music plane above which the "response surface" $\phi_1 z_{1i} + \phi_2 z_{2i}$ is to be estimated. It is convenient here to relabel the X_i's, calling that X_i with $(z_{1i}, z_{2i}) = (r, s)$ by the new name Y_{rs} $(7 \le r \le 10, 0 \le s \le 2)$. Specialize the answers of 5.17(b) (and optionally 17(c)) to the present setting, giving them in terms of the Y_{rs}'s. Note that, in practice, data are more likely to be presented in terms of the Y_{rs}'s than in terms of 12 X_i's.

5.20. *Numerical example of the preceding*: The data $\{Y_{rs}\}$ are:

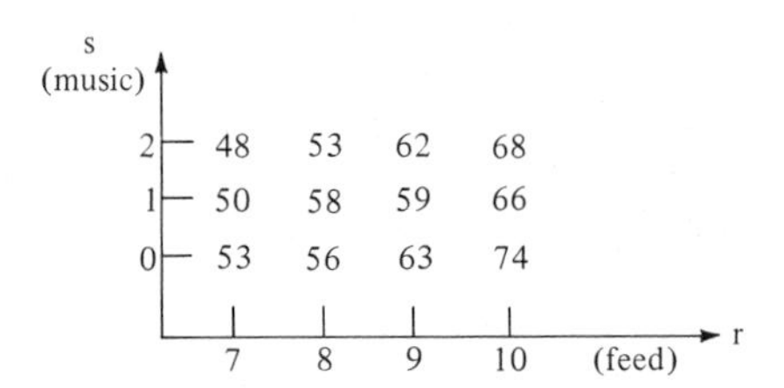

(a) Show that $B'B = \binom{882 \quad 102}{102 \quad 02}$ and that the unique solution of the normal equations is $t_1 = 7.14$, $t_2 = -1.68$.

(b) Make a 3×4 chart of the LS fit "surface" values Bt_{LS} (written this way rather than as a 12×1 vector), for comparison with the Y_{rs}'s, of whose expectations they are B.L.U.E.'s.

(c) As in (d) of Problem 5.15, compute $Q_{\min}$ as the sum of squares of the 12 differences of corresponding elements Y_{rs} and $(Bt_{\text{LS}})_{rs}$; and/or compute by another method, e.g.,

$$Q_{\min} = \sum_{r,s} Y_{rs}^2 - (7.14, -1.68)(B'B)\begin{pmatrix} 7.14 \\ -1.68 \end{pmatrix}.$$

[Note how one loses significant figures in computing such a difference of two large numbers!]

(d) Thus compute an *unbiased estimator* of $\text{var}_F(t_2)$. [Note that $\text{var}_F(t_2) = \sigma_F^2 \times$ [lower right element of $(B'B)^{-1}$].]

(e) Predict the average daily weight of milk from a cow that would be obtained from a daily diet of 5 units of feed and 4 hours of music. Estimate the variance of your prediction.

5.21. This problem is meant to illustrate that one can often find B.L.U.E.'s without going through the solution of the normal equations, just as in Problem 5.15(e). If you can *guess* a linear unbiased estimator and show that (i) it is in the estimation space, and (ii) its expectation is what you are trying to estimate, then you have found the B.L.U.E.. We return, for this example, to the setting of Problem 5.19, but with two levels 1 and 2 of each of r and s, instead of 4 and 3 as before. It is desired to estimate $\phi_1(F) - \phi_2(F)$ without going through the calculations of Problem 5.19. (This parametric function measures whether music or hay is more important—of doubtful practical value, but here we go!) The experimenter feels that Y_{11} and Y_{22} don't seem likely to convey much information about $\phi_1 - \phi_2$ since, for each of these cows, hay and music have identical values.

He feels, intuitively, that something of the form $t^* = \text{const.}\,(Y_{12} - Y_{21})$ should be appropriate since it depends on increasing hay and decreasing music in a symmetric way in going from one observation to the other. (i) If b_1 and b_2 are the two columns of B, show that t^* is a linear combination of $b_1'X$ and $b_2'X$ and is therefore in the estimation space. (ii) Find the choice of "const." in t^* to yield a B.L.U.E. of $\phi_1(F) - \phi_2(F)$. (iii) Now (and only now) check that the development of Problem 5.19 would have yielded the same result (but note how much more work it entailed!).

5.22 The results of Problem 5.19 may be generalized to the setting where $n = \alpha\beta$ (instead of 12), there being α levels $1, 2, \ldots, \alpha$ of feed and β levels $1, 2, \ldots, \beta$ of music. Once more, label that X_i with $(z_{1i}, z_{2i}) = (r, s)$ by the new name Y_{rs} $(1 \le r \le \alpha,\ 1 \le s \le \beta)$. Once more, $EY_{rs} = r\phi_1 + s\phi_2$. Specialize the results of Problem 5.17 to the present setting, giving them in terms of the Y_{rs}, by using the facts $\sum_{j=1}^{m} j = m(m+1)/2$ and $\sum_{j=1}^{m} j^2 = \dfrac{m(m+1)(2m+1)}{6}$. Note that data in practice are likely to be presented in terms of the Y_{rs}'s, not the X_i's, so it is useful to put the results in terms of the Y_{rs}'s.

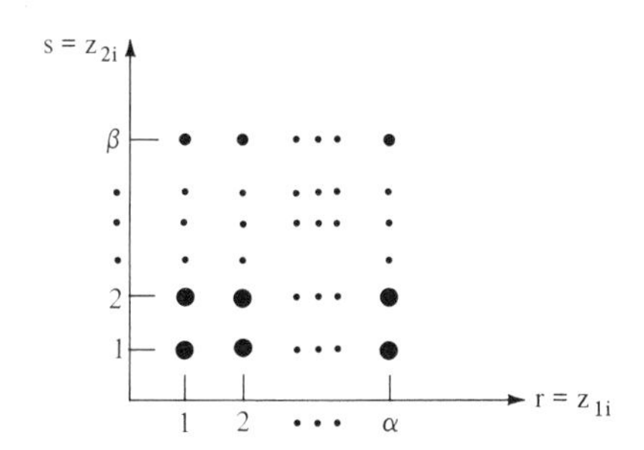

5.23. *Another numerical example* of Problem 5.22 (similar to Problem 5.20).

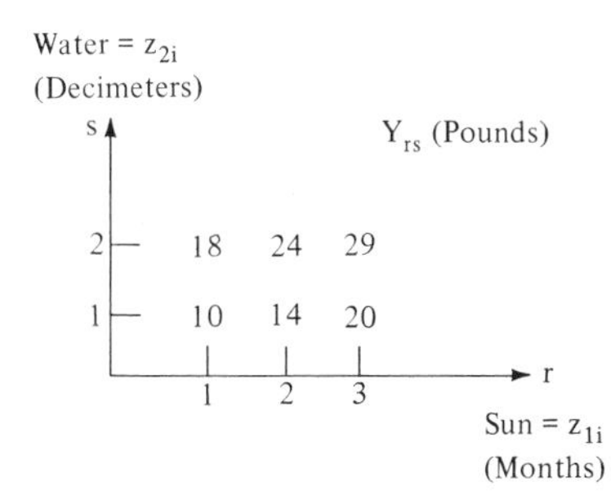

(a) Suppose $\alpha = 3$, $\beta = 2$, and the data are as shown for weight of 6 watermelons grown with various total amounts (r, s) of sun and water. The model is that of Problem 5.22. Show that the normal equations have unique solution

$$t_1 = 4.34, \qquad t_2 = 7.19.$$

[Here $B'B = \begin{pmatrix} 28 & 18 \\ 18 & 15 \end{pmatrix}$.] Show that LS fit "surface" is 4.34 (sun) + 7.19 (water), and evaluate it at the six given points (r, s) at which there were observations, to yield the following for Bt_{LS} (written as a 2×3 array instead of a 6×1 vector):

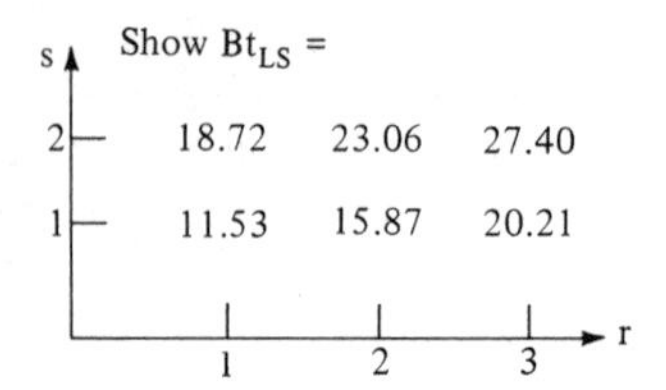

These should be compared with the Y_{rs}'s, of whose expectations they are B.L.U.E.'s.

In particular, $Q_{\min}$ is the sum of squares of the six differences:

$$(.72)^2 + (.94)^2 + (1.6)^2 + (1.53)^2 + (1.87)^2 + (.21)^2 \approx 10.$$

Try also another method of computing Q:

e.g.,

$$\sum_{r,s} Y_{rs}^2 - \left[(4.34, 7.19)(B'B)\begin{pmatrix}4.34\\7.19\end{pmatrix}\right]$$

or

$$\sum_{r,s} Y_{rs}^2 - [(18.72)^2 + (23.06)^2 + \cdots + (20.21)^2].$$

(Note how easy it is to lose a significant figure where the method involves finding a small difference of two large numbers!) Consequently, show that the estimator t_1, whose variance is $\sigma_F^2 \times$ [upper left element of $(B'B)^{-1}$], has an *unbiased estimator of that variance* approximately equal to .4.

(b) *Prediction problem:* How would you estimate the weight of a watermelon grown with 3 decimeters of water and 4 months of sunshine? What would you estimate the variance of your prediction to be?

5.24. A third application of the model of Problem 5.17 is to *curve-fitting in one variable* (of the type sketched in Section 5.2) where the interest is in a response to a single stimulus, but of a functional form involving a linear combination of two known functions of the single stimulus (the coefficients in the linear combination being

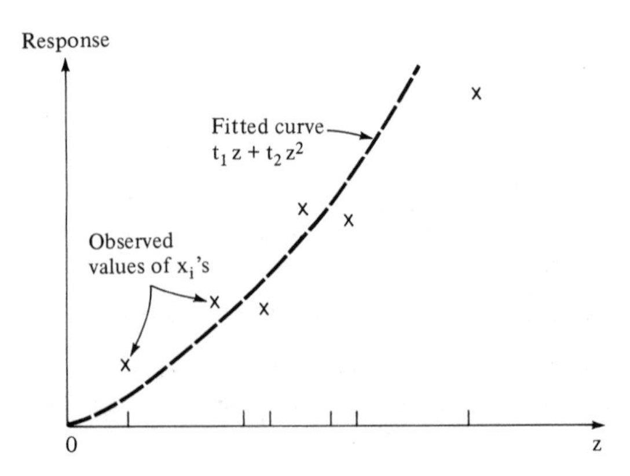

unknown). Suppose the ith plant of a given species is planted with amount of fertilizer z_i and that the dependence of total leaf weight X_i on fertilizer is given by $E_F X_i = \phi_1(F)z_i + \phi_2(F)z_i^2$. Specialize the expressions of Problem 5.17 to the present case by putting $z_{pi} = z_i^p$ for $p = 1, 2$.

5.25. *Numerical example of Problem 5.24*. Suppose $n = 5$ and the values are

grams per pot, z_i	*ounces per plant*, x_i
4	16
9	50
6	30
11	73
8	40

Make columns "z_i^2, z_i^3, z_i^4, $z_i x_i$, $z_i^2 x_i$." The sums of these are 318, 2 852, 26 850, 1 817, 16 779, which numbers appear in the normal equations and yield

$$t_1 = 2.26, \qquad t_2 = .38.$$

(a) Plot the fitted curve $2.26z_i + .38z_i^2$ along with the X_i's. Note that the height of the fitted curve at the values $z_i = 4, 9, 6, 8, 11$ *gives the components of* $\mathbf{B}\mathbf{t}_{\text{LS}}$. What do you estimate $\text{var}_F(t_1)$ to be?

(b) *Prediction:* With 20 grams of fertilizer per pot, how much weight of leaf per pot would you get? What do you estimate the *variance* of this prediction to be?

5.26. [Another example (similar to Problem 5.24) of curve-fitting in one variable, where interest is in a response to a single stimulus, but of a functional form involving a linear combination of two known functions of the single stimulus (the coefficients in the linear combination being unknown).] Suppose the ith hog is fed an amount z_i of Hogitol every day for a month, and that the dependence of total hog weight X_i on dosage of Hogitol is given by

$$E_F X_i = \phi_1(F)z_i + \phi_2(F)\sqrt{z_i}.$$

Specialize the expressions of Problem 5.17 to this case by putting

$$z_{pi} = z_i^{1/p} \quad \text{for} \quad p = 1, 2.$$

5.27. *Numerical example of* Problem 5.26. Suppose $n = 5$ and the values are

daily units of Hogitol, z_i	final hog weight, X_i
5	120
9	210
16	360
12	270
10	230

(a) Make columns similar to those in the table, labeled z_i^2, $(z_i^{1/2})^2$, $z_i z_i^{1/2}$, $z_i x_i$, $z_i^{1/2} x_i$. The sums of the numbers in these columns give the entries of the normal equations. Find t_1 and t_2. What do you estimate $\text{var}_F(t_1)$ to be?

(b) Plot the points (z_i, x_i) and also the fitted curve $t_1 z + t_2 z^{1/2}$ and especially the points on that curve corresponding to values $z = z_i$. (The ordinates of these points are the components of Bt_{LS}.)

(c) How much do you think a hog will weigh if it is fed 20 units of Hogitol daily? Estimate the *variance* of your guess.

5.28. [Another example (like Problems 5.24 and 5.26) of curve-fitting in one variable, where interest is in a response to a single stimulus, but of a functional form involving a linear combination of two known functions of the single stimulus (the coefficients in the linear combination being unknown).] Suppose that a certain economic indicator X_i is assumed to consist of a linear trend plus a seasonal variation (plus some chance deviation from this relationship), so that

$$E_F X_i = \phi_1(F)z_i + \phi_2(F)\sin(\pi z_i/2).$$

Specialize the expressions of 5.17(b) (and, optionally, 5.17(c)) to this case by putting $z_{1i} = z_i$ and $z_{2i} = \sin(z_i\pi/2)$. [There are 4 seasons per year, so the period of the sinusoidal function is taken to be 4. To make the problem simpler (two unknown parameters) the intercept of the linear part is assumed, and the latter has been subtracted off to yield the X_i's as given.]

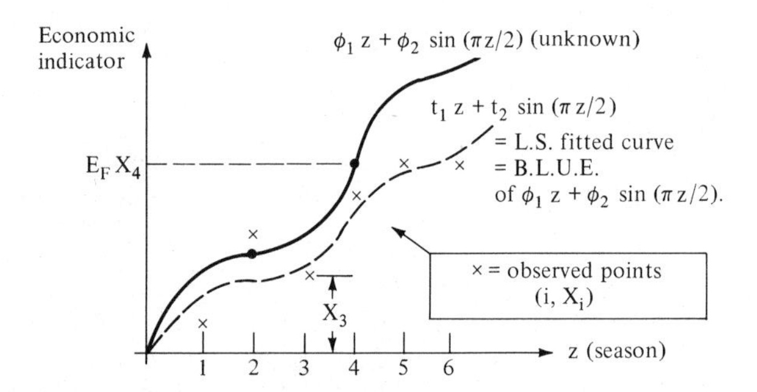

5.29. *Numerical example of Problem* 5.28. Suppose $n = 6$ and the values are

season, z_i (in order; 4 per year)	economic indicator, X_i
1	132
2	216
3	272
4	384
5	558
6	591

(a) Make columns next to those in the table, labeled z_i^2, $\sin^2(\pi z_i/2)$, $z_i \sin(\pi z_i/2)$, $z_i x_i$, $x_i \sin(\pi z_i/2)$. The sums of the numbers in these columns give the entries of the normal equations. Find t_1 and t_2. Estimate $\text{var}_F(t_1)$.

(b) Plot the points (z_i, x_i) and also the fitted curve $t_1 z + t_2 \sin(\pi z/2)$ and especially the points on that curve corresponding to values $z = z_i$. (The ordinates of these points are the components of Bt_{LS}.)

(c) What do you think the indicator will read after the first quarter of next year $(z = 9)$? Estimate the *variance* of your guess.

5.30. *Orthogonalization.* [*Note*: It is best to work the problem which follows from elementary vector space notions rather than from substituting into the subsection on orthogonal polynomials. Consult the latter to see how this problem fits in.] This will illustrate the simplification in arithmetic achieved by going through an appropriate change of variables before writing down the normal equations. The setup is that of Problem 5.17. Call b_j the original columns of B; that is, $b_j' = (z_{j1}, z_{j2}, \ldots, z_{jn})$ for $j = 1, 2$. Assume $h = k$ for simplicity in what follows.

(a) Let $\tilde{b}_1 = b_1$ and $\tilde{b}_2 = b_2 - db_1$ where d is a scalar. What value of d will make $\tilde{b}_1 \perp \tilde{b}_2$ (that is, $\tilde{b}_1'\tilde{b}_2 = 0$)? [Answer: $d = b_1'b_2/b_1'b_1$.]

We hereafter write $\tilde{B} = \begin{bmatrix} \tilde{b}_1 & \tilde{b}_2 \end{bmatrix}$ and $\tilde{b}_j' = (\tilde{z}_{j1}, \tilde{z}_{j2}, \ldots, \tilde{z}_{jn})$.

(b) What value must the scalar e take on, in terms of d, in order that one can write two new parametric functions

$$\tilde{\phi}_1 = \phi_1 + e\phi_2, \qquad \tilde{\phi}_2 = \phi_2, \qquad \tilde{\phi} = \begin{pmatrix} \tilde{\phi}_1 \\ \tilde{\phi}_2 \end{pmatrix}$$

for which (**) $B\phi = \tilde{B}\tilde{\phi}$ (that is, for which $E_F X_i = z_{1i}\phi_1(F) + z_{2i}\phi_2(F) = \tilde{z}_{1i}\tilde{\phi}_1(F) + \tilde{z}_{2i}\tilde{\phi}_2(F)$? [*Answer*: $e = d$.]

(c) We can, in view of (**), think of the model as $E_F X = \tilde{B}\tilde{\phi}(F)$ instead of $E_F X = B\phi(F)$. In terms of this newer way of writing the model, show that the normal equations $\tilde{B}'\tilde{B}\tilde{t} = \tilde{B}'X$ are in *diagonal form*, so that the B.L.U.E.'s $\tilde{t}_i$ of the $\tilde{\phi}_i$ are easier to compute than the estimators t_i of the ϕ_i in Problem 5.17. Write down the $\tilde{t}_i$'s as functions of the X_i's and $\tilde{z}_{ji}$'s.

(d) One can find tables of the vectors $\tilde{b}_j$ for certain special important cases. In the setting of Problem 5.24, with $n = 10$ and $z_i = i (1 \le i \le 10)$ (which is a common design of "uniformly spaced stimulus values"), use the facts that

$$\sum_{i=1}^{m} i^2 = m(m+1)(2m+1)/6$$

and

$$\sum_{i=1}^{m} i^3 = [m(m+1)]^2/4$$

and the conclusion of (a) to show that $d = 55/7$. Since $\tilde{b}_2$ will have fractional entries, books of tables would (instead of listing $\tilde{b}_1, \tilde{b}_2$) often give the vectors $\tilde{b}_1$ and $\tilde{\tilde{b}}_2 = 7\tilde{b}_2$ in this situation. Write down the explicit numerical values of the vectors $\tilde{b}_1$ and $\tilde{\tilde{b}}_2$ in the present case. Also write down the values of $\tilde{b}_1'\tilde{b}_1$ and $\tilde{\tilde{b}}_2'\tilde{\tilde{b}}_2$ (which books of tables would include) and state how they would be of use in solving the normal equations; here you will also use $\sum_{i=1}^{m} i^4 = \dfrac{m(2m+1)(m+1)(3m^2+3m-1)}{30}$.

(e) In the setting of (d) and Problem 5.24, if it is desired only to write down the "fitted curve," this can be done without going back to the original model ($B\phi$), since E_F {height of a plant with amount z of fertilizer} $= \tilde{\phi}_1 z + \tilde{\phi}_2(z^2 - dz)$; why is this so? What is the form of this function (numerical coefficients included) if $\tilde{t}_1 = 18$, $\tilde{t}_2 = 25$?

(f) If one wishes to obtain *not* the result of (e) but rather the explicit values t_1 and t_2 (B.L.U.E.'s of the original ϕ_1, ϕ_2), show how to obtain these by solving the equations of part (b) for ϕ to obtain $\phi_1 = \tilde{\phi}_1 - e\tilde{\phi}_2$, $\phi_2 = \tilde{\phi}_2$, and from these obtain the t_i's from the $\tilde{t}_i$'s. Note that the results are consistent with those of (e), from which fitted curve the t_i's could hence also have been obtained.

5.31. *Extension of model of Problems* 5.19 *and* 5.22 *to include a third parameter.* (*Nonhomogeneous* model with two independent controllable variables.) $X_1, X_2, \ldots, X_n$ are independent with common variance σ^2. The observation X_i is the height of a plant which has been grown with z_{1i} hours of sunshine and z_{2i} inches of rainfall, and it is assumed that $E_F X_i = \theta_0(F) + \theta_1(F)z_{1i} + \theta_2(F)z_{2i}$. The numbers (z_{1i}, z_{2i}) are known.

(a) Show that linear unbiased estimators of θ_0, θ_1, θ_2 exist, by showing that the normal equations have a solution which gives such estimators, *unless*

$$\sum_i (z_{1i} - \bar{z}_1)^2 \sum_i (z_{2i} - \bar{z}_2)^2 = \left[\sum_i (z_{1i} - \bar{z}_1)(z_{2i} - \bar{z}_2)\right]^2$$

where $\bar{z}_j = \sum_i z_{ji}/n$.

Hence (for example, by considering the correlation coefficient for a probability function on the plane which assigns probability $1/n$ to each of the points (z_{1i}, z_{2i})), show that this lack of estimability occurs precisely when all the points (z_{1i}, z_{2i}) fall on a line.

(b) In the case when L.U.E.'s exist, find the best L.U.E.'s of θ_0, θ_1, and θ_2; find the variances and covariances of the best L.U.E.'s.

5.32. *Orthogonalization in the three-parameter model of Problem* 5.31. In the setup of Problem 5.31 where the X_i's are uncorrelated with $\text{Var}_F(X_i) = \sigma_F^2$, and where $E_F X_i = \theta_0(F) + \theta_1(F)z_{1i} + \theta_2(F)z_{2i}$, go through the following steps of orthogonalization to simplify the computations:

(a) Find the numbers c, d, and e (depending on the z_{ji}) such that

$$\sum_i (z_{1i} - c \cdot 1) \cdot 1 = 0, \qquad \sum_i (z_{2i} - d \cdot 1) \cdot 1 = 0,$$

and

$$\sum_i [(z_{2i} - d) - e(z_{1i} - c)](z_{1i} - c) = 0.$$

(b) Writing $z'_{1i} = z_{1i} - c$ and $z'_{2i} = z_{2i} - d - e(z_{1i} - c) = z_{2i} - d - ez'_{1i}$, find numbers f, g, and h (depending on the z_{ji}, or on c, d, and e) such that

$$\theta_0(F) + \theta_1(F)z_{1i} + \theta_2(F)z_{2i} = \theta'_0(F) + \theta'_1(F)z'_{1i} + \theta_2(F)z'_{2i},$$

where

$$\theta'_0 = \theta_0 + f\theta_1 + g\theta_2 \quad \text{and} \quad \theta'_1 = \theta_1 + h\theta_2.$$

(c) Considering the problem $E_F X_i = \theta'_0(F) + \theta'_1(F)z'_{1i} + \theta_2(F)z'_{2i}$, find the best L.U.E.'s t'_0, t'_1, and t'_2 of θ'_0, θ'_1, and θ_2 by solving the normal equations (in terms of the z_{ji}'s). Find the variances of these best L.U.E.'s and state why the covariance between any two of them is zero.

(d) Using part (b), find linear functions of t'_0, t'_1, and t'_2 which are best L.U.E.'s t_0, t_1, and t_2 of θ_0, θ_1, and θ_2. Use (c) to find the variances of t_0, t_1, and t_2 (leave these in terms of the z_{ji}'s). Are t_0, t_1, and t_2 necessarily uncorrelated?

CHAPTER 6

Sufficiency

6.1. On the Meaning of Sufficiency

In the construction of statistical procedures using decision theory, one primary consideration in the search for good procedures is to limit the scope of the search as much as possible. It is the purpose of this chapter to describe sufficient statistics and sufficient partitions and to indicate why, from the point of view of decision theory, one need only use those procedures which are functions of a sufficient statistic or sufficient partition. (To do this it will sometimes be necessary to use randomized statistical procedures, defined in Section 4.3.) Since a sufficient statistic is often much simpler than X itself (often 1- rather than n-dimensional), it can be a considerable saving of effort only to look at functions of a sufficient statistic rather than at *all* procedures (functions of X).

The discussion of sufficient statistics for continuous random variables presents mathematical difficulties. The discussion of these difficulties is omitted in the presentation which follows. The following calculations will be limited to discrete random variables, although analogous results hold in absolutely continuous settings.

In the discrete case—that is, where S is a countable set, a function s on S is said to be a *sufficient statistic* (sometimes one says a *sufficient statistic for* Ω) if

$$\begin{aligned}&\text{whenever } P_F(s(X) = b) > 0, \text{ then}\\&P_F(X = x \mid s(X) = b) \text{ does not depend on } F.\end{aligned} \tag{6.1}$$

If X is a continuous random variable there is an analogous definition. Note that s can be any function on S and need not be real-valued. (This differs from some early usage, in which, if Ω were a 1-dimensional interval of possible values, a sufficient statistic meant a continuous real-valued function; such functions do not exist in all settings, whereas we shall *always* have sufficient partitions or statistics in the sense used here.)

For example, if we let $X = (X_1, X_2)$ be a pair of independently and identically distributed Bernoulli random variables, then $s(X) = X_1 + X_2$ is a sufficient statistic. It is easily seen that

$$P_\theta(X = (0,0) | X_1 + X_2 = 0) = 1;$$

$$P_\theta(X = (a,b) | X_1 + X_2 = 1) \begin{cases} = 1/2 & \text{if } (a,b) = (0,1) \text{ or } (1,0), \\ = 0 & \text{otherwise;} \end{cases}$$

$$P_\theta(X = (1,1) | X_1 + X_2 = 2) = 1.$$

Since these conditional probabilities do not depend on θ, i.e., the values do not change as θ changes, $s(X) = X_1 + X_2$ is a sufficient statistic.

To see that not every statistic is sufficient, consider

$$s^*(X_1, X_2) = \begin{cases} 1 & \text{if } X_1 = X_2 = 1, \\ 0 & \text{otherwise.} \end{cases}$$

Then $P_\theta\{(X_1, X_2) = (0,0) | s^*(X_1, X_2) = 0\} = P_\theta\{(X_1, X_2) = 0 |$ at least one $X_i = 0\} = (1-\theta)^2/[(1-\theta)^2 + 2\theta(1-\theta)] = (1-\theta)/(1+\theta)$, which is *not* independent of θ.

In most problems there will be many sufficient statistics. To see why this might be so, suppose γ is any 1-1 function and s is a sufficient statistic. Then $\gamma(s(X))$ is likewise a sufficient statistic. For

$$\begin{aligned} P_F(X = x | \gamma(s(X)) = b) &= P_F(X = x | s(X) = \gamma^{-1}(b)) \\ &= P(X = x | s(X) = \gamma^{-1}(b)). \end{aligned} \tag{6.2}$$

That is to say, the conditional probabilities in (6.2) do not depend on F; therefore $\gamma(s(X))$ is a sufficient statistic.

If γ is not a 1-1 function, the result stated in the preceding paragraph is false. The argument there fails because γ^{-1} is no longer a function. For example, $\bar{\gamma}(s(X)) \equiv 3$ is not sufficient in the Bernoulli example given since $P_\theta\{X = (0,0) | \bar{\gamma}(X) = 3\} = (1-\theta)^2$ since $\{x : \bar{\gamma}(s(x)) = 3\} = S$.

The fact that a 1-1 function of a sufficient statistic is again a sufficient statistic suggests that the essential point about a sufficient statistic is not the set of values it takes on, but rather the collection of disjoint sets $\{A_b\}$ into which the statistic partitions the set S, where $A_b = \{x | s(x) = b\}$. Note that $P(X = x | s(X) = b)$ is in fact a shorthand for $P(X = x | \{x | s(x) = b\})$, or $P(X = x | A_b)$. The conditional probabilities really only depend on the sets $\{A_b\}$.

We consider an example, $X = (X_1, X_2, X_3, X_4)$, where the X_i are independent Bernoulli variables with unknown mean θ. Here S consists of 16 points, and you can easily compute (6.1) to see that $s(X_1, \ldots, X_4) = \sum X_i$ is a sufficient statistic. The 16 points in S are broken up by s into five sets, as follows:

$$
\begin{array}{c|c|c|c|c}
S: & & (1,1,0,0) & & \\
 & (1,0,0,0) & (1,0,1,0) & (1,1,1,0) & \\
 & (0,1,0,0) & (1,0,0,1) & (1,1,0,1) & \\
(0,0,0,0) & & & & (1,1,1,1). \\
 & (0,0,1,0) & (0,1,0,1) & (1,0,1,1) & \\
 & (0,0,0,1) & (0,1,1,0) & (0,1,1,1) & \\
 & & (0,0,1,1) & &
\end{array}
\tag{6.3}
$$

The conditional law (6.1) is easily seen to be uniform on the points in each subset. This is also intuitively clear: $P_\theta\{\sum X_i = 1\}$ *will* depend on θ, but if someone tells you that $\sum X_i = 1$ and asks you what the conditional odds are that the single head occurred on the first, second, third, or fourth toss, you would assign equal probabilities of 1/4 to each possibility, independently of θ. (In other examples, the conditional probabilities might not of course be *equal*.) The five sets into which S has been divided are (from left to right in (6.3)) $A_i = \{(X_1, X_2, X_3, X_4) | i = X_1 + X_2 + X_3 + X_4\}$, $i = 0, 1, 2, 3, 4$. If we labeled these sets instead by the values 0, 1, 8, 27, 64 of $(\sum X_i)^3$, which is a 1-1 function of $\sum X_i$, only the labels on the five sets, but not the sets themselves, would change.

From (6.1) it is clear that $P_F(X = x | s(X) = b)$ is a conditional probability function for the random variable X, which, when $s(X)$ is a sufficient statistic, states that the conditional distribution of X given that X is in $\{x | s(x) = b\}$ does not depend on F in Ω. In terms of a sufficient partition $\{A_b\}$, the conditional distribution of X given that X is in A_b does not depend on F in Ω.

In most problems there will be a number of different ways of partitioning S, each partition being a sufficient partition. In the discrete case (i.e., S countable) each sufficient partition is given by a sufficient statistic, so a second reason there may be many different sufficient statistics is that there can be many different sufficient partitions.

One way that this arises is as follows. Suppose the partition of S, $\{A_b\}$, is sufficient, and that another partition $\{B_b\}$ of S has the property that to each B_b there is some A_{b^*}, with $B_b \subset A_{b^*}$. Then $\{B_b\}$ is also a sufficient partition. (Usually we say in this case $\{B_b\}$ is a *refinement* of $\{A_b\}$.) We show this to be the case by a direct calculation. Let B_b be in $\{B_b\}$ and $B_b \subset A_{b^*}$.

$$
\begin{aligned}
P_F(X = x | B_b) &= P_F(X = x \text{ and } B_b)/P_F(B_b) \\
&= \frac{P_F(X = x, B_b \cap A_{b^*})/P_F(A_{b^*})}{P_F(B_b \cap A_{b^*})/P_F(A_{b^*})} \\
&= \frac{P_F(X = x \text{ and } x \text{ in } B_b | X \text{ in } A_{b^*})}{P_F(X \text{ in } B_b | X \text{ in } A_{b^*})} \\
&= \frac{P(X = x, x \text{ in } B_b | X \text{ in } A_{b^*})}{P(X \text{ in } B_b | X \text{ in } A_{b^*})}
\end{aligned}
\tag{6.4}
$$

Therefore the conditional distribution of X given X in B_b does not depend on F; that is, $\{B_b\}$ is a sufficient partition.

If we consider the partition (6.3), it is clear how to subdivide the five sets there into refinements, and all such refinements will be sufficient partitions. The finest partition of all is the collection of sets $\{\{x\}, \quad x \in S\}$, which consists of every set $\{x\}$ consisting of a single point of S. This is always a sufficient partition of S.

Analytically, the preceding result says that if the original sufficient statistic s is a function of some other statistic s', that is, $s(x) = q(s'(x))$ for all x in S, then s' is also a sufficient statistic. For, if we define $A_b = \{x : s(x) = b\}$ and $B_b = \{x : s'(x) = b\}$, then $B_b \subset A_{q(b)}$ and

$$P_F(X = x | s'(X) = b) = P_F\{X = x | B_b\},$$

which, according to the result of (6.4), does not depend on F. It is easy to see that the partition of S induced by s' is a refinement of the partition of S induced by s. Thus, in our example, if $s(x_1, x_2, x_3, x_4) = \sum_1^4 x_i$, then the conditional probability law (6.1) assigns equal probabilities to each of the four points when $b = 1$, to each of the six points when $b = 2$, etc. If we write $s'(x_1, x_2, x_3, x_4) = (x_1 + x_2, x_3 + x_4)$, then the set $\sum_1^4 x_i = 1$ of the original partition is broken up into the two sets where $s' = (0, 1)$ and $s' = (1, 0)$; each consists of two points, and the conditional probability law, given that $s' = c$, assigns probability 1/2 to each of the two points in the set $s' = c$. Similarly, corresponding to $\sum_1^4 x_i = 2$, there are three sets, namely, $s' = (2, 0)$, $s' = (1, 1)$, and $s' = (0, 2)$, containing 1, 4, and 1 point, respectively. As another example of a finer sufficient partition (the finest one), the statistic defined by $s''(x) = x$ is *always* sufficient (no matter what Ω is), since (6.1) with $s = s''$ takes on the value 1 or 0 according to whether or not $b = x$, independently of F.

In our example, $\sum_1^4 x_i$ can be written as a function of s' or s'', but not conversely (e.g., the statement that $\sum_1^4 x_i = 2$ does not determine s', which could be (2, 0), (1, 1), or (0, 2)). From a practical (as well as theoretical) point of view, if it is only necessary to know the value taken on by a sufficient statistic, it is more convenient to record the value of $\sum_1^4 X_i$ than of s' or s'', and more convenient to consider only procedures which are functions of the single variable $\sum_1^4 X_i$ than procedures which are functions of the two components of s' or of the four components of s''. Similarly, in general, if a sufficient statistic can be expressed as a function of every other sufficient statistic s' (or, equivalently, if it yields a coarser partition than every sufficient s', in that each set of the s'-partition is a subset of the s-partition), then such a sufficient statistic, called a *minimal* sufficient statistic (or *necessary* statistic), is really what we seek. It can be shown that $\sum_1^4 X_i$ is a minimal sufficient statistic in the example. We do not prove this here. (Problem 6.19 gives a method for verifying minimal sufficiency.)

6.2. Recognizing Sufficient Statistics

Suppose that s is a sufficient statistic. If $P_F(X = x) > 0$, then

$$P_F(X = x) = P_F(s(X) = s(x))P_F(X = x \mid s(X) = s(x)). \tag{6.5}$$

The first factor on the right side of (6.5) depends on x only through the value of $s(x)$, so we write

$$g(F, s(x)) = P_F(s(X) = s(x)). \tag{6.6}$$

The second factor, $P_F(X = x \mid s(X) = s(x))$, does not depend on F, since s is a sufficient statistic; therefore, we write

$$h(x) = P(X = x \mid s(X) = s(x)). \tag{6.7}$$

Using (6.6) and (6.7), we may write (6.5) as

$$P_F(X = x) = g(F, s(x))h(x). \tag{6.8}$$

This expresses $P_F(X = x)$ as a product of two factors, the first of which can depend on F but depends on x only through the value of $s(x)$; the second factor can depend on x in any way but does not depend on F.

Conversely, suppose for some statistic s (i.e., a function of x) that (6.8) holds for suitable functions g and h. Then, from (6.8), we may sum over all x such that $s(x) = a$ to write

$$\begin{aligned} P_F(s(X) = a) &= \sum_{x\,:\,s(x)=a} P_F(X = x) = \sum_{x\,:\,s(x)=a} g(F, s(x))h(x) \\ &= g(F, a) \sum_{x\,:\,s(x)=a} h(x). \end{aligned} \tag{6.9}$$

(The sum in the last expression is finite even when S is infinite, for each a for which $P_F(s(x) = a) > 0$, from this development.)
Also

$$\begin{aligned} P_F(X = x, s(x) = a) &= 0 && \text{if} \quad s(x) \neq a \\ &= P_F(X = x) && \text{if} \quad s(x) = a. \end{aligned} \tag{6.10}$$

Therefore, using (6.9) and (6.10), if $P_F(s(x) = a) > 0$ we have

$$\begin{aligned} P_F(X = x \mid s(X) = a) &= 0 \quad \text{if} \quad s(x) \neq a, \\ &= h(x) \Big/ \sum_{x\,:\,s(x)=a} h(x) \quad \text{if} \quad s(x) = a. \end{aligned} \tag{6.11}$$

Since the value of this conditional probability does not depend on F in Ω, it follows from (6.1) that s is a sufficient statistic.

Although we have discussed the problem only when S is countable, even in the case of a density function, if $f_{F;X}(x)$ can be written in the form of the right side of (6.8) for some s, then that s is a sufficient statistic. This *Neyman Decomposition* (or *Factorization*) *Theorem* is the main tool for recognizing sufficient statistics in a given problem; we try to factor $p_{F;X}$ or $f_{F;X}$ into this form, which shows us that a particular s is sufficient. (With a little more effort, one could verify whether or not s is minimal. There is a method for computing minimal sufficient statistics which we shall not treat here.)

It is important to understand that it is only necessary to verify (6.8) for some g and h, in order to prove s sufficient. For we have proved that (6.8) implies (6.1).

EXAMPLE 6.1. In our Bernoulli example, we have

$$p_{X;\theta}(x_1,x_2,x_3,x_4) = \begin{cases} \prod_{i=1}^{4} \theta^{x_i}(1-\theta)^{1-x_i} & \text{if all } x_i \text{ are 0 or 1,} \\ 0 & \text{otherwise.} \end{cases}$$

Hence, with $h(x_1,x_2,x_3,x_4) = 1$ if all x_i are 0 or 1 and $h = 0$ otherwise, and with $g(\theta, y) = \theta^y(1-\theta)^{4-y}$, we obtain (6.8) with $s(x_1,x_2,x_3,x_4) = \sum_1^4 x_i$. If the four "observations" were replaced by n observations, $\sum_1^n x_i$ would still be sufficient and, in fact, minimal sufficient.

EXAMPLE 6.2. Let $X = (X_1,\ldots,X_n)$ where the X_i are independent and identically distributed normal random variables with variance one and unknown mean θ. We then have

$$\begin{aligned} f_{\theta;X}(x_1,\ldots,x_n) &= \prod_{i=1}^{n}\left(\frac{1}{\sqrt{2\pi}}e^{-(x_i-\theta)^2/2}\right) \\ &= (2\pi)^{-n/2}e^{-\sum_1^n (x_i-\theta)^2/2} \\ &= (2\pi)^{-n/2}e^{-\sum_1^n (x_i-\bar{x}_n)^2/2}e^{-n(\bar{x}_n-\theta)^2/2}. \end{aligned}$$

Hence, taking the last factor to be g, we see that $\bar{X}_n$ is a sufficient statistic. It was necessary to break up the term $\sum (x_i - \theta)^2$ in the next to last line in order to obtain this result; if we had let g be the exponential expression of that line, we would not have obtained $\bar{X}_n$ alone (which is minimal) as the sufficient statistic. In general, the Neyman decomposition theorem does *not* say that any way in which you happen to write down f (e.g., as in the second line) will exhibit the fact that a certain sufficient s is sufficient; rather, it says that there *is* a way of writing f (as in the third line) which exhibits the sufficiency.

Examples 6.1 and 6.2 are special cases of a family of probability laws which includes many important practical cases, known as the Koopman-Darmois family. A random variable (or vector) X_1 is said to have a *Koopman-Darmois (KD) law* if its discrete or absolutely continuous density function is of the form

$$f_{\theta;X_1}(x_1) = b(x_1)c(\theta)e^{k(x_1)q(\theta)}. \tag{6.12}$$

Here θ is real, Ω is some interval of possible θ-values, and b, c, k, q are real-valued functions. *We assume* q *is increasing in* θ; this makes different values of θ correspond to different laws of X_1; if instead $q(3) = q(7)$, one can see that $c(3) = c(7)$ (these are just constants that make the total probability unity) and X_1 would have the same law whether $\theta = 3$ or $\theta = 7$, an inconvenient labeling. [Of course, q could be *decreasing* in θ, instead.]

The family (6.12) for $\theta \in \Omega$ is sometimes referred to in books as a *law of exponential type*. This may be a little confusing, since (6.12) includes the exponential density $\theta e^{-\theta x_1}$ for $x_1 > 0$ and $\theta > 0$ (or $\theta^{-1}e^{-x_1/\theta}$ for $x_1 > 0$ and $\theta > 0$), but also many others, as the problems will illustrate. [Some books use *KD* for the more general form encountered in Problem 6.2.]

If q is also continuous, and we write $\phi = q(\theta)$, then if $\Omega = \{\theta : a < \theta < b\}$ and $q(a) = A$, $q(b) = B$, we can equally well think of the parameter set, in terms of ϕ, as $\{\phi : A < \phi < B\}$. Writing $c(\theta) = h_1(\phi)$, (6.12) then becomes $b(x_1)h_1(\phi)e^{k(x_1)\phi}$, which is the form we will encounter in Chapter 7.

Finally, writing $Y_1 = k(X_1)$, one sees that Y_1 is sufficient for Ω (considering only X_1) and has a density or probability function of the form

$$\bar{b}(y)c(\theta)e^{q(\theta)y} \quad \text{or} \quad \bar{b}(y)h_1(\phi)e^{\phi y}. \tag{6.13}$$

EXAMPLE 6.3 (Recall the notation introduced earlier: We denote by $\mathcal{N}(\mu, \sigma^2)$ the normal law with mean μ and variance σ^2.) X_1 is $\mathcal{N}(0, \theta^2)$, $\Omega = \{\theta : 0 < \theta < \infty\}$. Hence (6.12) is

$$\underbrace{\frac{1}{\sqrt{2\pi}}}_{b} \cdot \underbrace{\theta^{-1}}_{c} \cdot e^{\overbrace{(x_1^2)}^{k}\overbrace{(-1/2\theta^2)}^{q}}.$$

Or substitute $\phi = -1/2\theta^2$, $-\infty < \phi < 0$. Finally, with $Y_1 = X_1^2$, so that $dk(x_1)/dx_1 = 2x_1 = 2\sqrt{y}$, or $\left|\frac{dx_1}{dy}\right| = \frac{1}{2\sqrt{y}}$, (6.13) becomes the density of X_1^2, a gamma density, as discussed in Appendix A. When $\theta = 1$, this is called the *chi-square* (χ^2) *distribution with one degree of freedom*. The form of (6.13) is

$$\frac{1}{\theta\sqrt{2\pi y}}e^{(-1/2\theta^2)y} = \frac{\sqrt{-\phi}}{\sqrt{\pi y}}e^{\phi y}.$$

[If you derive this last from the $\mathcal{N}(0, \theta^2)$ density, recall that in calculating the density of $Y_1 = X_1^2$ we must consider the *two* values of X_1 which yield a given value of Y_1.]

EXAMPLE 6.4. Suppose the variance σ^2 is also unknown in Example 6.2, so that

$$\begin{aligned} f_{(\theta,\sigma);X}(x_1,\ldots,x_n) &= (2\pi\sigma^2)^{-n/2}e^{-\Sigma(x_i-\theta)^2/2\sigma^2} \\ &= (2\pi\sigma^2)^{-n/2}e^{-[\Sigma(x_i-\bar{x}_n)^2+n(\bar{x}_n-\theta)^2]/2\sigma^2} \\ &= g[(\theta,\sigma),(\bar{x}_n,\Sigma(x_i-\bar{x}_n)^2)]. \end{aligned}$$

Thus (setting $h = 1$), we have shown that the pair $(\bar{X}_n, \sum(X_i - \bar{X}_n)^2)$ is sufficient (it is actually minimal). Some books say in an example like this that "$\bar{X}_n$ is sufficient for θ, and $\sum(X_i - \bar{X}_n)^2$ is sufficient for σ^2," but this is somewhat misleading, since both components may be needed in inference about either parameter (as we shall see in testing hypotheses, in the "t-test"). For us, "sufficiency" always refers to *all* of Ω, that is, to the pair (θ, σ) in the present example.

EXAMPLE 6.5. Suppose $X = (X_1,\ldots,X_n)$, where the X_i's are independent and identically distributed with uniform density from 0 to θ $(0 < \theta < \infty)$. Then

$$f_{\theta;X}(x_1,\ldots,x_n) = \begin{cases} \theta^{-n} & \text{if} \quad 0 < x_i < \theta \quad \text{for all } i, \\ 0 & \text{otherwise.} \end{cases}$$

This can be written as $h(x_1,\ldots,x_n)g(\theta, \max_i x_i)$, where h is 1 if all x_i's are > 0 and h is 0 otherwise, and where $g(\theta, y) = \theta^{-n}$ or 0 according to whether or not $y < \theta$. Thus, $\max_i x_i$ is sufficient (it is actually minimal).

EXAMPLE 6.6. If $X = (X_1,\ldots,X_n)$ where the X_i's are independent and uniformly distributed from $\theta - 1/2$ to $\theta + 1/2$ $(-\infty < \theta < \infty)$, we obtain

$$\begin{aligned} f_{\theta;X}(x_1,\ldots,x_n) &= \begin{cases} 1 & \text{if} \quad \theta - 1/2 < x_i < \theta + 1/2 \quad \text{for all } i, \\ 0 & \text{otherwise} \end{cases} \\ &= \begin{cases} 1 & \text{if} \quad \theta - 1/2 < \min_i x_i \le \max_i x_i < \theta + 1 \\ 0 & \text{otherwise,} \end{cases} \end{aligned}$$

so that the pair $(\min_i X_i, \max_i X_i)$ is sufficient (it is actually minimal). Here we have a 1-dimensional parameter θ and a 2-dimensional minimal sufficient statistic. (It is possible to have examples of a real parameter θ and a minimal sufficient statistic of any number of real components $\le n$; we have obtained such examples with 1 and 2 components and in Examples 6.7 and 6.8 shall give one with n components.) It is important to keep in mind the fact that $(\min X_i + \max X_1)/2$, which is a good (e.g., minimax for squared error loss) estimator of θ, is nevertheless *not* a sufficient statistic: the pair $(\min x_i, \max x_i)$, and not merely their average, gives all the information about θ in the sample. But a point estimator of θ must take on real values (not vector values), and thus $(\min X_i, \max X_i)$, although sufficient, is not even allowable as an estimator. The lack of sufficiency of $t_1(X) = (\min X_i + \max X_i)/2$ manifests itself in the fact that it is *not* true that *any* estimator t can be replaced by a function of t_1 which is at least as good for all F; this does not contradict the fact that t_1 is minimax.

EXAMPLE 6.7. Suppose $X = (X_1, \ldots, X_n)$, where the X_i's are independently and identically distributed; Ω need not be more precisely specified at the moment. If, for any values $x_1, \ldots, x_n$, we let $y_1 \leq y_2 \leq \cdots \leq y_n$ be the ordered values of the x_i's, we have (e.g., in the continuous case)

$$f_{F;X}(x_1, \ldots, x_n) = \prod_{i=1}^{n} f_{F;X_1}(x_i) = \prod_{i=1}^{n} f_{F;X_1}(y_i),$$

since the terms in the last product are merely a permutation of those in the previous product. We conclude that the set of order statistics $(Y_1, \ldots, Y_n)$ constitutes a sufficient statistic. This is an n-dimensional statistic and does not represent much of a reduction from the original $X = (X_i, \ldots, X_n)$, which, as we have mentioned, is always sufficient. (The partition corresponding to $(Y_1, \ldots, Y_n)$ lumps into the same set with a given point $(x_1, \ldots, x_n)$ of S precisely those points obtained by permuting the coordinates of the given point; a procedure depending only on this sufficient statistic merely has the property that the decision reached for a given set of values of the observations is the same as that reached for any permutation of these values—the order in which the values were obtained is irrelevant because the X_i were independent and identically distributed.) The question naturally arises, is $(Y_1, \ldots, Y_n)$ ever a *minimal* sufficient statistic? In the examples considered previously, it is not. In most genuinely nonparametric problems, it is, and no coarser summarization of the data contains all the information present in the sample. There are also many simple parametric examples in which it is minimal, one of which we shall now mention without giving any details:

EXAMPLE 6.8. If $X_1, \ldots, X_n$ are independent Cauchy random variables with common density $1/\pi[1 + (x - \theta)^2]$, where $\Omega = \{-\infty < \theta < \infty\}$, then it can be shown that the full set of order statistics $(Y_1, \ldots, Y_n)$ is *minimal* sufficient.

Thus, Example 6.8 illustrates the "worst case" from the sufficiency viewpoint, in which a minimal sufficient s puts $n!$ points (if the observations take on distinct values) into each set of the sufficient partition, but s is still n-dimensional just as X was. Thus, it is almost as horrible to look at all procedures that depend on $s(X)$ as to look at all that depend on X in this case. At the other extreme, in Example 6.2 or (as shown in Problem 6.1) under (6.12), the dimensionality n of $(X_1, \ldots, X_n)$ is reduced to 1 for minimal $s(X)$, so sufficiency is very helpful in reducing the class of procedures we must examine. Example 6.5, in which the dimension is reduced to 2, and Problem 6.2, in which it is reduced to m (independent of n) even for 1-dimensional Ω, lie between these two extremes.

6.3. Reconstruction of the Sample

The notion of sufficiency is often described by saying that "given the value taken on by a sufficient statistic, we can reconstruct the sample." This is not to be taken literally; for example, in our Bernoulli Example 6.1, if we are told that $s(X) = 1$, there is no way for us to determine which of the four possible values of X (satisfying $s(X) = 1$) was actually taken on in the experiment. Rather, the statement in quotation marks is to be interpreted in a probabilistic sense: We can give a prescription for producing a random quantity Y, the prescription depending only on the value of $s(X)$, and such that Y has the same probability law as X for every F. The prescription is this: If s is sufficient, then (in the discrete case), for each b, the function

$$g(x|b) = P_F(X = x|s(X) = b)$$

does not depend on F and is a probability function on S (as a function of x); if we are told that $s(X) = b$, we construct a chance mechanism with possible outcomes Y taking on values in S and (conditional) probability law $P(Y = y|s(X) = b) = g(y|b)$. (Note again that sufficiency makes g not depend on the unknown F, so that this construction is possible.) Thus, in our example of Section 6.1 in the Bernoulli case, with $s(x_1, x_2, x_3, x_4) = \sum_1^4 x_i$, if $b = 0$ we would set $Y = (0, 0, 0, 0)$; if $b = 1$, we would perform a chance experiment with four equally likely outcomes, (0001), (0010), (0100), and (1000), and would set Y equal to the outcome; and so on. If we think of X being observed, $s(X)$ being computed, and Y being obtained by an additional chance trial (depending on the value of $s(X)$ but not on F) as outlined previously, it is clear that Y need not equal X; in our example, we may have $X = (0001)$, perform the indicated chance trial with $b = 1$, and obtain $Y = (0100)$. Thus, we do not "reconstruct the sample" in the sense of recreating the value taken on by X (and in fact, we cannot, as we have seen). But we *have* recreated the sample probabilistically, in the sense that

$$P(Y = x|s(X) = b) = P(X = x|s(X) = b)$$

for every x and b, so that $P_F(X = x) = P_F(Y = x)$ for all x and F. This gives us an intuitively evident method for constructing a (randomized) procedure depending only on s and which is equivalent to a given procedure t or δ. Suppose D and Ω are given, and s is sufficient; having computed $s(X)$, we could then produce (e.g., with a spinner) the random variable Y as previously, and then make decision $t(Y)$ (or, if we started with a randomized procedure δ and $Y = y$, make a decision according to the probabilities $\delta(y, d)$). Since Y has the same probability function as X for all F, this new procedure will have the same operating characteristic as the original one (although the actual decisions reached by the two procedures in any given experiment and subsequent production of Y may differ). In fact, you can check easily that P (make decision d, based on $Y|s(X) = b$) $= P$ (make decision d, based on $X|s(X) = b$). Multiplying both sides of this equality by $P_F(s(X) = b)$ and summing over b, we obtain P_F {make decision d, based on Y} $= P_F$ {make decision d, based on X}. Consequently, the rule t based on X and that based on $s(X)$ (through randomization producing Y, and using $t(Y)$) have the same risk function, no matter what the loss function W (a function of F and d) may be!

In the next section we formalize the construction of a δ^* depending only on $s(X)$ and which has the same operating characteristic as a given δ, without explicit "probabilistic reconstruction of the sample"; but the preceding development should help give an intuitive feeling for why sufficiency works in the way it does.

6.4. Sufficiency: "No Loss of Information"

Fisher's intuitive notion that no "information" is lost by restriction to procedures that depend on X only through sufficient $s(X)$ is made precise by the following decision-theoretic fact (sketched in the previous section in terms of the "reconstructed sample" notion), stated here for discrete S and D, although valid more generally.

Given S, D, Ω, if s is sufficient for Ω, let $s(S) \stackrel{\text{def}}{=} \{b : b = s(x) \text{ for some } x \text{ in } S\}$ be the "reduced sample space" of possible s-values (range of s). Then, for each randomized procedure δ for the original S and D, there is a randomized procedure Δ defined on $s(S) \times D$ (in the way δ was defined on $S \times D$ in Section 4.3)—so that $\Delta(b, d)$ is a probability function of d on D for each fixed b—such that use of δ based on X and use of Δ based on $s(X)$ result in identical operating characteristics (OC's), and consequently in identical risk functions for every loss function W (of F and d). Thus, for each such W, the set of all risk functions of such δ's is the same as the set of all risk functions of such Δ's. This makes explicit the sense in which one "loses nothing" by restricting consideration to procedures that depend only on a sufficient statistic. We shall prove the result in the next paragraph. There is a more difficult converse which we shall not prove: given S, Ω, and a statistic u on S, if for each D and δ there is a Δ on $u(S) \times D$ with the same oc as δ (or, alternatively, if for each D, W, and δ there is a Δ with the same risk function as δ), then u is sufficient for Ω. (The reason that the existence of a Δ corresponding to each δ must be postulated for a variety of D's will be indicated in Problem 6.24.

We now prove the assertion of the previous paragraph. Suppose S, D, Ω are given (with S, D discrete), that s is sufficient for Ω, and that δ is any procedure. Define

$$\begin{aligned}\Delta(b, d) &= \sum_x P\{X = x | s(X) = b\} \delta(x, d) \\ &= \sum_{x\,:\,s(x)=b} P(X = x | s(X) = b) \delta(x, d) \\ &= E\{\delta(X, d) | s(X) = b\},\end{aligned}$$

and then define, as the procedure on S that computes $s(X)$ and then uses Δ,

$$\delta^*(x, d) = \Delta(s(x), d).$$

We shall now show that δ^* satisfies the requirements of a randomized statistical procedure, and that the operating characteristics satisfy

$$q_{\delta^*}(F,d) = q_\delta(F,d).$$

That is, we must show

$$1 = \sum_d \delta^*(x',d) \quad \text{for all } x';$$

$$q_{\delta^*}(F,d) = q_\delta(F,d).$$

But

$$\begin{aligned}\sum_d \delta^*(x',d) &= \sum_d \sum_x P(X = x|s(X) = s(x'))\delta(x,d)\\ &= \sum_x P(X = x|s(X) = s(x'))\sum_d \delta(x,d)\\ &= \sum_x P(X = x|s(X) = s(x')) = 1,\end{aligned}$$

and

$$\begin{aligned}q_{\delta^*}(F,d) &= E_F\delta^*(X,d) = E_F\Delta(s(X),d)\\ &= \sum_y P_F(s(X) = y) \sum_{x:s(x)=y} P(X = x|s(X) = y)\delta(x,d)\\ &= \sum_y \sum_{x:s(x)=y} P_F(X = x)\delta(x,d)\\ &= E_F\delta(X,d) = q_\delta(F,d).\end{aligned}$$

It follows that, *for all possible loss functions* W, this same δ^* is equivalent to δ. Thus, we have shown that, for any δ, there is a δ^* depending only on the given sufficient statistic, and which is equivalent to δ for all W. (Of course, the form of δ^* depends on δ.)

Note that the sufficiency of s was essential to make Δ and δ^* as defined previously *not* depend on F.

Even if δ *had* been nonrandomized, the δ^* produced in this manner might be a randomized procedure; this is why we dealt with randomized procedures δ (which include nonrandomized procedures) from the beginning here. In the next section we shall see that such randomization can sometimes be dispensed with in finding a procedure that depends on X only through a sufficient statistic $s(X)$, and that is at least as good as a given δ.

6.5. Convex Loss

We digress momentarily to consider some general properties of convex functions. First, if q is a convex function of a real variable, then through each point $(z_0, q(z_0))$ on the graph of q we can draw at least one (tangent) line, say $y = L(z) = a + bz$, such that $q(z) \geq L(z)$ for all z:

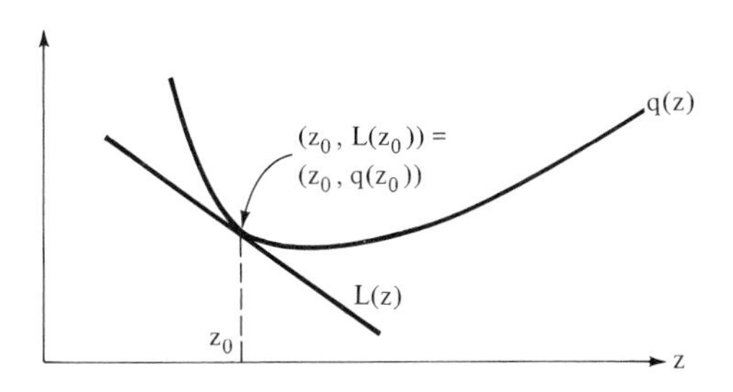

We shall not prove this fact here, but it should seem intuitively reasonable from the picture. Next, we shall prove *Jensen's inequality*: *If* q *is a convex function and* Z *is any random variable with finite expectation, then*

$$Eq(Z) \geq q(EZ). \tag{6.14}$$

To prove this, we construct the linear function $L(z)$ just described with $z_0 = EZ$. Thus, $q(EZ) = L(EZ)$ and $q(z) \geq L(z)$ for every value z. Multiplying both sides of this last inequality by the probability function of Z and summing over all z (with the obvious analogue in the continuous case), we obtain $Eq(Z) \geq EL(Z)$. Now, $EL(Z) = L(EZ)$, since $L(z) = a + bz$ is linear; moreover, $L(EZ) = q(EZ)$. Hence (6.14) is proved. (See Appendix C1.1 for a special case of (6.14) and more on convexity.)

Sometimes it is possible to replace a given statistical procedure with a nonrandomized statistical procedure which is as good as the original procedure. This is in particular true if the decision space D is an interval I of real numbers and the loss function $W(F, d)$ is a convex function of d.

In that setting, if Z is any random variable with finite expectation and which takes on values only in I, Jensen's inequality (6.14) implies that

$$EW(F, Z) \geq W(F, EZ). \tag{6.15}$$

The distribution of Z will be known (independent of F) in our use of (6.15).

Now suppose $D = I$ is a finite interval, $W(F, d)$ is a convex function of d in I, and that δ is a randomized statistical procedure. Relative to δ, d is a random variable. Using (6.15) we compute, letting Z be a random variable with law $P\{Z = d\} = \delta(x, d)$ for fixed x,

$$\sum_d W(F, d)\delta(x, d) \geq W\left(F, \sum_d d\delta(x, d)\right).$$

Since $\sum_d d\delta(x, d) = t(x)$ is a value in the decision space D, $t(x)$ is a nonrandomized statistical procedure and

$$r_t(F) = \sum_x W(F, t(x))p_{F;X}(x) \leq \sum_x \left(\sum_d W(F, d)\delta(x, d)\right) p_{F;X}(x) = r_\delta(F).$$

Therefore t is as good as δ. (The proof can be extended to infinite intervals.)

The result just proved is an example of a *complete class theorem.* We have shown that when D is a real number interval and $W(F, d)$ is a convex function of d, then every (randomized) statistical procedure δ may be replaced by a statistical procedure t as good as δ which is nonrandomized; the t's are said to be an *essentially complete class.* Such a result gets the name *essentially complete class theorem* because it says, in effect, every admissible procedure δ is equivalent to (has the same risk function as) some nonrandomized procedure t; and we say the nonrandomized procedures (admissible or not) form an essentially complete class. (Earlier we have noted that in certain problems every admissible procedure is a Bayes procedure. In such problems the Bayes procedures form a complete class. We have discussed such results further in Chapter 4, in particular in Section 4.4.) Note that the conclusion here differs in three ways from that of the previous section: we obtain, for a given δ, a nonrandomized t rather than a randomized Δ and δ^*; the procedure t may not have the same risk function as δ, but the difference is in a favorable direction in that t is *at least as good* as δ; and the conclusion is not for all W, but for convex (in d) W.

The preceding result has no relation to sufficiency. We now turn to the useful result that, for any procedure δ, there is a nonrandomized t^* depending only on the sufficient statistic s, and which is at least as good as δ for all convex W. (Of course, the form of t^* depends on δ.) This result follows at once upon using the result of the previous section to obtain a Δ (and δ^*) equivalent to the given δ, and then using the earlier result of the present section, with S and δ replaced by $s(S)$ and Δ, to obtain a t^* on $s(S)$ that is at least as good as Δ (and, hence, at least as good as δ). Thus, for convex W the nonrandomized procedures depending only on s form an essentially complete class, and we may restrict our study to such procedures in choosing the one to be used.

We can write down a particularly simple formula for t^* in the case when the original procedure δ is itself a nonrandomized procedure t: Suppose D is an interval and $W(F, d)$ is, for each fixed F, a convex function of d (for example, $(\phi(F) - d)^2$). Suppose $s(X)$ is a sufficient statistic. Then, for any nonrandomized estimator t for which $E_F t(X)$ is finite, we use Jensen's inequality

$$W(F, \text{average decision}) \leq \text{average over } d \text{ of } W(F, d),$$

to obtain (written in the discrete case)

$$\begin{aligned} W\left(F, \left[\sum_x t(x) P\{X = x | s(X) = s_0\}\right]\right) \\ \leq \sum_x W(F, t(x)) P\{X = x | s(X) = s_0\} \end{aligned} \tag{6.16}$$

for every possible value s_0 of $s(X)$. *Note sufficiency has been used here* in our being able to compute $P\{X = x | s(X) = s_0\}$ without knowing F; if s were *not* sufficient, this conditional probability would depend on F, and $\sum_x t(x) P_F\{X = x | s(X) = s_0\}$ would depend on F and thus would not be a "procedure." But it does *not* depend on F, only on s_0, so we define

$$t^*(s_0) = \sum_x t(x) P\{X = x | s(X) = s_0\}. \tag{6.17}$$

This t^* is a function on the space $s(S)$ of possible values of $s(X)$. Multiplying both sides of (6.16) by $P_F\{s(X) = s_0\}$ and summing on s_0, we obtain

$$\begin{aligned} E_F W(F, t^*(s(X))) &= \sum_{s_0} W(F, t^*(s_0)) P_F\{s(X) = s_0\} \\ &\leq \sum_{s_0} \sum_x W(F, t(x)) P\{X = x | s(X) = s_0\} P_F\{s(X) = s_0\} \quad (6.18) \\ &= E_F W(F, t(X)), \end{aligned}$$

the last equality since $\sum_{s_0} P\{X = x | s(X) = s_0\} P_F\{s(X) = s_0\} = P_F\{X = x\}$.

Equation (6.18) indicates that t^* is at least as good as t: Given any procedure t with finite expectation, there is a t^* *depending on* X *only through the values of* $s(X)$, such that t^* is at least as good as t. (The form of t^* depends on t.) In Example 6.9 this is used to find, from $t(x_1, x_2, x_3, x_4) = (x_1 + 2x_2 + 3x_3 + 4x_4)/10$, a t^* depending only on the sufficient statistic $\sum_1^4 x_i$, with t^* better than t.

Note that one t^* is at least as good as the given t, *simultaneously for every loss function* W which is convex in d.

If you compute $E_F t(x)$ by first summing on values x of X given that $s(X) = s_0$ (as on the right side of (6.18) with $W(F, t(x))$ there replaced by $t(x)$), you will see easily that $E_F t(X) = E_F t^*(s(X))$ in the preceding. In particular, *if* t *is unbiased*, so is t*. We shall use this result in Chapter 7.

EXAMPLE 6.9. In the Bernoulli Problem of Example 6.1, consider the estimator

$$t(x_1, x_2, x_3, x_4) = (x_1 + 2x_2 + 3x_3 + 4x_4)/10.$$

If θ is to be estimated and $W(\theta, d) = (\theta - d)^2$, we see that, since t is an unbiased estimator of θ,

$$r_t(\theta) = E_\theta(t(X) - \theta)^2 = 10^{-2} \operatorname{var}_\theta[X_1 + 2x_2 + 3x_3 + 4x_4] = .3\theta(1 - \theta).$$

Let $s(x) = \sum_1^4 x_i$. We recall that $P(X = x | s(X) = b)$ assigns equal probabilities to each of the possible points in the set of x for which $s(x) = b$. (*Warning*: In other examples, this conditional law need not be uniform!) For $b = 0$, there is just one point, so

$$t^*(0) = t(0, 0, 0, 0) = 0.$$

For $b = 1$, there are four points, and we have

$$\begin{aligned} t^*(1) &= \frac{1}{4}[t(1, 0, 0, 0) + t(0, 1, 0, 0) + t(0, 0, 1, 0) + t(0, 0, 0, 1)] \\ &= \frac{1}{4}\left[\frac{1}{10} + \frac{2}{10} + \frac{3}{10} + \frac{4}{10}\right] = \frac{1}{4}. \end{aligned}$$

Continuing in this way, one obtains $T(b) = b/4$ for each b, for the given t. Thus, $t^*(s(x)) = \bar{x}_4$ for the given t, and

$$r_{t^*}(\theta) = \text{var}_\theta(\bar{X}_4) = .25\theta(1-\theta).$$

Finally, we remark on circumstances in which the t defined after (6.15) is not just *as good as*, but *better than* δ, and similarly for the t^* of (6.17) compared with t. If q is *strictly* convex—i.e., its graph contains no line segments of positive length, or q'' is strictly *positive* for differentiable q—and if the law of Z does not assign all probability to a single point, then the figure shows that strict inequality holds in (6.14). Consequently, $r_t(F) < r_\delta(F)$ provided that W is strictly convex in d and F assigns positive probability to the set of x-values for which $\delta(x, d)$ *really randomizes* (i.e., does *not* assign probability one to a single d). Similarly, $r_{t^*}(F) < r_t(F)$ if W is strictly convex and when F is true t^* is at least sometimes different from t; i.e., F assigns positive probability to the set of values of $s(X)$ for which the conditional law of t does not concentrate all its probability at a single point.

Problems

Suggested: 6.1; at least a selection of examples such as 6.3–6.5 or 6.6–6.8; 6.13 and 6.16, or 6.14–6.15; at least read 6.2 and 6.23. All the sufficient statistics you find in these problems are "minimal." This is incidental intelligence; don't prove it, unless you work the following problem:

[*Optional*] Use the conclusion of Problem 6.19 (in particular, (6.20)) to prove in Problems 6.1–6.14 that the sufficient statistics obtained are minimal. In Problem 6.2 assume the q_j are linearly independent on Ω.

6.1. Suppose $X = (X_1, X_2, \ldots, X_n)$, where the X_i's are iid, each with density function (discrete or continuous case) given by (6.12) with Ω, b, c, k, q as described there. Show that $s(X_1, \ldots, X_n) = \sum_1^n k(X_i)$ is a sufficient statistic for Ω by using the Neyman factorization theorem.

6.2. The term *KD* or *exponential type* is used in some books to designate a form more general than (6.12), namely,

$$f_{\theta; X_1}(x_1) = b(x_1)c(\theta)e^{\sum_1^m k_j(x_1)q_j(\theta)}. \tag{6.19}$$

[Here θ may be a vector. Ω is some specified set of θ-values.]

Show that with n observations, a sufficient statistic in this case is the m-dimensional $s(X) = \left(\sum_1^n k_1(x_i), \ldots, \sum_1^n k_m(x_i)\right)$.

6.3. State which 5 of the following 10 laws of X_1 can be written in the form (6.12) and find the k (and thus the s of Problem 6.1) in each case. [In the case of (iv) and (v), see whether you can guess the result before writing down f_{θ, X_1}—not obvious!] Make sure you understand the difference between the family of exponential densities of (vii) and that of (viii)! Also, verify that (vi) is a density.

(i) normal, mean 3, unknown variance θ^2; $\Omega = \{\theta : \theta > 0\}$;
(ii) normal, variance 2, unknown mean θ; $\Omega = \{\theta : -\infty < \theta < \infty\}$;
(iii) normal, unknown mean θ_1 and unknown variance θ_2 (so that $\theta = (\theta_1, \theta_2)$); $\Omega = \{(\theta_1, \theta_2) : -\infty < \theta_1 < \infty, \theta_2 > 0\}$;
(iv) normal, unknown variance θ, and unknown mean $\theta + 1$; $\Omega = \{\theta : \theta > 0\}$;
(v) normal, unknown mean θ, and unknown variance θ^2; $\Omega = \{\theta : -\infty < \theta < \infty\}$;
(vi) the density $(1 + \theta x)/2$ for $-1 < x < 1$ (and 0 otherwise); here $-1 \leq \theta \leq 1$;
(vii) exponential density $\theta^{-1}e^{-x/\theta}$ for $x > 0$ (and 0 otherwise); here $\theta > 0$;
(viii) exponential density $e^{-(x-\theta)}$ if $x > \theta$ (and 0 otherwise); here $-\infty < \theta < \infty$;
(ix) exponential density $\theta_2^{-1}e^{-(x-\theta_1)/\theta_2}$ if $x > \theta_1$ (and 0 otherwise); here $\theta_2 > 0$, $-\infty < \theta_1 < \infty$;
(x) geometric law [see Appendix A] of X_1 = number of heads before first tail in iid flips of a coin with probability θ on each toss of heads. Here $0 < \theta < 1$.

6.4. Which *one* of the "failures" in Problem 6.3, although not of *KD* form (6.12), has a "1-dimensional sufficient statistic," namely, $s(X_1, \ldots, X_n) = \min(X_1, X_2, \ldots, X_n)$? (Use Neyman factorization.)

6.5. Of the four "failures" in Problem 6.3 not covered in Problem 6.4, two are of the form (6.19) with $m = 2$. Find them, and find $s(X)$ in each case. Note that θ is 1-dimensional in one case and 2-dimensional in the other, although s is 2-dimensional in both cases. (This and the other examples indicate that the relationship between the dimensionality of s and that of θ is not always obvious.) Of the remaining two "failures," one has a 2-dimensional sufficient statistic although it is not of the form (6.19). Find s in this case.

6.6. For parts of this problem, we will use the following notation for some of the densities considered:

$$f^*_{\theta_1,\theta_2}(x_1) = \begin{cases} \dfrac{1}{\theta_2^{\theta_1}\Gamma(\theta_1)} \cdot x_1^{\theta_1 - 1}e^{-x_1/\theta_2} & \text{if} \quad x_1 > 0, \\ 0 \quad \text{if} \quad x_1 \leq 0. \end{cases}$$

Here $\theta_1 > 0$, $\theta_2 > 0$. This is a family of Γ-laws with unknown index and scale (Appendix A).

$$f^{**}_{\theta_1,\theta_2}(x_1) = \begin{cases} \dfrac{\theta_1}{\theta_2}\left(\dfrac{x_1}{\theta_2}\right)^{-\theta_1 - 1} & \text{if} \quad x_1 > \theta_2, \\ 0 \quad \text{if} \quad x_1 \leq \theta_2. \end{cases}$$

Again, θ_1 and θ_2 are positive. *Verify that f^{**} is a density function.* Make sure you realize, in each of problems (i) through (viii), which of the parameters θ_i are fixed and which are unknown.

Each of the following 10 parts gives a family Ω of laws of X_1. State which 5 of these can be written in the form (6.12) and find the k (and thus the s) of Problem 6.1 in each case. In doing this, having worked (i), (ii), (iii), see whether you can guess the result in (iv) and (v) before writing down the family of densities; the results are not so obvious!

(i) $\Omega = \{f^*_{\theta_1,3} : 0 < \theta_1 < \infty\}$ (that is, "θ_2 is known to be 3").
(ii) $\Omega = \{f^*_{2,\theta_2} : 0 < \theta_2 < \infty\}$.

(iii) $\Omega = \{f^*_{\theta_1,\theta_2} : 0 < \theta_1, \theta_2 < \infty\}$ (both θ_1 and θ_2 unknown).
(iv) $\Omega = \{f^*_{\theta,\theta} : 0 < \theta < \infty\}$ (θ_1 and θ_2 are unknown but equal).
(v) $\Omega = \{f^*_{\theta,1/\theta} : 0 < \theta < \infty\}$ (θ_1 and θ_2 are unknown but reciprocals).
(vi) $\Omega = \{f^{**}_{\theta_1,1} : 0 < \theta_1 < \infty\}$.
(vii) $\Omega = \{f^{**}_{4,\theta_2} : 0 < \theta_2 < \infty\}$.
(viii) $\Omega = \{f^{**}_{\theta_1,\theta_2} : 0 < \theta_1, \theta_2 < \infty\}$.
(ix) $\Omega = \{$Poisson probability functions with mean $\theta, 0 < \theta < \infty\}$.
(x) $\Omega = \{$densities $f_\theta^{***}, 0 < \theta < \infty\}$,

where

$$f_\theta^{***}(x) = \begin{cases} 2\theta^2/(x+\theta)^3 & \text{if } x > 0, \\ 0 & \text{if } x \le 0. \end{cases}$$

(Verify that this *is* a pdf.)

6.7. Which *one* of the "failures" in Problem 6.5, although not of *KD* form (6.12), has a "1-dimensional sufficient statistic," namely, $s(X_1, \ldots, X_n) = \min(X_1, X_2, \ldots, X_n)$? (Use Neyman factorization.)

6.8. Of the four "failures" in Problem 6.6 not covered in Problem 6.7, two are of the form (6.19) with $m = 2$. Find them, and find $s(X)$ in each case. Note that θ is 1-dimensional in one case and 2-dimensional in the other, although s is 2-dimensional in both cases. (This and the other examples indicate that the relationship between the dimensionality of s and that of θ is *not always* obvious.) Of the remaining two "failures," one has a 2-dimensional sufficient statistic although it is not of the form (6.19). Find s in this case.

6.9. With $\Omega = \{\theta : \theta > 0\}$, suppose $X = (X_1, \ldots, X_N)$ with $N \ge 2$ and the X_i iid $\mathcal{N}(\theta, \theta^p)$ where p is a *known* real value. Thus, the unknown mean and variance of the normal law have the known given relation. Find the two values of p for which (6.12) is satisfied.

6.10. For the models of Problem 6.9 where p is not one of the two values you obtained there, show that $f_{\theta; X_1}$ is of the form (6.19) with $m = 2$. Note that θ is 1-dimensional in each case, although s is 2-dimensional. (This and other examples indicate that the relationship between the dimensionality of s and that of θ is not always obvious.)

6.11. A well-known random phenomenon has known probability density g (continuous case) on R^1. An ogre (perhaps a bureaucrat or polluter, but some sort of bad guy) cannot permit any random variable to take on a value larger than θ (his limit of tolerance). So if Y_j are iid random variables with density g, he annihilates any $Y_j > \theta$ and lets us observe only the Y_j's that are $\le \theta$. (In statistical literature, this is called *censoring*.) Call these observed random variables X_i. The density of X_1 is then the same as the conditional density of Y_1 *given that* $Y_1 \le \theta$, namely,

$$f_{\theta; X_1}(x_1) = \begin{cases} c(\theta) g(x_1) & \text{if } x_1 \le \theta, \\ 0 & \text{if } x_1 > \theta, \end{cases}$$

where $c(\theta) = 1/\int_{-\infty}^{\theta} g(x)\,dx$. If N iid rv's X_i are observed, show that $\max(X_1, \ldots, X_N)$ is sufficient for $\Omega = \{\theta : -\infty < \theta < \infty\}$.

6.12. Suppose $X_1, \ldots, X_n$ ($n \ge 3$) are iid rv's with absolutely continuous case pdf

$$f_{(\theta_1,\theta_2,\theta_3)}(x_1) = \begin{cases} H(\theta_1,\theta_2,\theta_3)x_1^{\theta_1-1}e^{-(x_1-\theta_2)/\theta_3} & \text{if} \quad x_1 \geq \theta_2, \\ 0 \quad \text{otherwise,} \end{cases}$$

where $H(\theta_1,\theta_2,\theta_3) = 1/\int_{\theta_2}^{\infty} x_1^{\theta_1-1}e^{-(x_1-\theta_2)/\theta_3}\,dx_1$ and $\Omega = \{(\theta_1,\theta_2,\theta_3): \theta_1 > 0, \theta_2 > 0, \theta_3 > 0\}$. Find a 3-dimensional sufficient statistic; note that f is not of the form (6.19), though.

6.13. If $X = (X_1, X_2)$ where the X_i are iid, each with geometric law $\mathscr{G}(\theta)$ of Appendix A, where $\Omega = \{\theta: 0 \leq \theta < 1\}$, show that $s(X) = X_1 + X_2$ is sufficient for Ω by using the original definition of sufficiency (6.1) (as illustrated after (6.1) in the Bernoulli case); do *not* use Neyman decomposition (6.8). (Familiarity with the law of $X_1 + X_2$ is unnecessary.)

6.14. Same as Problem 6.13 with $\mathscr{G}(\theta)$ replaced by Poisson law $\mathscr{P}(\theta)$ (Appendix A) and $\Omega = \{\theta: 0 \leq \theta < \infty\}$. (In this case $P\{X_1 = x_1, X_2 = b - x_1 | s(X) = b\}$ turns out to be binomial rather than uniform as in the Bernoulli example.)

6.15. In the setting of Problem 6.14, if $t(x_1, x_2) = .3x_1 + .5x_2 + 1$ is proposed as an estimator of θ, use (6.17) and the conditional binomial law obtained in Problem 6.14 to show that the t^* given by $t^*(x_1 + x_2) = 1 + .4(x_1 + x_2)$ is at least as good as t for all W convex in d. [Optional: Use the remark at the end of Section 6.5 to conclude t^* is *better than* t for W strictly convex in d.]

6.16 (a) In the setting of 6.13, if $t_1(x_1, x_2) = x_1$ is proposed as an estimator of $\phi_1(\theta) = \theta/(1-\theta)$ (of which it is an unbiased estimator—see Appendix A), use (6.17) and the conditional (uniform) law of X_1 given $X_1 + X_2 = b$ (found in Problem 6.13) to show that the t_1^* given by $t_1^*(x_1 + x_2) = .5(x_1 + x_2)$ is at least as good as t_1 for all W convex in d.

(b) If (a) is altered by proposing to estimate $\phi_2(\theta) \stackrel{\text{def}}{=} [\phi_1(\theta)]^2$ by $t_2 = t_1^2$ (reasonable intuitively if t_1 is proposed to estimate ϕ_1, but of course no longer unbiased), show that the corresponding t_2^* is given by $t_2^*(x_1 + x_2) = (x_1 + x_2)[2(x_1 + x_2) + 1]/6$. Note that, although $t_2 = t_1^2$, $t_2^* \neq t_1^{*2}$.

[*Optional*: Use the remark at the end of Section 6.5 to conclude that t_i^* is *better than* t_i for W strictly convex in d.]

6.17. Prove (by writing out the details) the assertion preceding Example 6.9 that $E_F t = E_F t^*$, so that t^* is an unbiased estimator of ϕ if t is.

6.18. This will illustrate that, in Section 6.5, if $W(F, d)$ is not convex in d, t^* need not be at least as good as t. Let X_1, X_2 be iid Bernoulli random variables with expectation θ, $\Omega = \{\theta: 0 \leq \theta \leq 1\}$; $D = \{d: 0 \leq d \leq 1\}$ and $W(\theta, d) = 0$ if $|\theta - d| \leq .2$, $= 1$ if $|\theta - d| > .2$. This familiar W is reasonable but not convex. Let $t(x_1, x_2) = x_1$, from which we verify that $t^*(x_1 + x_2) = .5(x_1 + x_2)$. Show that $P_\theta\{t(X_1, X_2) = t^*(X_1 + X_2) | X_1 + X_2 = 0 \text{ or } 2\} = 1$, and that the (conditional risk) functions $P_\theta\{|\bar{t} - \theta| > .2 | X_1 + X_2 = 1\}$, for $\bar{t} = t$ and t^*, are "crossing." Conclude that r_t and r_{t^*} are "crossing risk functions."

6.19. *Finding a minimal sufficient partition.*

(a) In the discrete case with $p_F(x) > 0$ throughout S for each F, show that if s is sufficient, then for x' and x'' in S

$$s(x') = s(x'') \Rightarrow p_F(x')/p_F(x'') \text{ does not depend on } F.$$

(b) A partition of S is defined in terms of an equivalence relation "$\sim$": $x' \sim x''$

means x' and x'' are in the same set of the partition. Make sure you understand that every partition of S gives an equivalence relation and conversely. Then show that

$$x' \sim x'' \Leftrightarrow p_F(x')/p_F(x'') \text{ does not depend on } F \tag{6.20}$$

defines an equivalence relation, which thus yields a partition of s.

(c) Conclude from (a) that the partition defined by (6.20) is sufficient, and that every other sufficient partition is a refinement of this one. Thus, (6.20) defines the *minimal sufficient* (or *necessary*) partition. [This conclusion also holds without the restriction that p_F be positive throughout S, and also in the absolutely continuous case.]

6.20. We saw in Example 6.7 that, whenever $X = (X_1, \ldots, X_N)$ and the X_i are iid, the set of "order statistics" $s(X) = (Y_1, \ldots, Y_N)$, where $Y_j = j^{\text{th}}$ largest of $X_1, \ldots, X_N$, is always sufficient. Consequently, the "worst cases" from the point of view of sufficiency (and, hence, from the point of view of finding analytically tractable "good" statistical procedures) are those where the set of order statistics is *minimal* sufficient, as mentioned after Example 6.8. It turns out that one of the density functions of Problem 6.3 whose minimal sufficient statistic was not found in Problems 6.3, 6.4, or 6.5 is of this "worst" variety. Use the result of Problem 6.19 in terms of (6.20) to demonstrate this, at least when $n = 2$, by showing that, if $\bar{\bar{\bar{x}}} = (\bar{\bar{\bar{x}}}_1, \ldots, \bar{\bar{\bar{x}}}_n)$ and $\bar{\bar{x}} = (\bar{\bar{x}}_1, \ldots, \bar{\bar{x}}_n)$ are two possible values of X, then we can have the ratio

$$\prod_1^n f_{\theta; X_1}(\bar{\bar{\bar{x}}}_i) \Big/ \prod_1^n f_{\theta; X_1}(\bar{\bar{x}}_i)$$

independent of the value of θ, if and only if $\bar{\bar{\bar{x}}}$ is a permutation of $\bar{\bar{x}}$. From this, conclude that Y is minimal sufficient.

6.21. Work Problem 6.20 with Problems 6.6–6.8 replacing Problems 6.3–6.5 in the statement of Problem 6.20.

6.22. Suppose $X = (X_1, \ldots, X_n)$ where the X_i's are iid random variables with

$$p_{\theta; X_1}(x_1) = \begin{cases} (\theta + x_1)/K\left(\theta + \dfrac{K+1}{2}\right) & \text{if} \quad x_1 = 1, 2, \ldots, K; \\ 0 \quad \text{otherwise}; \end{cases}$$

here K is a known integer >1 and $\Omega = \{\theta : 0 < \theta < \infty\}$. Use the conclusion of Problem 6.19 (in particular, (6.20) to show that, if $N_j(X) \overset{\text{def}}{=}$ number of X_i's equal to j $(1 \le i \le n)$, then $s^*(X) = (N_1(X), N_2(X), \ldots, N_K(X))$ is minimal sufficient. (This is a discrete case analogue of the absolutely continuous case phenomenon of Problems 6.20 and 6.21, as you can see by showing that $s^*(X)$ is a 1-to-1 function of Y. But in the discrete case, some X_i, and thus some Y_i, can be equal, and such ties occur with zero probability in the absolutely continuous case.)

6.23. *Another characterization of sufficiency.* Although this characterization holds more generally, we assume here S and Ω discrete and $p_F(x) > 0$ for all F and x. *Assertion*: s is sufficient for $\Omega \Leftrightarrow$ for each prior law ξ, the posterior law of F given $X = x$ (given as $p_\xi(F|x)$ in Section 4.1) depends on x only through the value of $s(x)$. That is, s is sufficient for $\Omega \Leftrightarrow p_\xi(F|x') = p_\xi(F|x'')$ for all ξ and F, whenever

$s(x') = s(x'')$. (a) Use (6.1) or (6.8) to prove the "$\Rightarrow$" part of the assertion. (b) To prove the "$\Leftarrow$" part, consider for fixed F_0 in Ω the particular *family* of prior laws $\{\xi_F : F \in \Omega, F \neq F_0\}$ where, for the chosen F_0 in Ω, ξ_F assigns probability 1/2 to each of F_0 and F. For a given s for which the assertion of the right side of $\Leftrightarrow$ holds, consider that assertion when $\xi = \xi_F$; vary F and use (6.8) or (6.20). (c) Although the proof of (b) shows that "$\Leftarrow$" still holds if the set Ξ of *all* ξ is replaced by a smaller set, that set cannot be too small. Note that "$\Leftarrow$" is trivially no longer true if Ξ is replaced by $\{\tilde{\xi}_F : F \in \Omega\}$ where $\tilde{\xi}_F(F) = 1$. Less trivially, if X is $\mathcal{N}(\theta, 1)$ with $\Omega = \{\theta : -\infty < \theta < \infty\}$ and Ξ is replaced by the set of all prior densities *symmetric about* 0, show that each resulting posterior density of θ, given that $X = x$, depends only on $\bar{s}(x) \stackrel{\text{def}}{=} |x|$; but $\bar{s}$ is not sufficient (as can be verified directly or from the fact that $s(x) = x$ is minimal sufficient).

6.24. In Section 5.4, as a converse to the decision-theoretic meaning of sufficiency, it was stated that if, for given S and Ω, there is a statistic u such that, for every D, W, and δ there is a procedure depending only on $u(X)$ with the same risk function as δ, then u is sufficient for Ω. It was also indicated that "for every D, W" cannot be altered to consideration of too small a set of D, W (although it can often be lessened some). Trivially, note that it does not suffice to consider only D, W for which D has only one element or W is constant. Only slightly less trivially, give an Ω and subset Ω_0 of Ω such that it does not suffice to consider only D, W for which $W(F, d)$ is the same constant for all d and all F in Ω_0, but is arbitrary for F in $\Omega - \Omega_0$.

6.25. Sometimes there are settings in which the loss incurred depends also on the outcome of the experiment and consequently is of the form $W(F, d, x)$ rather than the simpler $W(F, d)$ we consider. For example, upon inferring the quality of a box of bolts that remain after withdrawing a sample, we may decide whether or not to return some of the inspected items (whose quality is known from X) to the box, to be sold to a customer or returned to the manufacturing division. Give a simple example (not necessarily that of sampling inspection) to show that, if W can depend on the value taken on by X, it is no longer true that each δ may be replaced by a δ^* depending only on sufficient $s(X)$ and with the same risk function. [*Hint*: A trivial example can be obtained by noting that a constant s is sufficient if Ω has only one element. More complex examples can be constructed when $X = (X_1, X_2)$, the X_i iid Bernoulli, $\Omega = \{\theta : 0 < \theta < 1\}$.]

CHAPTER 7

Point Estimation

In previous chapters, particularly Chapters 1 through 4, the ideas of decision theory and various types of statistical problems have been introduced. Chapter 5 has developed these ideas in the context of squared error and linear theory, treating the subjects of best linear unbiased estimation and the analysis of variance. Related to the material of Chapter 5 is the subject of experimental design. Chapter 6 introduced the idea of sufficiency. All of these are ideas which have great applicability to statistical problems.

Chapter 7 treats a number of mathematical ideas and techniques (most of which were introduced in Chapter 4), using the statistical problem of estimation as a vehicle for the discussion. Many of these ideas find application in solving problems on testing of hypotheses, nonparametric statistics, etc. Section 7.1 introduces the idea of completeness and uses this idea in the study of unbiased estimation. Section 7.2 on the "information inequality" discusses a method sometimes useful in proving optimality of procedures. Section 7.3 is on the idea of invariance. Section 7.5 discusses the method of maximum likelihood. Section 7.6 discusses asymptotic theory. And Section 7.4 continues the discussion of finding minimax procedures.

In the terminology of this book the statistician is given a description $\{S, \Omega, D, W\}$ in which his or her problem is to estimate the value of $\phi(F)$, F in Ω. $\phi(F)$ may be real-valued, vector-valued, or function-valued (estimate the *df* F). D is the set of possible values of $\phi(F)$.

The problem of the statistician is to choose a decision procedure t. In this context we usually give t a less formidable name and call t an *estimator*. An estimator is a function defined on S taking values in D. If $X = x$ is observed, the *estimate* $t(x)$ is made. The risk of this procedure is

$$r_t(F) = E_F W(F, t(X)).$$

It is the aim of the statistician to choose a desirable estimator, that is, an admissible estimator which has other desirable properties in the behavior of its risk function compared with that of other procedures. The criterion of being unbiased or invariant or minimax is sometimes used to help select a procedure, but ultimately a more detailed examination of risk functions is desirable.

Thus, for example, if X is a binomial (n, p) random variable, then the estimator $t(X) = X/n$ of p is admissible for squared error loss. It is also invariant and best unbiased. These may seem to be sufficient grounds for using t, or they may not. A more careful look at possible risk functions of other estimators, as in Problem 2.1, will help decide.

7.1. Completeness and Unbiasedness

We now return to the subject of unbiased estimation without the restriction to linear estimation imposed in Chapter 5. Of course, our earlier remarks in Section 4.6, on the questionability of imposing the restriction to use only unbiased estimators, still apply.

Just as was the case in the example involving two X_i's near the end of Section 5.1 in the case of linear unbiased estimation, so in the general case of unbiased estimation it can often happen that Ω is such that no single unbiased estimator is uniformly better than all other unbiased estimators. It is important to realize this, since most standard texts discuss "(uniformly) best unbiased estimators" (with illustrations) without mentioning the fact that such estimators do not always exist, thus creating the misimpression among many people that such estimators always exist and engendering a futile hunt for them in many cases.

In cases in which uniformly best unbiased estimators do exist, there are two main ways of recognizing them. The more useful of the two methods will be discussed in the present section, and a less useful approach (which, however, contains some ideas which are also useful in estimation theory outside the realm of best unbiased estimation) will be discussed in Section 7.2.

A statistic s is said to be *complete* if the only real-valued function h of s for which

$$E_F h(s(X)) = 0 \quad \text{for all } F \text{ in } \Omega$$

is the function h which is itself zero with probability one for all F. Changing the definition of h on a set which has probability zero (for example, on a finite set in the continuous case) does not change $E_F h(s)$; this is why the definition cannot merely conclude with "is the function $h \equiv 0$." Thus, completeness essentially requires that the *only* unbiased estimator of the constant zero based upon s is the obvious estimator $t(s(X)) \equiv 0$.

For example, if X is a Bernoulli random variable the statistic $s(X) = X$ is complete. For, if $0 = E_\theta h(s) = \theta h(1) + (1 - \theta)h(0)$, $0 \leq \theta < 1$, then it clearly follows that $h(0) = h(1) = 0$. Other values, say $h(1/2)$, do not matter here, since

$P_\theta(X = 1/2) = 0$. We will discuss later additional examples where s is complete and others where it is not complete.

In this section we will be interested in statistics s which are *both* complete statistics (in the sense described here) and sufficient statistics (in the sense described in Chapter 6). Such statistics are called *complete sufficient statistics.* The main result (on unbiased estimation) is the following theorem.

Theorem. *Suppose in a statistical problem* $\{\mathrm{S}, \Omega, \mathrm{D}, \mathrm{W}\}$ *that a real parameter* $\phi(\mathrm{F})$ *is to be estimated, that* D *is an interval, and that* $\mathrm{W}(\mathrm{F}, \mathrm{d})$ *is, for each* F, *convex in* d. *If there exists a complete sufficient statistic* s *for* Ω *and if there exists some unbiased estimator of* ϕ, *then there exists a uniformly best unbiased estimator of* ϕ; *it is the unique* (*if* W *is strictly convex*) *function of* s *which is an unbiased estimator of* ϕ.

PROOF. (Refer to Chapter 6.) If the statistic s is a sufficient statistic then $P(X = x|s(X) = b)$ does not depend on F in Ω. Consequently if t is any unbiased estimator, we can define (as in Section 6.5)

$$t^*(b) = \sum_x t(x)P(X = x|s(X) = b). \tag{7.1}$$

t^* is defined on the space of values of s. As described in Section 6.5, t^* is used by making the decision $t^*(s(X))$ when X is observed. As mentioned there (and proved in Problem 6.17), it is easily verified that

$$E_F t^*(s(X)) = E_F t(X).$$

Since t is an unbiased estimator of $\phi(F)$, we conclude that t^* is, too. By virtue of the hypotheses of the theorem and the results of Section 6.5, we also know that

$$r_{t^*}(F) \le r_t(F) \quad \text{for all } F \text{ in } \Omega.$$

That is, t^* is at least as good as t.

Now suppose t_1^* and t_2^* are two functions on the space of s-values, such that $E_F t_1^*(s(X)) = E_F t_2^*(s(X))$ for all F in Ω. Then

$$0 = E_F(t_1^*(s(X)) - t_2^*(s(X))) \quad \text{for all } F \text{ in } \Omega.$$

By the assumption of completeness,

$$t_1^*(b) - t_2^*(b) = 0,$$

with probability one, for all F in Ω; that is,

$$t_1^*(b) = t_2^*(b)$$

with probability one, for all F in Ω.

Therefore if t_1 and t_2 are two unbiased estimators of $\phi(F)$, and if from these we derive t_1^* and t_2^* as previously, then from the completeness of s as a statistic it follows that $t_1^*(s(X)) = t_2^*(s(X))$ with probability one for all F in Ω. In other words, for *every* unbiased estimator t, the *same* t^* is at least as good as t. This t^* is thus uniformly best among unbiased estimators, which is what was to be proved. □

Note that the *same* t^* is the best unbiased estimator of ϕ for *all* W for which $W(F, d)$ is convex in d. Most books discuss only the classical case of squared error, but we have seen that this restriction is unnecessary.

If s is not complete, (7.1) will not yield the same t^* for every t. There will then not exist a uniformly best unbiased estimator.

As we saw in Section 4.6, not all functions ϕ possess unbiased estimators; hence, the assumption that ϕ does possess *some* unbiased estimator was made in the statement of the theorem.

An important aspect of the present approach compared with that of Section 7.2 is that here the existence of a complete sufficient statistic proves, once and for all, the existence of uniformly best unbiased estimators for *all* ϕ which can be estimated without bias, whereas the method of Section 7.2 is only capable of proving the result in a few special cases of Ω, for a single ϕ, and for $W(F, d) = [\phi(F) - d]^2$. (As we have mentioned, the approach of Section 7.2 has additional uses.)

The preceding theorem and its proof indicate two methods for obtaining a best unbiased estimator when there exists a complete sufficient statistic s:

Method 1: Hunt for that (essentially unique) function of s which is an unbiased estimator of ϕ.

Method 2: Find *any* unbiased estimator t of ϕ and construct the corresponding t^*. (This is the Blackwell-Rao method.)

The second method, as we shall see in an example, is often the easier one, since it may be very complicated analytically to find that function $q(s)$ which satisfies $E_F q(s) = \phi(F)$ for all F, whereas it may be obvious how to find an analytically simple unbiased estimator t (depending on X other than through s) which, although possibly poor in its own right, will yield the desired result in the form of the corresponding t^*.

Example 7.1. In the setting of Example 6.1 (with n Bernoulli observations), we saw in Section 4.6 that only polynomials in θ of degree $\leq n$ *can* possess unbiased estimators (and it is not difficult to see that all such polynomials *do* possess them). We shall show that $T_n = \sum_1^n X_i$ is a complete sufficient statistic. Thus, it follows that (for every convex W) the estimator $\bar{X}_n = T_n/n$ is the unique best unbiased estimator of θ, since it satisfies the two prerequisites of (i) being unbiased and (ii) depending only on T_n. To find a best unbiased estimator of θ^2 (which we can do only if $n \geq 2$), we may, using Method 1, first try $\bar{X}_n^2$. A direct computation shows us that

$$E_\theta \bar{X}_n^2 = \mathrm{Var}_\theta(\bar{X}_n) + (E_\theta \bar{X}_n)^2 = \theta(1-\theta)/n + \theta^2 = \theta^2(n-1)/n + \theta/n.$$

Thus $\bar{X}_n^2$ will not do the job, but $n\bar{X}_n^2/(n-1) - \bar{X}_n/(n-1)$ is now easily seen to have expectation θ^2; being a function only of T_n, it is therefore the best unbiased estimator of θ^2. Similarly, if we wanted to find the best unbiased estimator of θ^k (where $k \leq n$), we could compute $E_\theta \bar{X}_n^j$ for $1 \leq j \leq k$ and then find numbers a_j such that $E_\theta \sum_1^k a_j \bar{X}_n^j = \theta^k$. You will find that this becomes a slightly messy algebraic task if approached directly in this manner. Let us try Method 2. To that end, we must find *some* unbiased estimator of θ^k. Such an estimator is easy to guess: since the X_i's are independent and since $E_\theta X_i = \theta$, we clearly have $E_\theta(X_1 X_2 \cdots X_k) = \theta^k$. We must now apply equation (7.1), to which end we will compute

$$P_\theta(X_1 X_2 \cdots X_k = c \mid T_n = b),$$

which will actually not depend on θ. In the present example, each X_i is 0 or 1, so that $X_1 \cdots X_k$ is 0 or 1, and we need only consider $c = 0$ or 1 (not all examples will work out so easily). Since $X_1 X_2 \cdots X_k = 1$ if and only if all X_i's are 1, we have

$$\begin{aligned} P_\theta(X_1 \cdots X_k = 1 \mid T_n = b) &= P_\theta(X_1 = 1, \ldots, X_k = 1 \mid T_n = b) \\ &= \frac{P_\theta(X_1 = 1, \ldots, X_k = 1, T_n = b)}{P_\theta(T_n = b)}. \end{aligned}$$

The probability in the denominator is merely the binomial probability $\binom{n}{b}\ \theta^b(1-\theta)^{n-b}$. The event $X_1 = 1, \ldots, X_k = 1$ is synonymous with $X_1 + X_2 + \cdots + X_k = k$ (again because of the simplicity of the binomial case), and thus the event in the numerator can be rewritten as

$$\begin{aligned} &\{X_1 + \cdots + X_k = k, \qquad X_1 + \cdots + X_n = b\} \\ &\quad = \{X_1 + \cdots + X_k = k\} \cap \{X_{k+1} + \cdots + X_n = b - k\}. \end{aligned}$$

Thus, we have succeeded in rewriting the event in the numerator as the intersection of two independent events (since $X_1 + \cdots + X_k$ and $X_{k+1} + \cdots + X_n$ are independent), although it was not originally in this form ($X_1 + \cdots + X_k$ and T_n are *not* independent). This enables us to rewrite the numerator probability as a product of binomial probabilities, and we have

$$\begin{aligned} &P_\theta(X_1 \cdots X_k = 1 \mid T_n = b) \\ &\quad = \frac{P_\theta(X_1 + \cdots + X_k = k) P_\theta(X_{k+1} + \cdots + X_n = b - k)}{P_\theta(T_n = b)} \\ &\quad = \frac{\theta^k \binom{n-k}{b-k} \theta^{b-k}(1-\theta)^{n-b}}{\binom{n}{b} \theta^b (1-\theta)^{n-b}} = \frac{\binom{n-k}{b-k}}{\binom{n}{b}}, \end{aligned}$$

which of course turns out not to depend on θ. Thus, since $X_1 \ldots X_k = 0$ or 1, we have

$$E(X_1\cdots X_k|T_n=b)=\sum_j jP(X_1\cdots X_k=j|T_n=b)$$

$$=P(X_1\cdots X_k=1|T_n=b)=\frac{\binom{n-k}{b-k}}{\binom{n}{b}}$$

$$=\frac{b(b-1)\cdots(b-k+1)}{n(n-1)\cdots(n-k+1)}.$$

Hence, the unique best unbiased estimator of θ^k is $t^*(T_n)=T_n(T_n-1)\cdots(T_n-k+1)/n(n-1)\cdots(n-k+1)$. It might have been hard to guess (i.e., develop) this t^* using Method 1!

Example 7.2. Suppose $X=(X_1,\ldots,X_n)$ where the X_i's are independent and identically distributed Poisson random variables with unknown mean $\theta>0$. We have seen that $T_n=\sum_1^n X_i$ is a minimal sufficient statistic, and it can be shown to be complete. Suppose we want a best unbiased estimator (W being convex) of $\phi(\theta)=P_\theta\{X_1>0\}=1-e^{-\theta}$; there are many biological and physical examples, for which this quantity is important. Method 1 in this case is to find a function q for which, for all θ,

$$1-e^{-\theta}=E_\theta q(T_n)=\sum_{y=0}^{\infty}q(y)P_\theta(T_n=y)$$

$$=\sum_{y=0}^{\infty}q(y)e^{-n\theta}(n\theta)^y(y!)^{-1},$$

since T_n has the Poisson law with mean $n\theta$. One can try to find such a function q, but it is not even obvious what type of function to try; for example, no polynomial will work. Method 2 is particularly easy to use in examples like this (and Example 7.1), wherein the quantity to be estimated is the probability $P_\theta(A)$ of a specified event A; for, in such a case, the "counting" or "indicator" random variable I_A defined by

$$I_A=\begin{cases}1 & \text{if } A \text{ occurs}\\ 0 & \text{if } A \text{ does not occur}\end{cases}$$

obviously has $E_\theta Y=P_\theta(I_A=1)=P_\theta(A)$, so that I_A is an unbiased estimator of $P_\theta(A)$, which is all we need to apply Method 2. In the present case, then, we begin with

$$t(X_1,\ldots,X_n)=\begin{cases}1 & \text{if } X_1>0,\\ 0 & \text{if } X_1=0,\end{cases}$$

and compute (7.1). We assume $n>1$, since t is already best unbiased if $n=1$. It is easier to write down the conditional probability (given that $T_n=b$) that

$X_1 = 0$ rather than the conditional probability that $X_1 > 0$, since $\{X_1 = 0, T_n = b\}$ can be rewritten at once as $\{X_1 = 0\} \cap \{X_2 + \cdots + X_n = b\}$. This is the intersection of independent events, so we have

$$P_\theta(X_1 = 0 | T_n = b) = \frac{P_\theta(X_1 = 0)P_\theta(X_2 + \cdots + X_n = b)}{P_\theta(T_n = b)}$$
$$= \frac{e^{-\theta}e^{-(n-1)\theta}[(n-1)\theta]^b/b!}{e^{-n\theta}[n\theta]^b/b!} = \left(\frac{n-1}{n}\right)^b.$$

Hence,

$$E(t(X)|T_n = b) = \sum_j jP(t(X) = j | T_n = b)$$
$$= P(t(X) = 1 | T_n = b)$$
$$= 1 - \left(\frac{n-1}{n}\right)^b.$$

Thus, $t^*(T_n) = 1 - [(n-1)/n]^{T_n}$ is the unique best unbiased estimator of $\phi(\theta)$. Again, it might have been much harder to obtain this estimator by the "guesswork" or trial-and-error of Method 1.

We now turn to the question of which sufficient statistics are complete. We shall now (for simplicity, in the discrete case) verify that (i) *if some sufficient statistic is complete, then so is any minimal sufficient statistic.* On the other hand, we shall see that (ii) *a sufficient statistic which is not minimal cannot be complete.* (The idea is that the extra dependence on X makes it possible to find a function h whose expectation is zero.) To prove (i) suppose that s, s^* are sufficient statistics, s^* is minimal, and s is complete. Then with probability one, $s^*(X) = J(s(X))$ for some function J. Hence, if for some function h,

$$0 = E_F h(s^*(X)), \quad \text{all } F \text{ in } \Omega,$$

then $0 = E_F h(J(s(X)))$, all F in Ω, so that $h(J(s(X))) = 0$ with probability one, all F in Ω; that is, $h(s^*(X)) = 0$ with probability one, all F in Ω. This proves (i). On the other hand, if s is a sufficient statistic which is *not* minimal, then another sufficient statistic s^* and a function J may be found satisfying $s^*(X) = J(s(X))$, such that there is at least one value b_0 (say) of J for which $J(c) = b_0$ for each of (at least) two different values, $c = c_1$ and c_2, of $s(X)$, and such that $P_F\{s^*(X) = b_0\} > 0$ for some F and $P\{s(X) = c | s^*(X) = b_0\} > 0$ for $c = c_1$, c_2. (This last probability does not depend on F since s^* is sufficient; the assertion of the previous sentence is simply that s is not minimal, so that some s^* gives a coarser sufficient partition.) Define

$$h(c) = \begin{cases} (-1)^i/P\{s(X) = c_i | s^*(X) = b_0\} & \text{if} \quad c = c_i, i = 1 \text{ or } 2, \\ 0 \quad \text{otherwise.} \end{cases}$$

Then for all F

$$E_F h(s(X)) = \sum_{i=1}^{2} \frac{(-1)^i}{P(s(X) = c_i | s^*(X) = b_0)} \cdot P_F(s(X) = c_i),$$

which equals $(-1 + 1) \times P_F(s^*(X) = b_0) = 0$. But for some F, $P_F(s^*(X) = b_0) > 0$, so that the probability under F is positive that $s(X) = c_1$, in which case $h(s(X)) \neq 0$. Thus, s is not complete, which proves (ii).

Results (i) and (ii) together show that, in a given statistical problem, either no complete sufficient statistics exist, or else the class of complete sufficient statistics is identical with the class of minimal sufficient statistics. Thus, we can always answer the question of whether or not there is a complete sufficient statistic by seeing whether or not a minimal sufficient statistic is complete. (Since each minimal sufficient statistic is a function of every other minimal sufficient statistic, it does not matter which one we look at.)

In Example 6.1 with $n = 4$, the pair $s = (X_1 + X_2 + X_3, X_4)$ is sufficient but not minimal, and $h(s) = (X_1 + X_2 + X_3) - 3X_4$ has expectation zero without being identically zero itself. This illustrates result (ii).

The investigation of completeness is not elementary mathematically, and we can only indicate the idea in a couple of cases.

EXAMPLE 7.1 (Continued). (n Bernoulli observations). We want to show that, if $E_\theta h(T_n) = 0$, then $h(j) = 0$ for $j = 0, 1, \ldots, n$. We write out $E_\theta h(T_n)$ as a polynomial. That is, suppose

$$0 = E_\theta h(T_n) = \sum_{i=0}^{n} h(i) \binom{n}{i} \theta^i (1 - \theta)^{n-i}.$$

Then if $0 < \theta < 1$ we may divide both sides by $(1 - \theta)^n$ and obtain

$$0 = \sum_{i=0}^{n} h(i) \binom{n}{i} (\theta/(1 - \theta))^i.$$

If we set $\lambda = \theta/(1 - \theta)$ then we see that if $0 < \lambda < \infty$ then

$$0 = \sum_{n=0}^{n} h(i) \binom{n}{i} \lambda^i.$$

The only way a polynomial of degree n can be zero for more than n *distinct* values of the variable is for all coefficients to be zero. That is,

$$h(i) = 0, \qquad i = 0, 1, 2, \ldots, n.$$

Thus, we have shown that, if $E_\theta h(T_n) = 0$ for all θ, then $h(0) = h(1) = \cdots = h(n) = 0$; hence the sufficient statistic T_n is complete.

EXAMPLE 7.3. Suppose $X_1, \ldots, X_n$ are independent identically distributed random variables with common Koopman-Darmois law

$$c(\theta) b(x_1) e^{k(x_1) q(\theta)}$$

as in Chapter 6. The possible values of $q(\theta)$ are a nondegenerate interval. It can be shown that the minimal sufficient statistic $\sum_{1}^{n} k(X_i)$ is complete. However, the proof in general is more complicated than in the special Bernoulli case studied in Example 7.1. Keep in mind the many common distributions which are of the Koopman-Darmois form.

EXAMPLE 6.4 (Continued). The minimal sufficient statistic obtained in Example 6.4 can also be shown to be complete.

EXAMPLE 6.8 (Continued). For $n > 1$, the minimal sufficient statistic $(Y_1, Y_2, \ldots, Y_n)$ is not complete, since $E_\theta(Y_2 - Y_1) = E_\theta[(Y_2 - \theta) - (Y_1 - \theta)] = E_0[Y_2 - Y_1] = c_n$ is independent of θ, so that $E_\theta[Y_2 - Y_1 - c_n] = 0$ for all θ, but $Y_2 - Y_1 - c_n$ is not equal to zero with probability one for all θ.

EXAMPLE 6.7 (Continued). In most common genuinely nonparametric problems, the set of order statistics $(Y_1, \ldots, Y_n)$ *is* complete. Then $\bar{X}_n$, which is an unbiased estimator of $\phi(F) = E_F X_1$ and which depends only on $Y_1, \ldots, Y_n$ (since $\bar{X}_n = \sum_{1}^{n} Y_i/n$), is the best unbiased estimator of ϕ for all convex W. Note that linear unbiased estimators $\sum a_i X_i$ other than $\bar{X}_n$ do not depend on the order statistics alone, $\bar{X}_n$ being the only unbiased estimator of ϕ which is a function only of $(Y_1, \ldots, Y_n)$.

EXAMPLE 6.5 (Continued). The density function of $V_n = \max(X_1, \ldots, X_n)$ in the case of the rectangular density from 0 to $\theta (\theta > 0)$ is

$$f_{\theta; V_n}(v) = \begin{cases} nv^{n-1}/\theta^n & \text{if} \quad 0 < v < \theta, \\ 0 & \text{otherwise.} \end{cases}$$

If h is a real-valued function for which $E_\theta h(V_n) = 0$ for all θ, we have

$$0 = \theta^n E_\theta h(V_n) = \int_0^\theta h(v) n v^{n-1}\, dv$$

for all θ. If h were continuous, we could differentiate the integral with respect to the upper limit θ, obtaining $h(\theta) n \theta^{n-1} = 0$ for all $\theta > 0$, and hence $h(v) = 0$ for all $v > 0$. The analogous proof for general h is more difficult to carry out, but we can again conclude that $h(v) = 0$ for all $v > 0$, except possibly on a set of points which has probability zero for all θ. Thus, V_n is complete in this case.

EXAMPLE 6.6 (Continued). In the case of the rectangular density from $\theta - \frac{1}{2}$ to $\theta + \frac{1}{2}$, if $n \geq 2$ the minimal sufficient statistic is not complete, since (as in Example 6.8) there is a constant d_n such that $E_\theta[\max X_i - \min X_i - d_n] = 0$ for all θ. Even when $n = 1$, the minimal sufficient statistic X_1 is not complete, and $t(X_1) = X_1$ is not a uniformly best unbiased estimator of θ; this may seem surprising at first glance, since one is tempted to argue, intuitively, "What else but X_1 *could* one use to estimate θ?" To see that X_1 is *not* best unbiased when $n = 1$ for any loss function $W(\theta, d)$ for which $W(\theta, d) = 0$ or > 0 according to whether or not $d = \theta$, we consider the estimator t_1 defined by

$$t_1(x) = \text{closest integer to } x$$

(if $x = 1/2, 3/2$, etc., there are two integers equally close to x; the definition of t for such values of x, which occur with probability zero, does not matter). Whatever the true θ, there are integers N and $N + 1$ such that $N \leq \theta \leq N + 1$. If $\theta - \frac{1}{2} < X_1 < N + \frac{1}{2}$, which happens with probability $N + \frac{1}{2} - (\theta - \frac{1}{2})$, then N is the closest integer to X_1 and hence $t_1(X_1) = N$; similarly, if $N + \frac{1}{2} < X_1 < \theta + \frac{1}{2}$, we obtain $t_1(X_1) = N + 1$. (Drawing a picture may help you to see this.) Hence

$$\begin{aligned} E_\theta t_1(X_1) &= NP_\theta(t_1(X_1) = N) + (N + 1)P_\theta(t_1(X) = N + 1) \\ &= N[N + 1 - \theta] + (N + 1)[\theta - N] = \theta, \end{aligned}$$

and thus t_1 is an unbiased estimator of θ. Now, if θ is an integer, so that $|\theta - X_1| < 1/2$ with probability one, we have $t(X_1) = $ (closest integer to $X_1) = \theta$, with probability one; thus, $r_{t_1}(\theta) = 0$ whenever θ is an integer. (In Example 1 of Chapters 1 to 3, we considered estimators whose risk functions were zero for some θ with $0 < \theta < 1$, namely, $t(X) \equiv c$. However, those estimators were not unbiased, and in fact no unbiased estimator of θ can be constructed there for which the risk is 0 for some θ with $0 < \theta < 1$.) In the present example, if b is any real number, the estimator $t_b(X_1) = $ (closest number to X_1 of the form $\ldots, b - 1, b, b + 1, b + 2, \ldots$) can be shown in a way similar to the preceding to be an unbiased estimator of θ and to satisfy $r_{t_b}(\theta) = 0$ whenever $\theta = b, b \pm 1, \ldots$. Thus, considering all the estimators t_b for $0 \leq b < 1$, we see that any estimator t^* which is at least as good as all the t_b's must satisfy $r_{t^*}(\theta) = 0$ for *all* θ; that is, for all θ, $P_\theta(t^*(X_1) = \theta) = 1$. It is impossible to have a t^* for which this is so (which should seem at least intuitively clear to you). Thus, there is no uniformly best unbiased estimator in this case. Of course, the lack of completeness is easily seen on noting that $E_\theta[X_1 - t_1(X_1)] = 0$ for all θ.

More generally, in settings where there is no complete sufficient statistic and $\phi(\theta)$ can be estimated without bias, there will be several unbiased estimators of ϕ whose risk (variance) functions cross and which are admissible *within the class* $\mathscr{D}_{\text{UNB}}$ of unbiased estimators of ϕ. There is no uniformly best unbiased estimator in such circumstances, but one could use an additional criterion (e.g., minimax or Bayes) to select a procedure from $\mathscr{D}_{\text{UNB}}$ (which, for the respective criteria, would not coincide with the minimax or Bayes procedure in $\mathscr{D}$ if the latter procedure is biased). One possible choice that appears in the statistical literature is that of a *locally best* unbiased estimator at θ_0, an estimator in $\mathscr{D}_{\text{UNB}}$ which has smallest value of $\text{var}_{\theta_0}(t)$ among all t in $\mathscr{D}_{\text{UNB}}$. Thus, the estimator t_b in the previous paragraph is locally best unbiased at b (and at $b \pm 1, \ldots$). In other examples such a locally best (at θ_0) unbiased estimator will *not*, usually, have zero variance at θ_0. This choice of estimator, for those who believe in unbiasedness, is motivated by special concern about the risk in a neighborhood of θ_0, perhaps because the true θ is suspected to lie therein (see Problem 5.3 for an example).

More on Sample Moments

As a special case of remark (4) in Section 5.2, we found that if $X_1, \ldots, X_n$ are uncorrelated with common mean μ_F and variance $\sigma_F^2 < \infty$, then $(n-1)^{-1}\sum(X_i - \bar{X}_n)^2$ is an unbiased estimator of σ_F^2. Also, as discussed in conjunction with the method of moments (Section 4.8), if $\Omega = \{F : \mu'_{j,F}$ is finite$\}$, then the jth sample moment $n^{-1}\sum_1^n X_i^j = m'_{j,n}$ is an unbiased estimator of $\mu'_{j,F}$ (the jth moment of F, that is, $E_F X_i^j$, where the X_i's are independent with common distribution F). However, although $\frac{1}{n}\sum_i (X_i - \mu_F)^j$ (not an estimator since μ_F is unknown) has expectation $\mu_{j,F} \stackrel{\text{def}}{=} E_F(X_i - \mu'_{1,F})^j$, the random variable $m_{j,n} \stackrel{\text{def}}{=} \frac{1}{n}\sum_i (X_i - \bar{X}_n)^j$ is not an unbiased estimator of $\mu_{j,F}$ for $j > 1$, as we already know when $j = 2$ (beginning of this paragraph). We now illustrate the computations by showing that $[n^2/(n-1)(n-2)]m_{3,n}$ is an unbiased estimator of $\mu_{3,F} = E_F(X_i - \mu'_{1,F})^3$ for $n > 2$. By writing $X_i - \mu_F = Y_i$ we obtain, in terms of rv's Y_i with $E_F Y_i = 0$ and $E_F Y_i^3 = \mu_{3,F}$,

$$\begin{aligned}
n^3 E_F m_{3,n} &= n^2 E_F \sum_i (Y_i - \bar{Y}_n)^3 = n^3 E_F (Y_1 - \bar{Y}_n)^3 = E_F\left(nY_1 - \sum_1^n Y_i\right)^3 \\
&= E_F\left([n-1]Y_1 - \sum_2^n Y_i\right)^3 \\
&= (n-1)^3 E_F Y_1^3 - 3(n-1)^2 E_F\left\{Y_1^2 \sum_2^n Y_i\right\} \\
&\quad + 3(n-1)E_F\left\{Y_1\left(\sum_2^n Y_i^2\right)\right\} + E_F\left(\sum_2^n Y_i\right)^3.
\end{aligned}$$

The middle two terms of the last line are 0 because of independence of Y_1 from $\sum_2^n Y_i$ and the 0 expectation of each of these. The last term is, for similar reasons,

$$\begin{aligned}
E_F\left(\sum_2^n Y_i\right)^3 &= E_F \sum_2^n Y_i^3 + 3E_F \sum_{\left\{\substack{i \neq j \\ 2 \le i,j \le n}\right\}} Y_i^2 Y_j + 6E_F \sum_{\{1<i<j<k\le n\}} Y_i Y_j Y_k \\
&= (n-1)E_F Y_1^3.
\end{aligned}$$

Hence, $n^3 E_F m_{3,n} = [(n-1)^3 - (n-1)]\mu_{3,F} = n(n-1)(n-2)\mu_{3,F}$, which yields the stated result.

Similar calculations will be found in some of the problems. The resulting estimators, like $m_{3,n}$, will be best unbiased and useful only for nonparametric Ω's such as $\{F : \mu_{6,F} < \infty\}$ for $m_{3,n}$; this implies that $\text{var}_F m_{3,n}$ is finite and of order n^{-1} when n is large.

7.2. The "Information Inequality"

The information inequality is often called the *Cramér-Rao-Frechet-Darmois inequality*. We shall prove (in 7.10) that, under certain conditions, if $\Omega = \{\theta : a < \theta < b\}$ and t is any unbiased estimator of $\phi(\theta)$, then

$$\text{var}_\theta(t) \geq H(\theta), \tag{7.2}$$

where $H(\theta)$ is given explicitly. The value $H(\theta)$ depends on θ and on the form of ϕ, but in no way depends on t (this last is what makes the inequality useful). Suppose we can find a t^* which is an unbiased estimator of ϕ and such that $\text{var}_\theta(t^*) = H(\theta)$ for all θ. It follows from (7.2) that, if $W(\theta, d) = [\phi(\theta) - d]^2$ and t is any unbiased estimator of ϕ, then

$$r_t(\theta) \geq r_{t^*}(\theta) \quad \text{for all } \theta;$$

thus, we will have proved that t^* is a best unbiased estimator of ϕ. This is a method, different from that of completeness, for proving that a given estimator is best unbiased. (Of course, if no best unbiased estimator exists, no t^* that is unbiased can have $r_t(\theta) = H(\theta)$ for all θ.)

As mentioned in Section 7.1, the present method is less useful than that of Section 7.1 for proving such results. To start with, the regularity conditions needed to derive the inequality (7.2) are ones which rule out cases like that of Example 6.5. Second, only squared error loss is considered here. Finally, even if t^* *is* the best unbiased estimator, this method (unlike that of Section 7.1) may be incapable of proving the result, simply because the inequality (7.2) is not always a good one, in the sense that a best estimator t^* may have a variance which is *greater* than the lower bound $H(\theta)$. For example, in the setting of Example 7.1, the present method works *only* if $\phi(\theta) = a + b\theta$ where a and b are constants; if $\phi(\theta) = \theta^2$, one finds that $\text{var}_\theta(t^*) > H(\theta)$ for $0 < \theta < 1$, even though t^* was proved best unbiased by the method of completeness. These three objections to the present method can be eliminated only by considering a much more complicated development than that used later to obtain $H(\theta)$, in order to obtain a better bound $\bar{H}(\theta)$ (which is $\geq H(\theta)$) for which (7.2) holds, and the best $\bar{H}$ obtainable in this way will be extremely difficult to compute in any practical problem. Moreover, even with such an improved inequality, *any unbiased estimator which can be proved best by the present method could already have been proved best by the method of Section* 7.1.

Why, then, do we consider the present approach?

There are several reasons. One is that the approach will yield an inequality similar to (7.2) for the risk function of biased estimators as well, and this inequality will yield one method of proving that certain procedures are minimax in Section 7.4. It is possible to prove, sometimes, the admissibility of a procedure using (7.2). A second reason is that there are problems where no uniformly best unbiased estimator exists and where reasonably good (e.g., minimax) procedures are extremely messy functions of X. If data are cheap and the mathematical effort required to find "best" estimators is great, it may be of interest to consider an unbiased estimator t' which does *not* satisfy any criterion of optimality, but which is not too bad and which is easy to compute. In such a case, the ratio $H(\theta)/\text{var}_\theta(t')$, which is always ≤ 1 for unbiased t', gives a bound on how bad t' is; if this ratio is fairly close to 1 for all θ, then we can feel reassured that t' is not terribly bad, and that no other unbiased estimator can achieve a very great relative improvement in risk over t'. Finally, the type of consideration just mentioned is especially important when the sample size n is very large. We will see in Section 7.6 that when n is very large, it is "almost" true that there is a uniformly best procedure (without even any restriction to unbiased estimators) and that $H(\theta)$ is almost attainable as its risk function. Thus, the ratio $H(\theta)/E_\theta(t-\phi)^2$ discussed previously is particularly meaningful in this case, whether or not t is unbiased; for, the "almost uniformly best" procedure will often be exceedingly difficult to compute.

We now turn to the proof of our inequality. We assume that the (possibly vector-valued) random variable X has a probability (or density) function $p_{\theta;X}$ (where Ω is some interval of θ-values) which is such that $p_{\theta;X}(x) > 0$ for all x in S, for each θ (this rules out cases like Examples 6.5 and 6.6), and such that $p_{\theta;X}(x)$ is differentiable in θ and we can perform the steps of interchanging the order of differentiation and summation (or integration). This will consist of being able to write

$$\frac{\partial}{\partial\theta}\sum_x t(x)p_{\theta;X}(x) = \sum_x t(x)\frac{\partial}{\partial\theta}p_{\theta;X}(x),$$

which is immediate if the sum is over finitely many terms, but which requires some conditions if there are infinitely many terms or if the sum is replaced by an integral; the precise conditions, which will be satisfied in the most important practical examples, can be found in Cramér's book.

Before deriving our inequality, we make some definitions. Suppose ϕ is to be estimated. We assume ϕ to be differentiable. For any estimator t of ϕ, we write

$$\psi_t(\theta) = E_\theta t$$

and hereafter consider only estimators for which this is finite for all θ. We define the bias function of t:

$$b_t(\theta) = \psi_t(\theta) - \phi(\theta). \tag{7.3}$$

We also define

$$z_\theta(x) = \frac{\partial \log p_{\theta;X}(x)}{\partial\theta} = \frac{1}{p_{\theta;X}(x)}\frac{\partial}{\partial\theta}p_{\theta;X}(x). \tag{7.4}$$

Thus, by our assumption,

$$\begin{aligned} E_\theta z_\theta(X) &= \sum_x z_\theta(x)p_{\theta;X}(x) = \sum_x \frac{\partial}{\partial\theta}p_{\theta;X}(x) \\ &= \frac{\partial}{\partial\theta}\sum_x p_{\theta;X}(x) = \frac{\partial}{\partial\theta}1 = 0. \end{aligned} \tag{7.5}$$

Hence,

$$\operatorname{var}_\theta(z_\theta(X)) = E_\theta\{z_\theta(X)^2\} = \sum_x \left[\frac{\partial}{\partial\theta}\log p_{\theta;X}(x)\right]^2 p_{\theta;X}(x). \tag{7.6}$$

Assuming that we can differentiate a second time with respect to θ and can interchange this operation with summation, we have

$$\begin{aligned} 0 &= \frac{\partial}{\partial\theta}0 = \frac{\partial}{\partial\theta}E_\theta z_\theta(X) = \sum_x \frac{\partial}{\partial\theta}\{z_\theta(x)p_{\theta;X}(x)\} \\ &= \sum_x \left[\frac{\partial}{\partial\theta}z_\theta(x)\right]p_{\theta;X}(x) + \sum_x z_\theta(x)\frac{\partial}{\partial\theta}p_{\theta;X}(x) \\ &= \sum_x \left[\frac{\partial^2}{\partial\theta^2}\log p_{\theta;X}(x)\right]p_{\theta;X}(x) + \sum_x z_\theta(x)^2 p_{\theta;X}(x) \\ &= E_\theta\left\{\frac{\partial^2}{\partial\theta^2}\log p_{\theta;X}(x)\right\} + E_\theta\{z_\theta(X)^2\}. \end{aligned}$$

Hence, under this additional assumption we can also write

$$\operatorname{var}_\theta(z_\theta(X)) = -E_\theta\left\{\frac{\partial^2}{\partial\theta^2}\log p_{\theta;X}(X)\right\}. \tag{7.7}$$

We hereafter write, for brevity,

$$I(\theta) = \operatorname{var}_\theta(z_\theta(X)).$$

We shall see that (7.7) is often more convenient than (7.6) for computing $I(\theta)$. This quantity is often referred to, especially by R. A. Fisher and his school, as the "information about Ω available in the experiment when θ is true." We shall see that, for $\phi(\theta) = \theta$ and t unbiased, the function $H(\theta)$ of (7.2) is precisely $1/I(\theta)$, hence, the importance of this quantity. However, you should not be swayed by the use of the word *information* into thinking that this quantity has any deeper meaning.

We now prove our inequality. We have

$$\begin{aligned} \frac{\partial}{\partial\theta}\psi_t(\theta) &= \frac{\partial}{\partial\theta}\sum_x t(x)p_{\theta;X}(x) = \sum_x t(x)\frac{\partial}{\partial\theta}p_{\theta;X}(x) \\ &= \sum_x t(x)z_\theta(x)p_{\theta;X}(x) = E_\theta\{t(X)z_\theta(X)\}. \end{aligned} \tag{7.8}$$

By (7.5), $E_\theta(z_\theta(X))E_\theta t(X) = 0$. Hence, the quantity (7.8) is the covariance between $t(X)$ and $z_\theta(X)$. Since the square of the correlation coefficient between two random variables is always ≤ 1 (see also Appendix C.1.2, Schwarz's inequality), we have

$$\left[\frac{\partial}{\partial\theta}\psi_t(\theta)\right]^2 \leq \operatorname{var}_\theta t(X)\operatorname{var}_\theta(z_\theta(X)) = I(\theta)\operatorname{var}_\theta(t),$$

and hence, denoting the derivative of a function g by g',

$$\operatorname{var}_\theta(t) \geq \frac{\left[\frac{\partial}{\partial\theta}\psi_t(\theta)\right]^2}{I(\theta)} = \frac{[\phi'(\theta) + b_t'(\theta)]^2}{I(\theta)}, \tag{7.9}$$

which is the general "information inequality." In this form the inequality is not always easy to make use of, since the right side depends on t (however, the inequality in this form will be used in Section 7.4). If t is an *unbiased* estimator of ϕ, we have $b_t \equiv 0$ and hence $b_t' \equiv 0$, and thus

$$\operatorname{var}_\theta t(X) \geq \frac{[\phi'(\theta)]^2}{I(\theta)}. \tag{7.10}$$

This is the inequality (7.2), wherein the right side does not depend on t. In the particular case for which $\phi(\theta) = \theta$, we have $\phi'(\theta) = 1$, and thus (7.9) and (7.10) reduce to

$$\operatorname{var}_\theta t(X) \geq \frac{[1 + b_t'(\theta)]^2}{I(\theta)} \tag{7.11}$$

and

$$\operatorname{var}_\theta t(X) \geq \frac{1}{I(\theta)}, \tag{7.12}$$

respectively.

Before considering examples, we consider a simplification in the computation of $I(\theta)$ in the case where $X = (X_1, \ldots, X_n)$ with the X_i's independent and identically distributed. In that case we have

$$\begin{aligned} z_\theta(x) &= \frac{\partial}{\partial\theta} \log \prod_{i=1}^{n} p_{\theta;X_1}(x_i) = \frac{\partial}{\partial\theta} \sum_{i=1}^{n} \log p_{\theta;X_1}(x_i) \\ &= \sum_{i=1}^{n} \bar{z}_\theta(x_i), \end{aligned}$$

where $\bar{z}_\theta(x_1) = \frac{\partial}{\partial\theta} \log p_{\theta;X_1}(x_1)$. Exactly as in (7.5), we obtain $E_\theta \bar{z}_\theta(X_i) = 0$. The random variables $\bar{z}_\theta(X_i)$, $1 \leq i \leq n$, are independently and identically distributed. Hence,

$$I_n(\theta) = \sum_{i=1}^{n} E_\theta\{\bar{z}_\theta(X_i)^2\} = n \operatorname{var}_\theta(\bar{z}_\theta(X_1)) \stackrel{\text{def}}{=} n I_1(\theta), \tag{7.13}$$

where we have used $I_n(\theta)$ in this iid setup to denote *information from* n *observations*. This fact is often expressed by saying that "the information in n observations is n times the information in one observation" (more generally, the information in a vector of independent random variables, whether or not they have the same distribution, is the sum of the quantities of information in the components). This "additive" nature of I, as well as other superficial similarities of I to a quantity called *information* in communication theory (e.g., there is a logarithm of a probability present in both definitions), has led some people to stretch the analogy to a greater degree than it will bear. For example, if "information" is to have meaning as (roughly) the inverse of the variance of a good estimator for large sample size n, then in the setting of Example 6.5 this quantity would go up roughly as n^2 rather than linearly with n, the latter being the way information theorists think a quantity should behave if it is to be called information. Thus, it is at most in the case in which $p_{\theta;X}$ or $f_{\theta;X}$ satisfies our differentiability assumptions, called the *regular case*, that there is any meaningful definition of information which bears even a remote resemblance to that of communication theory. Also, the problems of communications and statistics usually differ. Moreover, although there is an extensive literature on the derivation of estimators or tests through the use of an "intuitively appealing" information concept, there is no theorem whatsoever which says that this approach yields good procedures. The situation is exactly that described for the maximum likelihood method in Sections 7.5 and 7.6: the method can lead to very poor procedures in general, and only some argument external to that method (like the arguments of Sections 7.1, 7.2, 7.3, or 7.4) will show that the procedure has some good property (if it does have any); the only properties of being "almost good" which such procedures will have for general $p_{\theta;X}$ are asymptotic ones which hold for very large sample sizes under certain regularity conditions, and which are discussed in Section 7.6.

Example 7.4. Suppose $X = (X_1, \ldots, X_n)$, where the X_i are independent and normal with mean 0 and variance θ; here $\Omega = \{\theta : \theta > 0\}$. We have

$$\begin{aligned}\bar{z}_\theta(x_1) &= \frac{\partial}{\partial\theta}\log\{(2\pi\theta)^{-1/2}e^{-x_1^2/2\theta}\}\\ &= -\frac{1}{2\theta} + \frac{x_1^2}{2\theta^2}.\end{aligned}$$

Since X_1 is normal with mean 0 and variance θ, we have $E_\theta X_1^2 = \theta$ and $E_\theta X_1^4 = 3\theta^2$; thus,

$$I_1(\theta) = E_\theta[\bar{z}_\theta(X_1)]^2 = E_\theta\left[\frac{X_1^2}{2\theta^2} - \frac{1}{2\theta}\right]^2 = 1/2\theta^2.$$

(The alternate computation of (7.7) in the present example is

$$\begin{aligned}I_1(\theta) &= -E_\theta\left\{\frac{\partial^2}{\partial\theta^2}\log f_{\theta;X_1}(X_1)\right\} = -E_\theta\left\{\frac{1}{2\theta^2} - \frac{X_1^2}{\theta^3}\right\}\\ &= 1/2\theta^2;\end{aligned}$$

this involves an extra differentiation but requires only a knowledge of $E_\theta X_1^2$ rather than of $E_\theta X_1^4$; this is typical of what one encounters in other examples.)

Thus, we obtain $I(\theta) = n/2\theta^2$. Suppose first that we want to estimate θ by an unbiased estimator t. By (7.12) we have

$$\text{var}_\theta(t) \geq 2\theta^2/n$$

for any such unbiased estimator. On the other hand, the estimator $t_n^*(X_1, \ldots, X_n) = n^{-1} \sum_1^n X_i^2$ is an unbiased estimator of θ, and $\text{var}_\theta(t_n^*) = n^{-1} \text{var}_\theta(X_1^2) = n^{-1}\{E_\theta X_1^4 - (E_\theta X_1^2)^2\} = 2\theta^2/n$. Thus, we have proved that t_n^* is a best unbiased estimator of θ (for squared error loss).

Suppose that we wanted, instead, to estimate the standard deviation, $\phi(\theta) = \theta^{1/2}$, of this normal density, by an unbiased estimator. According to (7.10), for any unbiased estimator t of ϕ we have

$$\text{var}_\theta(t) \geq \frac{[\phi'(\theta)]^2}{I(\theta)} = \theta/2n.$$

A reasonable unbiased estimator to consider might be one of the form $t_n^{**} = c_n(t_n^*)^{1/2}$, where t_n^* is the estimator of the previous paragraph and c_n is chosen to make t_n^{**} unbiased. To find c_n, we must compute $E_\theta(t_n^*)^{1/2}$. Now $X_1^2/2\theta$ has density

$$\begin{cases} y^{-1/2}e^{-y}/\Gamma(1/2) & \text{if} \quad y > 0, \\ 0 & \text{otherwise;} \end{cases}$$

and thus (Appendix A, on Γ-laws) $Y_n = \sum_1^n X_i^2/2\theta$ has density

$$\begin{cases} y^{n/2-1}e^{-y}/\Gamma(n/2) & \text{if} \quad y > 0, \\ 0 & \text{otherwise.} \end{cases}$$

Hence,

$$\begin{aligned} E_\theta(t_n^*)^{1/2} &= E_\theta\left(\frac{2\theta}{n} Y_n\right)^{1/2} \\ &= \left(\frac{2\theta}{n}\right)^{1/2} \int_0^\infty y^{1/2} y^{n/2-1} e^{-y}\, dy/\Gamma(n/2) \\ &= \theta^{1/2} \left\{\frac{2^{1/2}\Gamma[(n+1)/2]}{n^{1/2}\Gamma(n/2)}\right\}. \end{aligned}$$

Hence, the choice $c_n = n^{1/2}\Gamma(n/2)/2^{1/2}\Gamma[(n+1)/2]$ makes $t_n^{**} = c_n(t_n^*)^{1/2}$ an unbiased estimator of $\theta^{1/2}$. For example, when $n = 1$ we obtain $t_1^{**}(X_1) = c_1|X_1| = (\pi/2)^{1/2}|X_1|$ and, since $t_1^{**} = c_1(2\theta)^{1/2}Y^{1/2}$,

$$\begin{aligned} \text{var}_\theta(t_1^{**}) &= c_1^2(2\theta)\,\text{var}(Y_1^{1/2}) \\ &= 2\theta c_1^2[EY_1 - (EY_1^{1/2})^2] \\ &= \pi\theta[1/2 - 1/\pi] \\ &= .5708\theta. \end{aligned}$$

This is strictly greater than the lower bound $.5\theta$ obtained previously, and so we cannot conclude by this argument that t_1^* is best unbiased (for squared error loss), *although the conclusion that* t_n^{**} *is best unbiased is immediate from the considerations of Section* 7.1, *since* t_n^{**} *is an unbiased estimator of* $\theta^{1/2}$ *and it depends on* X *only through the value of the complete sufficient statistic* $\sum_1^n X_i^2$. A similar result holds for $n > 1$: the variance of t_n^{**} is always $> \theta^2/2n$. Incidentally, it can be shown (as indicated earlier) that

$$\mathrm{var}_\theta(t_n^{**})/[\theta^2/2n] \to 1 \quad \text{as} \quad n \to \infty.$$

The preceding example illustrates something about which you should be warned. The word *efficiency* is unfortunately used in many different ways in statistics, and you must be careful to see exactly which definition is being used at any time. The most common usage is that of Section 7.6, but another common usage is to call an unbiased estimator t of ϕ *efficient* if its variance equals the right side of (7.10) and to call the ratio of this right side of (7.10) to $\mathrm{var}_\theta(t)$ the *efficiency* $e_t(\theta)$ of the estimator t when θ is true. Thus $e_t(\theta) \leq 1$, with equality for all θ if and only if t is efficient in the preceding sense. In our example the efficiency of t_1^{**} according to this definition is .88 for all θ. You can see how misleading this use of the term is: t_1^{**} actually is the best unbiased estimator of ϕ, but the figure .88 could easily lead you to think otherwise. The reason why this confusion occurs is that this definition of efficiency compares $\mathrm{var}_\theta(t)$ *not* with the best *attainable* bound, but with the bound $H(\theta)$ of (7.2) and (7.10), which is often unattainable! Thus, a much more meaningful definition of the efficiency of an unbiased estimator t in a case in which there exists a best unbiased estimator t^* would be $\mathrm{var}_\theta(t^*)/\mathrm{var}_\theta(t)$.

When is the lower bound on the right side of (7.10) attainable by some t for all θ, so that the preceding method of proving an estimator best unbiased will work? This is not difficult to see. If equality holds for all θ in (7.10) for some estimator t, it must hold in the preceding inequality (7.9); that is, the correlation coefficient between $t(X)$ and $z_\theta(X)$ must be ± 1 when θ is true. We have seen (Appendix C.1.2) that this is possible only if these two random variables are linearly related for each θ, that is, if there are numbers a_θ and b_θ (depending on θ but not on X) such that, on a set of x-values of probability one when θ is true,

$$\frac{\partial \log p_{\theta;X}(x)}{\partial \theta} = z_\theta(x) = a_\theta + b_\theta t(x).$$

Assume we are in the discrete case, for simplicity. Integrating both sides with respect to θ for fixed x, we obtain

$$\log p_{\theta;X}(x) = A(\theta) + B(\theta)t(x) + c(x),$$

where $A(\theta)$ and $B(\theta)$ are the functions obtained by integrating a_θ and b_θ, and $c(x)$ is the *constant of integration* (which may depend on x, which was held fixed). Thus, finally

$$p_{\theta;X}(x) = e^{A(\theta)}e^{c(x)}e^{B(\theta)t(x)}. \tag{7.14}$$

This is almost the form of the Koopman-Darmois laws. In fact, if $X = (X_1, \ldots, X_n)$ where the X_i's are independent and identically distributed, we can use the fact that $p_{\theta;X}(x)$ is of the form $\prod_1^n p_{\theta;X_1}(x_i)$ to show that (7.14) must be of the Koopman-Darmois form with $t(x)$ a linear function of the complete minimal sufficient statistic $\sum_1^n k(x_i)$. Thus, $\phi(\theta)$ must be a linear function of $E_\theta k(X_1)$ in order for the preceding method of proof to be capable of showing that some unbiased estimator of ϕ (namely, the corresponding linear function of $n^{-1}\sum_1^n k(X_i)$) is best unbiased for squared error loss. In Example 7.4, $k(x) = x^2$. Since $E_\theta X^2 = \theta$, the preceding method of proof can be used in that example only to prove that $h_1 + h_2 n^{-1} \sum_1^n X_i^2$ is a best unbiased estimator of $\phi(\theta) = h_1 + h_2\theta$ where h_1 and h_2 are constants; it fails in all other cases, one example of which ($\phi(\theta) = \theta^{1/2}$) we went through in detail.

In addition to the possible modifications of (7.2) or (7.9) whose existence we mentioned near the beginning of this subsection, there are also useful modifications to cover the case in which θ is a vector (rather than being real) and we are interested in estimating one or more components of θ, or functions thereof. For a reference, see Lehmann's book.

A remarkable error in the treatment of this subject can be found in some books. An example like Example 6.5, in which one *cannot* interchange the steps of integration and differentiation, is treated as if one could perform this operation, and $I(\theta)$ is thus computed to be linear in n. The variance of the best unbiased estimator $(n + 1)\max_i X_i/n$ of θ goes down roughly as n^{-2} rather than n^{-1}, and this leads the authors to the mystical conclusion that, in this case, the variance of the best unbiased estimator is miraculously smaller than a lower bound on that variance!

In Section 7.4, Method B, the information inequality is used to prove that a certain estimator with constant risk is admissible, hence minimax. That discussion completes the discussion of lower bounds in this book. Generalizations have been made in a number of directions and these are treated in an extensive literature on the subject. A discussion of these generalizations would take us too far afield.

7.3. Invariance

[Some of the material of this section is more difficult than the general level of this book and can be omitted. The principle of using an invariant procedure is a simple idea, but the computational details are often complicated and not to be worried about. We will try to work on computationally simple examples to illustrate the derivation of best invariant estimators. The invariance principle will arise again in hypothesis testing, where we will sometimes refer to a procedure as "best among invariant procedures" without going through the computational justification of this description.]

We now discuss another criterion, *invariance*, which can be imposed on a statistical problem to limit the procedures among which we look for the one actually to be used (see Chapter 4). In many respects, this is one of the most natural criteria; in typical examples it merely states that the meaning of the decision reached at the end of an experiment should not depend on the particular units used to record the observations. [Another, more recent, terminology uses *equivariance* and *equivariant* in place of *invariance* and *invariant* (see Lehmann's book on estimation).]

For example, suppose a customer came in with $S = \{x : x > 0\}$, $\Omega = \theta : \theta \geq 2\}$,

$$f_{\theta;X}(x) = \begin{cases} \theta^{-1} e^{-x/\theta} & \text{if} \quad x > 0, \\ 0 & \text{otherwise,} \end{cases}$$

$D = \{d : d \geq 2\}$, $W = \left(\dfrac{d}{\theta} - 1\right)^2$. This customer measured the (exponentially distributed) lifetime X of a single light bulb, known to have expected lifetime θ at least equal to 2 weeks, and wants to estimate θ, the loss being the square of the relative error $\dfrac{d}{\theta} - 1$. The statistician considers this problem and perhaps finds this $\{S, \Omega, D, W\}$ listed as Problem 12345 in his *Golden Statistical Cookbook*, in which there is listed each possible statistical problem together with a recommended procedure for it. Thus, he would find procedure t_{12345} recommended for use in the present problem. Now suppose the same experimenter (or one of his colleagues) had decided to describe the problem in terms of days instead of weeks. He observes the same light bulb but records $Y = 7X$. His sample space is $S' = \{y : y > 0\}$, his parameter space is $\Omega' = \{\psi : \psi \geq 14\}$ where $\psi = 7\theta$ and Y has density

$$f_{\psi;Y}(y) = \begin{cases} \psi^{-1} e^{-y/\psi} & \text{if} \quad y > 0, \\ 0 & \text{otherwise} \end{cases}$$

when ψ is true; his decision space is $D' = \{c : c \geq 14\}$, and his loss function is again the square of relative error, which does not depend on the units of measurement, namely $W'(\psi, c) = \left(\dfrac{c}{\psi} - 1\right)^2$. The statistician finds this is Problem 67890 in his book, and hence tells the experimenter to use procedure t_{67890}.

Without any other specification regarding the criterion used by the statistician to select his procedures, the one specification which it seems reasonable for us to make is that of an internal consistency: if, as in the preceding problem, the *same data*, as well as the final estimate, are recorded in units of weeks and then in units of days, then the decision in units of days should be 7 times the decision in terms of weeks. In other words, if $X = x$ and the decision $t_{12345}(x)$

in units of weeks is written as $7t_{12345}(x)$ in units of days, this decision should be the same as that obtained if we had at the outset transformed the data into units of days, so that we recorded $Y = 7x$ and thus reached decision $t_{67890}(7x)$. Thus, we insist that, for each x,

$$7t_{12345}(x) = t_{67890}(7x). \tag{7.15}$$

Now suppose instead that there had been no lower bound on the expected lifetime θ of the bulb, so that Ω had been $\{\theta : \theta > 0\}$ and D had been $\{d : d > 0\}$. The statistician finds this is Problem 10101 in his *Golden Book*, so he tells the customer to use procedure t_{10101}. Now when the problem is presented in units of days instead of weeks, we obtain $\Omega' = \{\psi : \psi > 0\}$ and $D' = \{c : c > 0\}$. Thus, we see that S' is the same as S, Ω' is the same as Ω (this of course means not merely that the set $\{\theta : \theta > 0\}$ of positive reals is the same as the set $\{\psi : \psi > 0\}$, but also that the family of (exponential) probability laws for which these sets of parameter values stand are identical!), D' is the same as D, and W' is the same as W. Only the dummy variables used to describe these four elements of the statistical problem have changed, but the elements themselves are the same; for example, D and D' are both the set of positive reals, and $W(u, v) = W'(u, v)$ for all positive u and v. That is, *the statistical problem in terms of days has exactly the same description* $\{\mathrm{S}, \Omega, \mathrm{D}, \mathrm{W}\}$ *as that in terms of weeks*. The statistician thus also says to use t_{10101} when the data are presented in units of days, and our demand of consistency, in place of (7.15) becomes

$$7t_{10101}(x) = t_{10101}(7x) \quad \text{for all } x > 0. \tag{7.16}$$

Similarly, if the units were changed from weeks to hours, we would require (7.16) with 7 replaced (in both places) by 168. In order to obtain the type of consistency which we have been discussing with respect to *every possible* change of units, we would have to require

$$kt_{10101}(x) = t_{10101}(kx) \quad \text{for all } x > 0 \quad \text{and} \quad k > 0. \tag{7.17}$$

(7.15) does not give us any information about the nature of t_{12345}. On the other hand, (7.17) tells us quite a bit about t_{10101}. In fact, since (7.17) must hold for all $x > 0$ and $k > 0$, it must in particular hold when $k = 1/x$, for every positive x. Substituting $k = 1/x$ in (7.17) and letting $b = t_{10101}(1)$, we obtain

$$t_{10101}(x) = bx. \tag{7.18}$$

Thus, our simple demand of internal consistency in decision making under a change of units has greatly reduced the possible forms of t_{10101} from all possible functions from S into D to only the homogeneous linear functions of the form (7.18).

The argument that led us to state (7.17) depended on the notion that, for any given problem $\{S, \Omega, D, W\}$, the statistician could uniquely specify $t_{\{S,\Omega,D,W\}}$, the statistical procedure to be used; and, were we to demand that this specification be internally consistent in the sense described previously, then (7.17) would follow. If the statistician attempts to specify these procedures with the aid of a criterion treated earlier, he or she should realize that we have not even exhibited any criterion which always specifies an admissible statistical procedure (except the Bayes criterion, under certain restrictions). And of the criteria listed in Chapter 4, many do not yield a unique procedure in general; some of the criteria such as Bayes, minimax, and maximum likelihood will, in certain classes of problems, specify a unique statistical procedure. Finally, if one used such a criterion, he or she would have to verify that the procedures yielded by it satisfied (7.17) (or the corresponding invariance condition, in another problem). It is possible instead, if we accept (7.17), to approach the subject by simply searching among the procedures (7.18) for our choice of a procedure.

The reader should therefore realize that our argument is as follows. Should we decide that for each problem $\{S, \Omega, D, W\}$ we will pick *one* statistical procedure $t_{\{S,\Omega,D,W\}}$ depending only on the specification of the problem, and if we require that this statistical procedure be consistent with respect to scale changes, as described previously, then (7.17) follows. Otherwise (7.17) does not follow. (A Bayesian would not believe in (7.17); it can be shown in this example that no Bayes procedure can satisfy (7.18), which is obvious if $P_\xi\{\theta < c\} = 1$, $c < \infty$.)

In Chapter 4 in our discussions of maximum likelihood and the method of moments we indicated that these methods specified procedures without reference to the way loss is measured. Consequently it was not surprising, we suggested, to discover inadmissible procedures resulting from application of the criteria. These remarks apply equally well to the criterion of invariance discussed here. It should not be surprising to find examples of inadmissible statistical procedures which in every other way meet the requirements of our discussion.

Let us suppose that in measuring the lifetime θ of a light bulb the statistician decides in favor of using t_{10101} with the form (7.18). Then among all possible constants b there will be certain values which give *uniformly best invariant procedures*. Just as with unbiased estimation where we looked for a uniformly best unbiased estimator, so here we seek a uniformly best invariant procedure. To find this procedure in our example we consider all procedures

$$\bar{t}_b(x) = bx \tag{7.19}$$

and compute the risk function of each such procedure:

$$\begin{aligned} r_{\bar{t}_b}(\theta) &= E_\theta W(\theta, \bar{t}_b(X)) \\ &= E_\theta\left[\frac{bX}{\theta} - 1\right]^2 \\ &= E_1[bX - 1]^2 \qquad (7.20) \\ &= \int_0^\infty (bx - 1)^2 e^{-x}\, dx \\ &= 2b^2 - 2b + 1. \end{aligned}$$

Thus, the risk function is, for each b, a constant, independent of θ. (The important step to note here, since it is the one which must be duplicated in other examples, is that of seeing that, θ being a scale parameter, the probability law of X/θ_0 when X has the law labeled by $\theta = \theta_0$ is the same as that of X when $\theta = 1$, so that $E_{\theta_0} q(X/\theta_0) = E_1 q(X)$ for any function q.)

Since the graph of each risk function $r_{\bar{t}_b}$ is a horizontal line, it is easy to find a uniformly best procedure among the $\bar{t}_b$'s: we have only to minimize $2b^2 - 2b + 1$ with respect to b. By setting the derivative with respect to b equal to zero, the minimum is found to be attained when $b = 1/2$. Thus, the statistician would naturally set $t_{10101}(x) = x/2$ in the preceding example.

Thus using only the natural principle that the meaning of our decision should not depend on the scale of units used in recording the data and decision, and the principle that we should use the uniformly best procedure (if one exists) among the class of invariant procedures to which we are restricted, our estimator was uniquely determined!

Not all statistical problems yield such a satisfactory answer. In the original statement of the light bulb testing problem, $\Omega = \{\theta : \theta \geq 2\}$. We saw that a change of units from weeks to days leads to a problem with a different specification. Consequently our argument fails and there is no analogue of (7.18). We have only the consistency relationship (7.16). In this case our principles yield no reduction of the problem. We shall see in other examples later that the analogue of (7.18) in other cases may not reduce the problem enough to yield a uniformly best procedure; in such cases, we would have to use some additional criterion to select that procedure which satisfies the analogue of (7.18) and which will be used.

Before looking at some additional examples, we shall introduce the general nomenclature and notation which will be used in the remainder of this section. Suppose we have a statistical problem $\{S, \Omega, D, W\}$ and collections $\{g_k\}$, $\{h_k\}$, $\{j_k\}$ of functions, where k ranges over some set of indices K, and where g_k is a function from S onto S, h_k is from Ω onto Ω, and j_k is from D onto D. (Reminder: A function from A to B is *onto* if its range is all of B.) Suppose further that, for each k, F, and d,

(i) When X has (probability) distribution F, the random variable $g_k(X)$ has (probability) distribution $h_k(F)$;

(ii) $W(F, d) = W(h_k(F), j_k(d))$.

Then we shall say that *the collection of transformations* $\{g_k, h_k, j_k; k \in K\}$ *leaves the problem* $\{S, \Omega, D, W\}$ *invariant.*

Let us see how our example is described in these terms. The index set K is the positive reals: $K = \{k : k > 0\}$. The functions g_k, h_k, j_k are defined by

$$g_k(x) = kx,$$

$$h_k\,(\text{exponential law with mean } \theta) = (\text{exponential law with mean } k\theta),$$

$$j_k(d) = kd.$$

When we changed units from weeks to days we multiplied the value of the observation, the mean of the exponential law, and the estimate of the mean by the same number $k = 7$, in this case. Note also that (i) and (ii) are satisfied here; in fact, (ii) would be satisfied in this example as long as $W(\theta, d)$ depended only on d/θ, since if $W(\theta, d) = w(\theta/d)$ we have $W(h_k(\theta), j_k(d)) = W(k\theta, kd) = w(k\theta/kd) = W(\theta, d)$.

A statistical procedure t is said to be *invariant under the set of transformations* $\{g_k, j_k; k \in K\}$ if

$$j_k[t(x)] = t[g_k(x)] \quad \text{for all } x \text{ and } k. \tag{7.21}$$

In our example you can easily check that (7.21) is precisely (7.17).

Put abstractly, our earlier argument would have said, suppose instead of X we observe $g_k(X)$. The random variable $g_k(X)$ has sample space S since g_k is a function from S *onto* S. The distribution of $g_k(X)$ is $h_k(F)$, which is in Ω. Further, as h_k is a function from Ω onto Ω, every F in Ω is a possible distribution for $g_k(X)$. The set of decisions from which we are to choose does not change, nor does the measure of loss W. Therefore the specification of the problem remains $\{S, \Omega, D, W\}$.

In the example of light bulb testing there was a natural relationship between the distribution F of X and the decision d. Our decision was, "d is the expected value of X." Consequently if we used instead kX it was appropriate to make the decision kd. In the abstract statement of the problem *we must assume a natural relationship* of the following form. The experimenter knows that instead of observing the random variable X he is observing the random variable $Y = g_k(X)$. If, then, t_{12345} is the statistical procedure he would use when observing X, and he now observes $Y = g_k(X)$ and makes the decision $t_{67890}(Y) = t_{67890}(g_k(X))$ (where t_{67890} is the statistical procedure he has specified for use when observing Y), then we ask, on grounds of consistency, that

$$j_k(t_{12345}(x)) = t_{67890}(g_k(x)) \quad \text{for each } k \text{ in } K \text{ and } x \text{ in } S. \tag{7.22}$$

This is the analogue of (7.15).

If we suppose the problem in terms of $Y = g_k(x)$ is the same as that in terms of X, for each k in K, and suppose there to be a procedure $t_{\{S,\Omega,D,W\}}$ uniquely determined by the specification of the problem, and if this choice of statistical procedure is to be consistent in the sense described, then

$$t_{\{S,\Omega,D,W\}}(g_k(x)) = j_k(t_{\{S,\Omega,D,W\}}(x)) \quad \text{for all } k \text{ in } K \text{ and } x \text{ in } S. \tag{7.23}$$

This is exactly (7.22) with $t_{\{S,\Omega,D,W\}} = t_{12345} = t_{67890}$. From (7.21) we see that (7.23) says that $t_{\{S,\Omega,D,W\}}$ should be invariant under the set of transformations $\{g_k, j_k; k \in K\}$. The *principle of invariance* as usually expressed, states that, if a statistical problem is left invariant by a family of transformations, then one should use an invariant (with respect to this family) procedure for the problem, in the sense that (7.21) and (7.23) describe invariant procedures. We hope the reader will now see that using the principle of invariance as a criterion for reducing a problem is quite different from concluding that a procedure uniquely specified on other grounds must be invariant because of consistency.

In order to use invariance in a given problem the idea is to find all possible transformations (g_k, h_k, j_k) leaving the problem invariant. It is surprising, but in some problems the resulting set of (g_k, h_k, j_k) may be so large that the resulting invariant statistical procedures are not, from the viewpoint of decision theory and risk functions, good statistical procedures. The best that can be said is that one may want to use some subset or perhaps all the (g_k, h_k, j_k) to determine procedures invariant relative to the chosen family of transformations. Then, using (7.21), one will find an analogue of (7.19). One then investigates the risk functions of the invariant procedures to find a uniformly best invariant procedure if one exists, or a good (according to some additional criterion) invariant procedure if there is no uniformly best invariant procedure.

In discussing the problem of light bulb testing it was shown that by (7.20) the risk function of each invariant procedure was constant-valued. Whether this is true for a given problem depends on the problem. It may happen that the only (g_k, h_k, j_k) leaving a problem invariant are the identity transformations of S, Ω, D, respectively. Consequently every statistical procedure will be invariant. And in general the risk function will not be constant. If, however, there are several (g_k, h_k, j_k) leaving the problem invariant, then use of an invariant t does limit the way in which r_t can vary. For if t is invariant, in the light bulb testing, then for each θ and k we have, using (ii), (7.21) and (i) (which last says that $E_F q(g_k(X)) = E_{h_k(F)} q(X)$, just as $E_\theta q(kX) = E_{k\theta} q(X)$ in our example),

$$\begin{aligned} r_t(F) &= E_F W(F, t(X)) \\ &= E_F W(h_k(F), j_k[t(X)]) \\ &= E_F W(h_k(F), t[g_k(X)]) \\ &= E_{h_k(F)} W(h_k(F), t(X)) \\ &= r_t(h_k(F)). \end{aligned} \tag{7.24}$$

Thus, the space Ω can be broken up into subsets on each of which r_t (we are considering invariant t!) is constant. Each of these sets, called an *orbit of the set of transformations*, includes with any F_0 in the orbit all Fs of the form $h_k(F_0)$ for k in K. If there is only one orbit, we see that r_t is constant on Ω if t is invariant. In this case $\{h_k\}$ is said to be transitive on Ω, and for each pair F_1 and F_2 of elements in Ω there is some h_k (depending on F_1 and F_2) for which $h_k(F_1) = F_2$. It is in this case (example, light bulb testing) that a uniformly best invariant procedure will exist, it being an invariant procedure for which the constant risk is least.

We have suggested that in some known examples if one takes all (g_k, h_k, j_k) then one has too many transformations, and although one may find a best invariant procedure, this turns out not to be a good procedure. In the known examples of this, $\{h_k\}$ is transitive on Ω in the sense described in the preceding paragraph, but there are proper subsets of $\{h_k\}$ that are also transitive. See Example 7.5(a), (v). It turns out in these problems that there is naturally a smallest subset of $\{h_k\}$ corresponding to indices $k \in K_0 \subset K$ such that $\{h_k\}$ with $k \in K_0$ is transitive on Ω. If one considers statistical procedures invariant relative to $\{(g_k, h_k, j_k), k \in K_0\}$ and finds the uniformly best such procedure, then this procedure has smallest risk among all procedures having constant risk and is a minimax procedure, but is not admissible. Unfortunately we do not have enough mathematical machinery to describe many of these examples in greater detail.

At the other extreme, if one uses too few transformations, then invariance may not sufficiently reduce the problem. For example, if we had considered only the transformation from weeks into days ($k = 7$) in our problem, we would have obtained only (7.16) instead of (7.17). This would not have yielded anything so restrictive as (7.18). In fact, if $t(x)$ is defined *in any manner whatsoever* for $1 \le x < 7$, we can then define $t(x)$ for $7 \le x < 49$ by

$$t(x) = 7t(x/7), \tag{7.25}$$

using the previously defined (since $1 \le x/7 < 7$) value $t(x/7)$. We can repeat this process, next using (7.25) to define $t(x)$ for $49 \le x < 343$ in terms of the values just obtained (or, more directly, by using $t(x) = 49t(x/49)$, where $1 \le x/49 < 7$), etc. Similarly, for $1/7 \le x < 1$, we define $t(x) = 7^{-1}t(7x)$, etc. In this manner, we define $t(x)$ for every $x > 0$, in such a way that t satisfies (7.16). Since $t(x)$ could be chosen arbitrarily for $1 \le x < 7$, there are many such t's, and no uniformly best one exists among them; thus, we see the importance of trying to find sufficiently many transformations leaving the problem invariant, as exhibited by (7.17) rather than merely (7.16) in our example.

We now consider some additional examples.

EXAMPLE 7.5(a). *Location parameter family (one observation).* Suppose that S is the reals (we have a single real observation), that $\Omega = \{\theta : -\infty < \theta < \infty\}$, where

$$f_{\theta;X}(x) = f^*(x - \theta),$$

f^* being a given function, and that (the aim being to estimate θ) W is a function only of the difference $\theta - d$ between parameter value and estimate, so that $W(\theta, d) = w(d - \theta)$ for a suitable function w. This problem is invariant under the transformations $\{g_k, h_k, j_k; -\infty < k < \infty\}$ where

$$g_k(x) = x + k,$$

$$h_k(\theta) = \theta + k,$$

$$j_k(d) = d + k.$$

(For example, if we had a normal random variable X with known variance 1 and unknown mean θ, $-\infty < \theta < \infty$, so that $f^*(x) = (2\pi)^{-1/2}e^{-x^2/2}$, then $X + k$ still has one of this class of distributions, namely, that corresponding to the value $\theta + k$.) Equation (7.21) now states that, if t is invariant, then

$$t(x) + k = t(x + k) \quad \text{for all real } x \text{ and } k. \tag{7.26}$$

Thus, writing $t(0) = b$, and setting $k = -x$ in (7.26), we obtain

$$t_b(x) = x + b \tag{7.27}$$

as the form of an invariant estimator. Since $\{h_k, k \in K\}$ is transitive on Ω, each invariant t has constant risk, so that $r_t(\theta) = r_t(0)$ if t is invariant. Hence,

$$r_{t_b}(\theta) = r_{t_b}(0) = E_0 w(X + b) = \int_{-\infty}^{\infty} w(x + b) f^*(x)\, dx, \tag{7.28}$$

and the best choice of b is that which minimizes this last integral. Five important examples are as follows:

(i) If $w(\theta - d) = (\theta - d)^2$, so that $w(x + b) = (x + b)^2$, and if the mean $E_\theta X$ of $f_{\theta;X}$ is $\theta + c$ so that f^* has mean c, then $E_0(X + b)^2$ is minimized by taking $b = -c$ (since the second moment of a random variable is a minimum if taken about its mean). Hence, $t(x) = x - c$ is the best invariant estimator. In particular, if $E_\theta X = \theta$ so that $c = 0$, we obtain $t(x) = x$. Note the difference between this case and the light bulb problem for which the best invariant estimator (for a *different* problem and set of transformations leaving the problem invariant) was *not* the unbiased estimator X of θ.

(ii) If f^* is any unimodal function which is symmetric about 0 (that is, $f^*(x) = f^*(-x)$ and $f^*(x)$ is nonincreasing for $x \geq 0$), and if $w(x)$ is symmetric about 0 and nondecreasing for $x \geq 0$ (e.g., $w(\theta - d) = |\theta - d|^r$ for $r > 0$, or $w(\theta - d) = 0$ or 1 according to whether or not $|\theta - d| < B$, where B is a constant), it is not hard to show that $b = 0$ is a best choice, so that $t(x) = x$ is a best invariant estimator. We shall not prove this here, but it should not be difficult for you to convince yourself of the reasonableness of this result (especially in the case of the 0-1 w mentioned) by superimposing the graphs of f and w, thinking of integrating (as in (7.28)) the product, and thinking of what happens to this integral if you shift the graph of w to the left by b units (compare the amount gained by the integral on the left side of the x axis to the amount lost on the right).

(iii) If f^* is symmetric about 0 (not necessarily unimodal as in (ii)) and w is symmetric about 0 and convex (which is more than assumed in (ii)—for example, of the illustrations cited there, $|\theta - d|^r$ for $r < 1$ and the 0-1 loss function are not convex), then $E_0 w(X + b) = E_0 w(-X + b)$ by the symmetry of f^* (X and $-X$ have the same probability law when $\theta = 0$), so that $E_0 w(X + b) = E_0[w(X + b) + w(X - b)]/2$, where we have written $w(-X + b) = w(X - b)$ because of the symmetry of w. Since w is convex, $[w(x + b) + w(x - b)]/2 \geq w(x)$ for each x, and thus, finally, $E_0 w(X + b) \geq E_0 w(X)$; that is, $b = 0$ is a best choice (unique if w is strictly convex).

(iv) (*Optional.*) Let us suppose as in (ii) that $W(\theta, d) = w(d - \theta)$, where w is symmetric and w is a strictly increasing continuous function of $x \geq 0$. As we have seen in the preceding examples, to find the best invariant estimator, $t_b(x)$, we must find a value b_0 such that

$$E_0 w(X + b_0) = \min_b E_0 w(X + b) = r. \tag{7.29}$$

It may happen in certain problems that there are two (or more) values, say b_0 and b_1, which give the minimum in (7.29). For example, if $w(x) = |x|$ and there are several values a such that $\int_{-\infty}^{a} f^*(x)\,dx = 1/2$, then (7.29) will not have a unique solution. Any median of f^* will then be a value of B that minimizes $E_0|X - B|$. In this case there is no admissible invariant procedure. If there were admissible invariant procedures then $t_{b_0}(x) = x + b_0$ and $t_{b_1}(x) = x + b_1$ would both be admissible. We now exhibit a procedure that is better than either of these procedures. Suppose $b_0 < b_1$. Let b be a number. Construct a new estimator as follows. If $X = x$ is observed and $x \leq b$, estimate $t(x) = x + b_1$; if $x > b$ is observed, estimate $t(x) = x + b_0$.

To show t is a better estimator we calculate the risk function of t

$$\begin{aligned} r_t(\theta) &= \int_{-\infty}^{\infty} w(t(x) - \theta) f^*(x - \theta)\,dx \\ &= \int_{-\infty}^{b} w(x + b_1 - \theta) f^*(x - \theta)\,dx \\ &\quad + \int_{b}^{\infty} w(x + b_0 - \theta) f^*(x - \theta)\,dx \end{aligned}$$

and changing variable of integration

$$\begin{aligned} &= \int_{-\infty}^{b-\theta} w(x + b_1) f^*(x)\,dx + \int_{b-\theta}^{\infty} w(x + b_0) f^*(x)\,dx \\ &= \int_{-\infty}^{\infty} w(x + b_1) f^*(x)\,dx + \int_{b-\theta}^{\infty} [w(x + b_0) - w(x + b_1)] f^*(x)\,dx. \end{aligned}$$

Since the first integral is the (constant) value of $r_{t_{b_1}}(\theta)$, we will have justified our claim if we can show the second integral is always ≤ 0 and is actually negative for some θ. We note the following. By (7.29),

$$\begin{aligned} r &= \int_{-\infty}^{\infty} w(x + b_1) f^*(x)\,dx = \int_{-\infty}^{\infty} w(x + b_0) f^*(x)\,dx \\ &= \lim_{\theta \to +\infty} \int_{b-\theta}^{\infty} w(x + b_1) f^*(x)\,dx \\ &= \lim_{\theta \to +\infty} \int_{b-\theta}^{\infty} w(x + b_0) f^*(x)\,dx. \end{aligned} \tag{7.30}$$

Let us define

$$g(\theta) = \int_{b-\theta}^{\infty} [w(x + b_0) - w(x + b_1)] f^*(x)\, dx. \tag{7.31}$$

By (7.30) it follows that

$$\lim_{\theta \to +\infty} g(\theta) = 0. \tag{7.32}$$

It is easily seen that

$$\lim_{\theta \to -\infty} g(\theta) = 0. \tag{7.33}$$

The difference $w(x + b_0) - w(x + b_1)$ clearly satisfies

$$\begin{aligned} &\text{if} \quad x < -b_1 \quad \text{then} \quad x + b_0 < x + b_1 < 0 \quad \text{and} \\ &\quad w(x + b_0) - w(x + b_1) > 0; \\ &\text{if} \quad x > -b_0 \quad \text{then} \quad 0 < x + b_0 < x + b_1 \quad \text{and} \\ &\quad w(x + b_0) - w(x + b_1) < 0. \end{aligned} \tag{7.34}$$

The >0 or <0 holds since w is a strictly increasing function of $x \geq 0$. If $-b_1 \leq x \leq -b_0$ then $x + b_0 \leq 0$ so that $w(x + b_0)$ is a strictly decreasing function of x; and $x + b_1 \geq 0$ so that $-w(x + b_1)$ is a strictly decreasing function of x. Therefore,

$$\begin{aligned} &\text{if} \quad -b_1 \leq x \leq -b_0 \quad \text{then} \\ &\quad w(x + b_0) - w(x + b_1) \text{ is a strictly decreasing function of } x. \end{aligned} \tag{7.35}$$

Since w is a continuous function it follows from (7.34) and (7.35) that there is a number a such that

$$\begin{aligned} &\text{if} \quad x < a \quad \text{then} \quad w(x + b_0) - w(x + b_1) > 0; \\ &\text{if} \quad x > a \quad \text{then} \quad w(x + b_0) - w(x + b_1) < 0. \end{aligned} \tag{7.36}$$

(7.36) shows $x = a$ is the only value of x such that

$$0 = w(x + b_0) - w(x + b_1).$$

The derivative of $g(\theta)$ is

$$[w(-\theta + b + b_0) - w(-\theta + b + b_1)] f^*(b - \theta)$$

so that from (7.36) it follows that $g(\theta)$ is a decreasing function if $\theta < b - a$ and is an increasing function if $\theta > b - a$; moreover, $g(\theta)$ is not constant since $g'(\theta)$ is not always zero, and $\lim_{\theta \to -\infty} g(\theta) = \lim_{\theta \to \infty} g(\theta) = 0$. From this it follows that $g(\theta) \leq 0$ for all θ, and for some θ, $g(\theta) < 0$. Therefore t is a better estimator than t_{b_0} or t_{b_1} (which have the same risk function).

(v) (*Optional.*) We have not considered the possibility that, in Example 7.5(a), some more complicated transformations also leave the problem invariant. The general question of how to find all transformations which leave a given problem invariant is quite complicated. To mention one example of a simple additional transformation which works in some cases, if f^* and w are symmetric about 0 in Example 7.5(a), (for example, if f^* is normal and $W(\theta, d) = |\theta - d|^r$), it is easy to see that the transformation

$$g(x) = -x$$
$$h(\theta) = -\theta$$
$$j(d) = -d$$

leaves the problem invariant. (The invariance of the problem under this transformation indicates that which direction is called positive and which negative does not matter.) Hence, any invariant t for the larger set of transformations must satisfy $-t(x) = t(-x)$. Using the fact that t must also satisfy $t(x) = x + b$, for some b, we have $t(x) = -t(-x) = -(-x + b) = x - b$; hence we see that (since $x + b = t(x) = x - b$) we must have $b = 0$: the *only* invariant procedure is $t(x) = x$. The presence of more transformations leaving the problem invariant, as we have stated before, reduces the number of invariant procedures.

In order to give a concrete example let us suppose f^* has the following definition: $f^*(x) = 1/2$ if $1 \le |x| \le 2$; otherwise $f^*(x) = 0$. Further let $w(\theta, d) = |d - \theta|^{1/2}$. Then, letting $v(b)$ denote the (constant) risk of t_b,

$$\begin{aligned} 2v(b) = 2r_{t_b}(\theta) &= \int_1^2 |x + b|^{1/2}\,dx + \int_{-2}^{-1} |x + b|^{1/2}\,dx \\ &= \tfrac{2}{3}\{(2 + b)^{3/2} - (1 + b)^{3/2} + (2 - b)^{3/2} - (1 - b)^{3/2}\}. \end{aligned}$$

We suppose here that $|b| < 1$. As a function of b we may compute the derivative and obtain

$$2\frac{d}{db}v(b) = \{(2 + b)^{1/2} - (1 + b)^{1/2} - (2 - b)^{1/2} + (1 - b)^{1/2}\},$$

$$2\frac{d^2}{db^2}v(b) = \frac{1}{2}\{(2 + b)^{-1/2} - (1 + b)^{-1/2} + (2 - b)^{-1/2} - (1 - b)^{-1/2}\}.$$

It is easily checked that if $|b| < 1$ then

$$(2 + b)^{-1/2} - (1 + b)^{-1/2} < 0.$$

Therefore if $|b| < 1$, $\dfrac{d^2}{db^2}v(b) < 0$ and v is a concave function of b, $|b| < 1$. Since the first derivative of v is zero at $b = 0$, it follows that the invariant estimator $t_0(x) = x$ is uniformly worse than the (translation invariant) estimator $t_{1/2}(x) = x + 1/2$.

Therefore in this example the optimal choice for the experimenter is to ignore the additional transformation $g(x) = -x$, $h(\theta) = -\theta$, $k(d) = -d$, and simply to find the best invariant estimator under the smaller set of transformations which are merely the translations. In view of Example 7.5(a)(iv), even this procedure necessarily yields an inadmissible statistical procedure (which can be shown to be minimax).

You should think of the various examples which fall within the preceding framework. Familiar examples of location parameter families are the normal with known variance, the Cauchy with known scale, the exponential with unknown location (density $e^{-(x-\theta)}$ for $x > \theta$ and 0 otherwise), and the rectangular from $\theta - 1/2$ to $\theta + 1/2$.

EXAMPLE 7.5(b). *Location parameter family (more than one observation).* (*Optional*: The computations here are more involved than in Example 7.5(a).) Suppose $X = (X_1, \ldots, X_n)$ with $n > 1$, where the X_i's are independently and identically distributed, $f_{\theta;X_1}$ being of the same form as in the preceding Example 7.5(a). Thus, $\Omega = \{\theta: -\infty < \theta < \infty\}$, and we again assume that $D = \{d: -\infty < d < \infty\}$ and $W(\theta, d) = w(d - \theta)$. The transformations leaving the problem invariant are the same as before, except that g_k is now defined by

$$g_k[(x_1, x_2, \ldots, x_n)] = (x_1 + k, x_2 + k, \ldots, x_n + k),$$

and $(X_1 + k, \ldots, X_n + k)$ now has the law labeled $\theta + k$ when $(X_1, \ldots, X_n)$ has the law labeled θ. In place of (7.26), we now have

$$t(x_1, x_2, \ldots, x_n) + k = t(x_1 + k, \ldots, x_n + k), \tag{7.37}$$

and, setting $k = -x_1$ and $t(0, y_2, \ldots, y_n) = -q(y_2, \ldots, y_n)$, we have, in place of (7.27),

$$t_q(x_1, \ldots, x_n) = x_1 - q(x_2 - x_1, \ldots, x_n - x_1) \tag{7.38}$$

as the general form of an invariant procedure. Since any function q of $n - 1$ variables yields such a procedure, we see that the variety of invariant procedures here is much greater than in (7.27). Nevertheless, the important point is that $\{h_k, k \in K\}$ is still transitive on Ω, so that, despite the great variety of procedures satisfying (7.38) there still exists a uniformly best invariant procedure. Since $r_t(\theta) = r_t(0)$ for t invariant, we again have only to compute $r_t(0)$ and to choose the function q so as to minimize it. Let us write $Y_i = X_i - X_1$ and $Y = (Y_2, Y_3, \ldots, Y_n)$. Let $f^*(x_1 | y)$ denote the conditional density, when $\theta = 0$, of X_1, given that $Y = y$, and let $f^{**}(y)$ denote the (joint) density of $Y = (Y_2, \ldots, Y_n)$ when $\theta = 0$. We can then write

$$r_{t_q}(0) = E_0 w(X_1 - q(Y)) = \int\int w(x_1 - q(y)) f^*(x_1 | y) f^{**}(y) \, dx_1 \, dy,$$

where dy stands for $dy_2 \cdots dy_n$ and the outer integral is $(n - 1)$-fold. If $q(y)$ is chosen for each fixed y to be the value c (depending on y) which minimizes the inner integral

$$\int w(u - c) f^*(u | y) \, du, \tag{7.39}$$

then $r_{t_q}(0)$ will be a minimum. We now illustrate how this computation can be carried out in an important special case. First we note that, using the method of changing variables (from $(X_1, \ldots, X_n)$ to $(X_1, Y_2, \ldots, Y_n)$) as discussed in probability theory, it is not hard to show that, when $\theta = 0$, the joint density of $X_1, Y_2, Y_3, \ldots, Y_n$ is $f^*(x_1)f^*(x_1 + y_2)f^*(x_1 + y_3)\cdots f^*(x_1 + y_n)$, where, as before, f^* is the density of X_1. Hence, we have

$$f^*(u|y) = \frac{f^*(u)f^*(u + y_2)\cdots f^*(u + y_n)}{\int_{-\infty}^{\infty} f^*(u)f^*(u + y_2)\cdots f^*(u + y_n)\,du} \tag{7.40}$$

and this can be substituted into (7.39) to compute the minimizing c (which equals $q(y_2, \ldots, y_n)$) for each fixed $y_2, \ldots, y_n$.

Suppose now that we consider the important special case $W(\theta - d) = (\theta - d)^2$. Then (7.39) is just $E_0\{(X_1 - c)^2 | Y = y\}$. As recalled in Example 7.5(a), the second moment of any density function is a minimum about the mean; that is, (7.40) achieves its unique minimum $c^* = q^*(y)$ (say) for $q^*(y) = c^* = E_0\{X_1 | Y = y\}$. Hence, the unique best invariant procedure t^* is given, for each fixed set of numbers $x_1, \ldots, x_n$, by

$$\begin{aligned} t^*(x_1, x_2, \ldots, x_n) &= x_1 - q^*(y) = x_1 - E_0\{X_1 | Y = y\} \\ &= E_0\{x_1 - X_1 | Y = y\}, \end{aligned}$$

since x_1 is a fixed real number. Writing $V = x_1 - X_1$ and $v = x_1 - u$, we see easily that the conditional density of $x_1 - X_1$, given that $Y = y$, is $f^*(-v + x_1 | y)$ (where v is the variable and x_1 is still held fixed). Hence, using (7.40) and substituting $y_i = x_i - x_1$, we have

$$\begin{aligned} t^*(x_1, \ldots, x_n) = E_0\{V | Y = y\} &= \int_{-\infty}^{\infty} v f^*(-v + x_1 | y)\,dv \\ &= \frac{\displaystyle\int_{-\infty}^{\infty} v \prod_{i=1}^{n} f^*(x_i - v)\,dv}{\displaystyle\int_{-\infty}^{\infty} \prod_{i=1}^{n} f^*(x_i - v)\,dv}. \end{aligned} \tag{7.41}$$

The use of this estimator was first suggested by Pitman. It is to be noted that it is rarely the same as the commonly used estimator obtained by the method of maximum likelihood (discussed in Section 7.5) (it is the same estimator $\bar{x}_n$ in the normal case). It is also to be noted that the procedure (7.41) will be very difficult to write down in many cases (e.g., the Cauchy), the ratio of integrals often being difficult to write down as a simple expression which is usable in practice; this lends support to earlier remarks made on the usefulness of more easily computable procedures which are fairly good but not optimum.

EXAMPLE 7.5(c). *Location parameter; sufficiency and invariance.* With the same setup as in Example 7.5(b), suppose there is (as in the normal case with known variance or the case of the density $e^{-(x-\theta)}$ for $x > \theta$) a 1-dimensional minimal sufficient statistic $s(x_1,\ldots,x_n)$. It can then be shown that some minimal sufficient statistic (e.g., $\bar{x}_n$ but not $e^{\bar{x}_n}$ in the normal case) has a density $\bar{f}^*(s-\theta)$ depending on θ only as a location parameter. In such a case, we can use the simpler computation of Example 7.5(a) instead of the more complicated computation of Example 7.5(b); it can be shown that the final result is the same with either method of computation. Similarly, in the case of the rectangular density from $\theta - 1/2$ to $\theta + 1/2$ (Example 6.6) the joint density of the minimal sufficient statistic $(Z_{1n}, Z_{2n}) = (\max X_i, \min X_i)$ can easily be seen to be of the form $\bar{f}(z_1-\theta, z_2-\theta)$; although Z_{1n} and Z_{2n} are not independent, this does not affect the application of the invariance principle as in Example 7.5(b) (with appropriate modification of $f_{\theta;X}$) with X replaced by (Z_{1n}, Z_{2n}). For example, in the case of squared error loss, in place of (7.41) we obtain

$$\frac{\int_{-\infty}^{\infty} v\bar{f}(z_1-v, z_2-v)\,dv}{\int_{-\infty}^{\infty} \bar{f}(z_1-v, z_2-v)\,dv}$$

which equals $(z_1+z_2)/2$; exactly the same result would have been obtained if we computed (7.41) directly.

Naturally, an analogue of this approach of simplifying computations by invoking sufficiency before using invariance is useful in cases other than those of a location parameter.

It is important to keep in mind that, if W had depended on θ and d other than through $\theta - d$ in Example 7.5 (or other than through d/θ in the light bulb example), then the problem would not have been invariant under all of the transformations we have considered previously, and invariance would thus not have reduced the problem (or would have reduced it less, if *some* transformations still left the problem invariant).

EXAMPLE 7.6. *Discrete location parameter*. An analogue of Example 7.5 in the discrete case is obtained when X_i and θ can only take on integer values, so that $p_{\theta;X_1}(x) = p_{0;X_1}(x-\theta) = p^*(x-\theta)$ (say), where $p^*(y)$ is different from 0 only if y is an integer, and $\Omega = \{\theta : \theta = \ldots, -2, -1, 0, 1, 2, \ldots\}$. The analysis is now almost exactly like that of Example 7.5, except that k is now also restricted to the integers, and all integrals are replaced by sums. An example is the geometric distribution with unknown starting point:

$$p_{\theta;X_1}(x) = \begin{cases} (1-\alpha)\alpha^{x-\theta} & \text{if} \quad x = \theta, \theta+1, \theta+2, \ldots, \\ 0 & \text{otherwise,} \end{cases}$$

where α is known and $0 < \alpha < 1$. This is the discrete analogue of the exponential density $ae^{-a(x-\theta)}$ for $x > \theta$, where a is known.

EXAMPLE 7.7(a). To cite some examples for which invariance is of little help even though Ω contains "all possible values" of a natural parameter θ, suppose first that $S = \{0, 1\}$, X being a single Bernoulli variable with unknown mean θ, so that $\Omega = \{\theta : 0 \leq \theta \leq 1\}$. We want to estimate θ, so $D = \{d : 0 \leq d \leq 1\}$. Suppose also that $W(\theta, d) = W(1 - \theta, 1 - d)$ (for example, $|\theta - d|^r$). Then there is one transformation (other than the identity) which leaves the problem invariant; it is given by

$$g(x) = 1 - x,$$

$$h(\theta) = 1 - \theta,$$

$$j(d) = 1 - d.$$

In terms of this transformation, the invariance principle essentially says that the answer should not be affected by whether $\{X = 1\}$ is chosen to mean "heads" or "tails" for a coin; that is, the estimate of the probability of getting a head for a given coin should not depend on whether X represents the number of heads or the number of tails which appear on a single flip. In this case (7.21) becomes

$$1 - t(x) = t(1 - x),$$

which is unfortunately not much of a restriction. (In Problem 2.1(b) involving this Bernoulli setup with one observation, only t_6 of the six procedures considered is ruled out.) The orbits in Ω are now pairs of points $(\theta, 1 - \theta)$, so that $r_t(\theta)$ is symmetric about $\theta = 1/2$ for t invariant. Thus, invariance helps very little in this case.

EXAMPLE 7.7(b). Suppose $S = \{0, 1, 2, \ldots\}$ and X is a single Poisson variable with mean θ; $\Omega = \{\theta : \theta > 0\}$. It can be shown that no transformation (other than the identity) satisfies criterion (i) of the definition of a transformation leaving a problem invariant. Thus, no matter what form W may have, invariance is of no help here.

Use of invariance as a mathematical tool is a help in obtaining minimax procedures; this will be discussed in Section 7.4.

We shall return to the subject of invariance (and examples of invariant procedures) in our discussion of hypothesis testing in Chapter 8.

Additional Remarks on Invariance

1. In most examples in which invariance is helpful (e.g., Examples 7.5 and 7.6 and the light bulb problem), Bayes procedures will never be invariant. Intuitively, knowledge of an a priori distribution eliminates the equal stature that all Fs have in the invariance approach. For example, in the testing of light bulbs, if an a priori law were known which assigned probability one to the set of values $\{\theta : 3 \leq \theta \leq 9\}$, then the Bayes procedure t could only take on values between 3 and 9 and hence could not be of the

form bx (which can take on all positive values). In Example 7.4(a), the Bayes procedure (if unique) relative to any a priori law which is *symmetric about* $\theta = 1/2$ (i.e., which, for each z, assigns the same a priori probability or density to $\theta = z$ as to $\theta = 1 - z$) is invariant. In Example 7.7(b) in which invariance is of no help, every procedure is trivially invariant, as is therefore every Bayes procedure.

2. A complete discussion of invariance using randomized procedures is possible but will not be given here. Whenever a best invariant randomized procedure exists in a problem like Example 7.5, an equivalent best invariant nonrandomized procedure exists.
3. The transformations leaving a problem invariant form a group, and a knowledge of elementary group theory is necessary for a more theoretical treatment of the subject.
4. Unfortunately, the best invariant procedure (if one exists) need not be admissible among *all* procedures, although in many important practical cases (e.g., in Example 7.5(a), in which $W(\theta, d) = (\theta - d)^2$ and $E_0|X|^3 < \infty$) it is. Examples of inadmissible best invariant procedures: (a) In Example 7.5(a), in the Cauchy case with $W(\theta, d) = (\theta - d)^2$, every invariant procedure $t_b = b + x$ has $r_{t_b}(\theta) = \infty$ for all θ. The *noninvariant* procedure $\bar{t}(x) \equiv 0$ has risk function $r_{\bar{t}}(\theta) = \theta^2 < \infty$; it is thus uniformly better than any invariant procedure. (b) Example 7.5(a)(iv) shows that sometimes lack of uniqueness implies inadmissibility. (c) If f^* is normal but $W(\theta, d) = 1/(|\theta - d| + 1)$ (i.e., the closer your guess is to the true θ, the more the loss—not too realistic!), the lone invariant estimator $t(x) = x$ *is uniformly worst* among *all* procedures! *Any* other procedure, e.g., $t(x) = x + 10^6$, is better! (d) In Example 7.6 if $p^*(1) = p^*(-1) = 1/2$ and $p^* = 0$ otherwise (so that, when θ is true, X is either $\theta + 1$ or $\theta - 1$, each with probability 1/2), the invariant procedures are of the form $t_b(x) = x + b$ with b an integer, and you can check that, if $W(\theta, d) = 0$ or 1 according to whether or not $d = \theta$, the procedures t_1 and t_{-1} are best invariant with constant risk 1/2 (any other t_b has constant risk 1). It is easy to compute the risk function of the *noninvariant* procedure

$$\bar{t}(x) = \begin{cases} x - 1 & \text{if } x \geq 0, \\ x + 1 & \text{if } x < 0, \end{cases}$$

and one obtains $r_{\bar{t}}(\theta) = 1/2$ except when $\theta = -1$ or 0, when $r_{\bar{t}}(\theta) = 0$. Thus, $\bar{t}$ is better than t_1 or t_{-1}. See also Example 7.5(a)(iv).

The general question of which best invariant procedures are admissible is quite complicated.
5. Examples can be given wherein all randomized (or nonrandomized) invariant procedures are poor. For example, let S, Ω, and D each be the set of all real numbers, let $P_\theta\{X = \theta\} = 1$ (no errors of observation), and let $W(\theta, d) = 1$ if $\theta = d$ and $W(\theta, d) = 0$ if $\theta \neq d$ (the object is *not* to estimate θ correctly). Let $K = \{(a, b) : a \neq 0, b \text{ real}\}$. It is easy to see that this problem is invariant under the transformations

$$g_{a,b}(x) = ax + b,$$
$$h_{a,b}(\theta) = a\theta + b,$$
$$j_{a,b}(d) = ad + b,$$

and that $t(x) = x$ is the only invariant procedure, randomized or nonrandomized. This procedure has risk identically one, and $t'(x) = x + 5$ has risk identically zero. The example of 4(c) is based on the same idea. Other (more realistic) examples which take longer to describe can also be given; many of these are in multivariate normal theory.

Thus, invariance does not always reduce a problem enough to yield a unique specification of a procedure; and, where it does, that procedure need not have good risk properties. Still, it is an appealing criterion, and in certain settings such as the location parameter problem with (usual) conditions that avoid the anomalies of remark 4, it is known to produce a procedure with satisfactory properties (for example, minimaxity, as discussed in Section 7.4, and admissibility).

7.4. Computation of Minimax Procedures (Continued)

The minimax criterion was discussed in Section 4.2; in Section 4.5 we introduced the first of three methods for computing minimax procedures, the method that makes use of Bayes procedures in the manner described there. We called that approach *Method A*. We now discuss the other two approaches. Although the mathematical details are optional reading, the main ideas, especially of Method C and the tool of Remark 1 at the end of this section, are important.

Method B: The Information Inequality Method

Suppose Ω is an interval of real numbers θ, that we are to estimate $\phi(\theta)$, and that $W(\theta, d) = q(\theta)[\phi(\theta) - d]^2$, where q is a given positive function of θ (this generalizes the squared error loss function). According to (7.9) for any procedure t we have

$$\begin{aligned} r_t(\theta) &= q(\theta)E_\theta[t(X) - \phi(\theta)]^2 \\ &= q(\theta)\{b_t(\theta)^2 + \text{var}_\theta(t)\} \geq q(\theta)\left\{b_t(\theta)^2 + \frac{[\psi_t'(\theta)]^2}{I(\theta)}\right\} \qquad (7.42) \\ &= H_t(\theta)\,(\text{say}). \end{aligned}$$

We can compute $H_t(\theta)$ explicitly for each t. Suppose we have a procedure t^* with constant risk $r_{t^*}(\theta) = c$. According to the theorem at the end of Section

4.5, we can conclude that t^* is minimax if we can prove that t^* is admissible. We would accomplish this if we could show that for any procedure t which is *at least as good as* t*, it is also true that t^* is *at least as good as* t, so that t and t^* are actually *equivalent*; for then there is no t better than t^*, and t^* is therefore admissible. In symbols, we must prove that

$$\begin{aligned} &\text{if} \quad r_t(\theta) \le c \quad \text{for all } \theta, \\ &\text{then} \quad r_t(\theta) \ge c \quad \text{for all } \theta. \end{aligned} \tag{7.43}$$

Suppose we can prove that

$$\begin{aligned} &\text{if} \quad H_t(\theta) \le c \quad \text{for all } \theta, \\ &\text{then} \quad H_t(\theta) = c \quad \text{for all } \theta. \end{aligned} \tag{7.44}$$

Since (according to (7.42)) $r_t(\theta) \ge H_t(\theta)$, the premise of (7.43) implies the premise of (7.44), and the conclusion of (7.44) implies the conclusion of (7.43). If (7.44) is proved, the premise of (7.43) thus implies the conclusion of (7.43). Thus, t^* will be proved admissible if we can prove (7.44).

Of course, (7.43) may be (and often is) satisfied without (7.44) being satisfied. However, (7.43), which is just the admissibility of t^*, is hard to prove in general, and the seemingly stronger result (7.44), since it is analytically more explicit, can be proved in certain special cases.

The cases in which it has been possible to prove the minimax character of certain t^*'s by proving (7.44) are ones in which $X = (X_1, \ldots, X_n)$, the X_i's have a Koopman-Darmois law, $\phi(\theta)$ is $E_\theta k(X_1)$ (where $\sum_1^n k(X_i)$ is the minimal sufficient statistic we have discussed before), and $q(\theta)$ is of a very special form. We shall demonstrate the method in detail in one special case:

Theorem. *Suppose* $\mathrm{X} = (\mathrm{X}_1, \ldots, \mathrm{X}_\mathrm{n})$ *where the* X_i's *are independently and identically distributed normal random variables with variance one and unknown mean* θ, $\Omega = \{\theta : -\infty < \theta < \infty\}$, $\mathrm{D} = \{\mathrm{d} : -\infty < \mathrm{d} < \infty\}$, *and* $\mathrm{W}(\theta, \mathrm{d}) = (\theta - \mathrm{d})^2$. *Then* $\bar{\mathrm{X}}_\mathrm{n}$ *is an admissible (and in fact, the unique) minimax estimator.*

Proof. In this case we have $I(\theta) = n$ and $\psi_t'(\theta) = \partial(\theta + b_t(\theta))/\partial\theta = 1 + b_t'(\theta)$. Also, since $t^* = \bar{X}_n$, we have $c = E_\theta(\bar{X}_n - \theta)^2 = 1/n$. Hence, the premise of (7.44) is

$$b_t(\theta)^2 + [1 + b_t'(\theta)]^2/n \le 1/n \quad \text{for all } \theta. \tag{7.45}$$

Since both terms on the left side of (7.45) are nonnegative, we must have

$$b_t'(\theta) \le 0 \quad \text{for all } \theta, \tag{7.46}$$

since if $b_t'(\theta) > 0$ for any θ the second term on the left side of (7.45) would by itself be greater than the right. By (7.46), $b_t(\theta)$ is a nonincreasing function of θ. We shall prove in the next paragraph that

$$\lim_{\theta \to \infty} b_t(\theta) = 0. \tag{7.47}$$

In a similar fashion, it can be shown that

$$\lim_{\theta \to -\infty} b_t(\theta) = 0. \tag{7.48}$$

Clearly, the only function $b_t(\theta)$ which is nonincreasing and which satisfies (7.47) and (7.48) is the function

$$b_t(\theta) = 0 \quad \text{for all } \theta. \tag{7.49}$$

But for this function, the corresponding $H_t(\theta)$ is $1/n = c$ for all θ, proving the conclusion of (7.44).

In order to prove (7.47) we shall first show that there is a sequence of numbers $\{\theta_j\}$ with $\theta_j \geq j$ and such that

$$b_t'(\theta_j) \geq -1/j \quad \text{for } j = 1, 2, \ldots. \tag{7.50}$$

If no such sequence exists, then there must exist a positive integer j_0 such that

$$b_t'(\theta) < -1/j_0 \quad \text{for all } \theta \geq j_0, \tag{7.51}$$

since if no such j_0 exists we can for each j find a $\theta_j \geq j$ for which (7.50) is satisfied. Now (7.51) implies that, for $\theta > j_0$,

$$b_t(\theta) = b_t(j_0) + \int_{j_0}^{\theta} b_t'(x)\,dx < b_t(j_0) - \frac{\theta - j_0}{j_0},$$

so that

$$\lim_{\theta \to \infty} b_t(\theta) = -\infty.$$

This last is impossible, since by (7.45) we have that $b_t(\theta)^2 \leq 1/n$ for all θ. Thus, we have proved that there does exist a sequence $\{\theta_j\}$ satisfying (7.50) and with $\lim_{j \to \infty} \theta_j = \infty$ (since $\theta_j \geq j$). From (7.45) we thus have

$$b_t(\theta_j)^2 \leq n^{-1}[1 - (1 + b_t'(\theta_j))^2] \leq n^{-1}[1 - (1 - j^{-1})^2]. \tag{7.52}$$

As $j \to \infty$, the last expression approaches zero, as does therefore the nonnegative left side. Thus, $b_t(\theta_j) \to 0$ on a sequence of points $\{\theta_j\}$ tending to infinity. Since $b_t(\theta)$ is nonincreasing in θ, we must therefore have $b_t(\theta) \to 0$ as $\theta \to \infty$, which is just (7.47). Thus, our result that $\bar{X}_n$ is admissible (hence, minimax) is proved.

Also, we have shown that any minimax t must have $b_t(\theta) \equiv 0$. But then t is unbiased, and it follows from the results of Section 7.2 (or from the results of Section 7.1) that only $\bar{X}_n$, among all unbiased estimators, has risk $1/n$ for all θ. This proves that $\bar{X}_n$ is indeed the unique minimax estimator. □

Method C: The Invariance Method

Under certain conditions it can be proved that, for each procedure t, there is an invariant procedure t' for which

$$\max_F r_{t'}(F) \leq \max_F r_t(F). \tag{7.53}$$

Hence, if t is minimax, so is t'. Thus, under these conditions *there is an invariant procedure which is minimax among all procedures; in particular, if a best invariant procedure exists, it is minimax.* (This is called the *Hunt–Stein Theorem.*)

This result is not always true, as is illustrated in Example 7.5(a), (v), and remark 5 where, by considering transformations in addition to those considered in the location-parameter problem, we obtained a single (hence, best) invariant procedure $t(x) = x$ which is not minimax. However, the result stated at the end of the preceding paragraph is true in many important settings (e.g., in most location-parameter or scale-parameter problems) and is of great use when it is true.

The use of the result is simple: in a problem falling into the framework of Example 7.5(a), for example, we have only to choose b so as to minimize (7.28), and we then know that $t^*(x) = b + x$ is minimax. There is no subtlety of having to guess a sequence $\{\xi\}$ and to compute $\min_t R_t(\xi_n)$ as in Method A, and no restriction to squared error loss or Koopman-Darmois laws as in Method B. Where a best invariant procedure exists and the problem satisfies conditions ensuring that the Hunt–Stein Theorem's result is valid, the present method is by far the easiest for obtaining minimax procedures with a minimum of computation.

Where no best invariant procedure exists, as in Examples 7.7(a) and (b), invariance is of less use in finding a minimax procedure, unless (as in certain problems we shall discuss under hypothesis testing) it markedly reduces the class of procedures among which we must look for a minimax procedure (even though no uniformly best procedure exists among these invariant procedures).

Keep in mind that, as in Example 7.5(c), it is possible to use the present method by first invoking sufficiency and then invariance; this will often reduce the computations even more.

Proving the Hunt–Stein Theorem (Optional)

The idea of a proof that, for any t, there is an invariant t' for which (7.53) is satisfied, will now be sketched briefly in the location-parameter problem of Example 7.5(a). Suppose t is a given procedure. For each fixed real number $p(-\infty < p < \infty)$, the procedure

$$t'_p(x) = t(p) - p + x$$

is an invariant procedure (being just the t_b of (7.27) with $b = t(p) - p$). Hence, each t'_p has constant risk $r_{t'_p}(0)$. If we could show that (i) a probabilistic *average* of the numbers $r_{t'_p}(0)$ is $\leq \max_\theta r_t(\theta)$, it would follow that (ii) at least one of the individual numbers $r_{t'_p}(0)$ is $\leq \max_\theta r_t(\theta)$. If $r_{t'_{p^*}}(0)$ is such a number, it follows that (7.53) is satisfied with $t' = t'_{p^*}$.

Since a probabilistic average (over θ) of values $r_t(\theta)$ is $\leq \max_\theta r_t(\theta)$, the result (i) of the previous paragraph would follow if we could show that (iii) some probabilistic average (over p) of values $r_{t_p}(0)$ is $\leq$some probabilistic average (over θ) of values $r_t(\theta)$. For example, for $A > 0$ we may try to show that

$$\frac{1}{2A}\int_{-A}^{A} r_{t_p}(0)\,dp \le \frac{1}{2A}\int_{-A}^{A} r_t(\theta)\,d\theta. \tag{7.54}$$

Writing out $r_{t_p}(0) = E_0 w(t(p) - p + X)$ and $r_t(\theta) = E_\theta w(t(X) - \theta)$, the inequality (7.54) becomes

$$\begin{aligned} &\frac{1}{2A}\int_{-A}^{A}\int_{-\infty}^{\infty} w(t(p) - p + y)f^*(y)\,dy\,dp \\ &\le \frac{1}{2A}\int_{-A}^{A}\int_{-\infty}^{\infty} w(t(X) - \theta)f^*(x - \theta)\,dx\,d\theta. \end{aligned} \tag{7.55}$$

If we can interchange the order of integrating over x and θ in the last expression and we make the substitution $y = x - \theta, p = x$, this last expression becomes

$$\begin{aligned} &\frac{1}{2A}\int_{-\infty}^{\infty}\int_{-A}^{A} w(t(x) - \theta)f^*(x - \theta)\,d\theta\,dx \\ &= \frac{1}{2A}\int_{-\infty}^{\infty}\int_{-A+p}^{A+p} w(t(p) - p + y)f^*(y)\,dy\,dp. \end{aligned} \tag{7.56}$$

It seems intuitively reasonable that, as $A \to \infty$, the left side of (7.55) and the right side of (7.56) tend to the same limits, since the integrands and limiting ranges of integration (the whole (y, p) plane) are identical. The details of the proof establish this fact if w is not too unruly. Thus, one can not quite prove (7.54) but can prove that, as $A \to \infty$, the left and right sides of (7.54) approach the same limit. This limiting version of (7.54) suffices to show that some t'_p does the desired job.

In some cases the proof is much easier; if there are only finitely many transformations leaving the problem invariant (as in Example 7.7(a) and other examples discussed in hypothesis testing), the integration over p is replaced by a finite sum, and no limiting operation is required. However, the general proof in this case entails the use of randomized procedures if $\{h_k, k \in K\}$ is not transitive on Ω, and we shall not take the space to give details here.

Summary of Methods of Minimax Proof

The Bayes Method A with its modifications, works in almost all cases, but entails the guessing of a ξ or sequence $\{\xi_n\}$, and the computations are often messy. Sometimes Method A also proves admissibility. The information inequality Method B works for only a few Ω's and Ws; but, where it does work, it also yields the admissibility of the procedure, unlike Method C and (sometimes) A. The invariance Method C does not always yield much, but in many practical cases, such as scale-parameter or location-parameter problems, it yields the minimax result with a minimum of computation, and without the preliminary guesswork (of ξ or of the t^* to be proved minimax) needed in Method A or B. (The cases in which Methods B and C work to prove the minimax result are overlapping subsets of the cases where A works. However, Method B also proves admissibility in some cases where Method A does not.)

Additional Remarks on Minimax Procedures

1. *A minimax extension method*: A useful tool in many applications, particularly nonparametric ones, is that *a procedure* t* *is minimax for the problem* $\{S, \Omega, D, W\}$ *if, for some* Ω' *which is a subset of* Ω,
   (a) t* *is minimax for the problem* $\{S, \Omega', D, W\}$, *and*
   (b) $\max_{F \in \Omega} r_{t^*}(F) = \max_{F \in \Omega'} r_{t^*}(F)$.

   The proof is simply that for any procedure t the maximum of $r_t(F)$ for F in Ω is greater than or equal to the maximum for F in Ω', which by (a) is at least the (Ω') maximum of $r_{t^*}(F)$; and by (b) the Ω-maximum for t^* is the same, so that t^* was Ω-maximum risk less than or equal to that of t.

   The generalization of the Bayes method, given just before Example 4.4 in Section 4.5, can be viewed as a special case of the preceding result, Ω' being a set of F's on which $r_{t^*}(F)$ is constant and attains its maximum over Ω and to which the corresponding least favorable ξ assigns probability one and the original theorem of Section 4.5 applies. Thus, in Example 4.4 of Section 4.5, you could think of Ω' as being the set of two points $\{1/3, 2/3\}$, of t^* as having been proved minimax by using the original theorem of that section, with $r_{t^*}(\theta) = \text{constant}$ on Ω', and of t^* as having then been proved minimax by using the result of the previous paragraph, proving (b) by showing analytically that $r_{t^*}(\theta) < r_{t^*}(1/3)$ for $1/3 < \theta < 2/3$.

   In other applications, too, Ω' can be thought of intuitively as a subset of Ω which is "hardest for the statistician to handle" (and on which a malevolent nature would concentrate its least favorable ξ, in the spirit of the discussion in Section 4.5).

   As a nonparametric example of the use of this technique, suppose $X = (X_1, \ldots, X_n)$, the X_i's being uncorrelated with unknown common mean $\phi(F)$ and known common variance σ^2, Ω consisting of *all* F's with these properties. It is desired to estimate $\phi(F)$, with $W(F, d) = (\phi(F) - d)^2$. If we let $t^*(x) = \bar{x}_n$ and let Ω' consist of the *normal* F's, then (a) is satisfied, as was proved in the first theorem of this section, and (b) is satisfied since $r_{t^*}(F)$ is constant (σ^2/n) on Ω. Thus, t^* is minimax. This proves the statement in Section 5.1 justifying the use of linear estimators in this simplest nonparametric linear estimation problem. Best linear estimators can similarly be proved minimax in the more complicated general linear estimation problem considered in Section 5.2.
2. If σ^2 were unknown ($0 < \sigma^2 < \infty$) in the example of the last paragraph, then, even if we considered only the parametric (normal) problem, the

minimax criterion would yield little; for, every procedure t would have an *unbounded* risk function on Ω (since, for fixed σ, $\max_\mu r_t(\mu, \sigma^2) \geq \sigma^2/n$ by the previous result, and σ is now unbounded), and hence *every procedure is minimax*. There are many other problems of this sort: e.g., any scale-parameter problem, or the estimation of a Poisson mean, *if* W *is taken to be squared error* (not, as we have seen, if W is squared *relative* error in the scale-parameter problem). In an example like that of the previous paragraph, there is still a satisfactory solution: for, there is a single procedure t^* (equals to $\bar{x}_n$) which, for each $c > 0$, is the unique minimax procedure for the problem in which Ω is replaced by $\Omega_c = \{(\mu, \sigma^2): -\infty < \mu < \infty, \sigma^2 = c\}$. This is a very strong property. For, ordinarily, if we break up Ω into several subsets Ω_j, the minimax procedure for one Ω_j will not be the same as for another. That the *same* procedure t^* is the unique minimax procedure for each Ω_c in our example, where $\Omega = \bigcup_c \Omega_c$, makes t^* a satisfactory solution to our problem, even though every other t is also minimax for Ω (for no other t is minimax on Ω_c). Unfortunately, such a satisfactory approach by means of a modified minimax criterion is not always possible in problems in which every t has unbounded risk, since such a partition of Ω into subsets Ω_j, on each of which the same t^* is the unique minimax procedure, is not always possible.

3. There are problems for which the same procedure t^* is a good one for many different W's. Where this is so, the practical implications are pleasing: the customer does not have to know the exact form of W, but only that it is one of those in a broad class, in order to be told which procedure to use.

   An important example is that in which $X = (X_1, \ldots, X_n)$, the X_i's being independent and identically distributed with normal density, known variance σ^2, and unknown mean θ $(-\infty < \theta < \infty)$. If θ is to be estimated and $W(\theta, d)$ is *any symmetric function of* $|\theta - \mathrm{d}|$ *which is nondecreasing in* $|\theta - \mathrm{d}|$, then by part (iii) of Example 7.5(a) and the discussion of Example 7.5(b), $\bar{X}_n$ *is best invariant*; by the use of Method C of the present subsection, we see that it is *minimax*. (In addition, it is also *best unbiased*.) Thus, as long as the customer feels that underestimation and over-estimation are equally important and that the larger the magnitude of error of estimation, the greater the loss, he or she need know nothing more precise about W in order to justify using $\bar{X}_n$.

   If σ^2 is unknown, remark 2 indicates that the use of $\bar{X}_n$ is still well justified in this example.

7.5. The Method of Maximum Likelihood

The most commonly employed estimation procedure among practical people is the maximum likelihood procedure. Popularized by R. A. Fisher at the time of his criticism of the method of moments (see Section 4.8), this procedure has been so widely accepted that most non-Bayesian books on statistics present it as the *only* estimation procedure. This is unfortunate, since the procedure, although intuitively appealing, need not be a good one, as we shall see in examples later. In fact, there is no theorem which says that the procedure has any good properties other than the asymptotic (large sample) ones mentioned in Section 7.6; and where the maximum likelihood estimator *is* a good one in some sense, this goodness is not provable by any theorem which is a part of the maximum likelihood theory itself. Usually if the maximum likelihood estimator is good, the maximum likelihood estimator happens to coincide with the estimator obtainable by other methods like those of Sections 7.1–7.4. The definition of a maximum likelihood estimator makes no reference to any W, which automatically means that the estimator cannot possibly be satisfactory in all problems.

The maximum likelihood (M.L.) method is a prescription for producing an estimator $\hat{t}$ of F (or θ), called the *M.L. estimator*. The prescription in the discrete case is to consider, when $X = x$, the function

$$p_{F;X}(x),$$

thought of for fixed x as a function (the *likelihood function*) of F, and to let $\hat{t}$, an *M.L. estimator of* F, be any estimator such that $\hat{t}(x)$ is a value of the argument F which maximizes $p_{F;X}(x)$. Intuitively, this may seem reasonable, since "$\hat{t}(x)$ is a value of F for which the probability $p_{F;X}(x)$ that the observed outcome x would occur when F is true, is largest." In the continuous case, the definition is the same, with $f_{F;X}$ replacing $p_{F;X}$.

Note that we have defined "*an* M.L. estimator" and not "*the* M.L. estimator." This is because there need not be a unique maximizer of the likelihood. In fact there need not be any maximizer of the likelihood. This point is often not understood, with the result that (as in Example 7.10) one can find many examples in the literature in which a precise estimate (decision) is not even given, but instead it is stated that "the M.L. estimate is any value between .73 and .85," etc. The M.L. method simply does not always yield a unique estimator.

The definition of an M.L. estimator reminds one slightly of the Bayes approach, and there is a relationship. Suppose S, Ω, and D are finite with d_j the decision that F_j is the distribution function of X, for $F_j \in \Omega$. If W_{ij} is the loss when F_i is the case and d_j is taken, suppose $W_{ii} = 0$ and $W_{ij} = 1$ if $i \neq j$, $i, j = 1, 2, \ldots, k$. It is easy to check that an M.L. estimator of F is a Bayes procedure relative to the uniform a priori law on Ω. An analogous result for Ω an interval is treated in Problem 5.10.

Therefore, for the measure of loss described in the preceding paragraph, for a k decision problem, an M.L. procedure will be a good procedure but only because it is Bayes relative to an a priori distribution which places positive mass on each $F \in \Omega$ (see Section 4.4). If the measure of loss or if the a priori probability is changed, then a given M.L. procedure may no longer be Bayes.

Since every admissible procedure, when S, Ω, D are finite, is a Bayes procedure, this again suggests that an M.L. procedure cannot always be satisfactory.

In the literature, the M.L. estimator of θ is often denoted by $\hat{\theta}$. We shall denote it by $\hat{t}$ instead, to eliminate possible confusion between parameter values and estimates of them.

EXAMPLE 7.8. Let $X = (X_1, \ldots, X_n)$, where the X_i's are independent, identically distributed, and normal.

(i) If the variance v is known and the mean μ is unknown with $\Omega = \{\mu : -\infty < \mu < \infty\}$, we have

$$f_{\mu;X}(x) = (2\pi v)^{-n/2} e^{-\sum (x_i - \mu)^2/2v}.$$

As in many examples, it is easier to work with the logarithm of this quantity than with the quantity itself; since $\log f$ is strictly increasing in f for $f > 0$, any value of F which maximizes $\log f_{F;X}(x)$ also maximizes $f_{F;X}(x)$, and conversely. In the present case we have

$$\log f_{\mu;X}(x) = -\frac{n}{2}\log(2\pi v) - \sum (x_i - \mu)^2/2v.$$

Since $\sum (x_i - c)^2$ is a minimum for $c = \bar{x}_n$ (if this is not familiar, you can check it by differentiating and setting equal to zero and making sure that this yields a minimum), the preceding expression is a maximum for $\mu = \bar{x}_n$. Thus, $\hat{t}(x_1, \ldots, x_n) = \bar{x}_n$ is the unique M.L. estimator of μ in this case.

(ii) If μ is known but v is unknown ($\Omega = \{v : v > 0\}$), then differentiating the preceding logarithm of the density with respect to v yields

$$-\frac{n}{2v} + \sum (x_i - \mu)^2/2v^2,$$

and setting this equal to zero yields $v = \sum (x_i - \mu)^2/n$. The second derivative is easily checked to be negative at this value, so $\hat{t}(x) = \sum (x_i - \mu)^2/n$ is the unique M.L. estimator of v.

(iii) If both μ and v are unknown ($\Omega = \{(\mu, v) : -\infty < \mu < \infty, v > 0\}$), then whatever the value of v, the argument of (i) shows that $\mu = \bar{x}_n$ maximizes $f_{\mu,;X}(x)$ for fixed v. Substituting this value, the computation of (ii) then shows that $v = \sum (x_i - \bar{x}_n)^2/n$ maximizes $f_{\bar{x}_n, v; X}(x)$. Thus, $\hat{t}(x) = (\bar{x}_n, \sum (x_i - \bar{x}_n)^2/n)$ is the unique M.L. estimator of (μ, v) in this case.

(iv) If $X_1, \ldots, X_n$ are independent and identically distributed with common Koopman-Darmois density (or probability function)

$$f_{\theta;X_1}(x_1) = b(x_1)c(\theta)e^{k(x_1)q(\theta)},$$

then the minimal sufficient statistic $s(X) = \sum_1^n k(X_i)$ also has a (different)

Koopman-Darmois law. The proof in the discrete case is simply to write out

$$\begin{aligned} P_\theta\{s(X) = y\} &= \sum_{\{x:s(x)=y\}} \prod_{i=1}^{n} P_\theta\{X_i = x_i\} \\ &= \sum_{\{x:s(x)=y\}} [c(\theta)]^n \left[\prod_{i=1}^{n} b(x_i)\right] e^{q(\theta)\sum k(x_i)} \\ &= C(\theta)B(y)e^{q(\theta)y}, \end{aligned}$$

where $C(\theta) = [c(\theta)]^n$ and $B(y)$ is the sum of $\prod_1^n b(x_i)$ over the set $\left\{(x_1, \ldots, x_n) : \sum_1^n k(x_i) = y\right\}$. Thus, $p_{\theta; s(X)}(y)$ has a K-D form. Consequently, if we know how to compute an M.L. estimator for an *arbitrary* K-D law with $n = 1$, we know how to compute it for any K-D family, for each n. (See Problem 5.1 for that computation.)

We see that M.L. estimators need not be unbiased: referring to (iii) we have seen previously that $\sum(x_i - \bar{x}_n)^2/n$ is *not* an unbiased estimator of v. Similarly, in (ii) if we had written $v = \sigma^2$ and asked for the M.L. estimator of σ, we would have obtained $[\sum(x_i - \mu)^2/n]^{1/2}$, which is not an unbiased estimator of v. [Of course, our discussion of unbiasedness indicates that biasedness of an estimator is not too serious a criticism to raise against it, anyway.]

The last illustration, wherein we changed parameters from v to $\sigma = v^{1/2}$ and found that (M.L. estimator of σ) = (M.L. estimator of v)$^{1/2}$, is an example of a more general phenomenon: Suppose Ω is parametrized in terms of a parameter θ and we change to a labeling of elements of Ω in terms of $\phi = q(\theta)$, where q is a 1-to-1 function of θ. Thus, if $f^*_{\phi;X}$ is the density in terms of this new parametrization, we have

$$f^*_{q(\theta);X}(x) = f_{\theta;X}(x).$$

If, for fixed x, the right side is a maximum for $\theta = \theta_0$, then so is the left side; i.e., $f^*_{\phi;X}(x)$ is maximized by $\phi = q(\theta_0)$; and conversely. Thus, if $\hat{t}(x)$ is any M.L. estimator of θ, the estimator $q(\hat{t}(x))$ is an M.L. estimator of ϕ, and conversely. This property of M.L. estimators makes them attractive to practical workers; for example, compare the computation involved in obtaining an M.L. estimator of σ from one of $v = \sigma^2$ in the preceding example, to that involved in the corresponding problem of unbiased estimation (i.e., the computation of c_n in Section 7.2). Moreover, this property of M.L. estimators makes them satisfy the same appealing relationship that was stated in (7.15): changing the parametrization changes the decision in a consistent manner. Best unbiased estimators do not have this property (the best unbiased estimator of σ^2 not being the square of that of σ in our normal example).

Thus far we have considered only M.L. estimators of F (of θ), the space D being the same as Ω. We may instead want to estimate some function $\phi(F)$.

Motivated by the discussion of the previous paragraph (where ϕ is 1-to-1), for general ϕ (not necessarily 1-1), we define $\hat{u}$ (taking on values in the space D of all possible values of $\phi(F)$ for F in Ω) to be an *M.L. estimator of* $\phi(F)$ if $\hat{u}(x) = \phi(\hat{t}(x))$ where $\hat{t}$ is an M.L. estimator of F. For example, in Example 7.8(iii) we would call $\sum(x_i - \bar{x}_n)^2/n$ the M.L. estimator of $v = \phi(\mu, v)$ and $|\bar{x}_n|/[\sum(x_i - \bar{x}_n)^2/n]^{1/2}$ would be the M.L. estimator of $|\mu|/v^{1/2}$.

EXAMPLE 7.9. Suppose $X = (X_1, \ldots, X_n)$ where the X_i are independent and identically distributed with common density

$$f_{\theta; X_1}(x_1) = \begin{cases} 1/\theta & \text{if } 0 < x_1 \leq \theta, \\ 0 & \text{otherwise.} \end{cases}$$

Here $\Omega = \{\theta : \theta > 0\}$. Thus, for fixed $x = (x_1, x_2, \ldots, x_n)$, we have

$$f_{\theta; X}(x) = \begin{cases} 0 & \text{if } \theta < \max_i x_i, \\ \theta^{-n} & \text{if } \theta \geq \max_i x_i. \end{cases}$$

This function of θ is 0 up to the value $\theta = \max_i x_i$, where it jumps to the value $(\max_i x_i)^{-n}$. Thereafter it decreases continuously. The unique maximum is thus at $\theta = \max_i x_i$, so that $\hat{t}(x) = \max_i x_i$ is the unique M.L. estimator of θ in this case. Again, $\hat{t}$ is not unbiased. More important, $\hat{t}$ always *underestimates* θ, since $P_\theta\{\hat{t}(X) < \theta\} = 1$. For almost any reasonable W (which penalizes for underestimation as well as for overestimation), there will be estimators t' satisfying $t'(x) > \hat{t}(x)$ for all x which are uniformly better than $\hat{t}$.

Remark. Note from the preceding example that it is *not* always true that $\hat{t}(x)$ can be obtained as the solution θ of the equation $\partial f_{\theta; X}(x)/\partial\theta = 0$. This phenomenon occurs in some "regular" cases, too: For x Bernoulli, $p_\theta(0) = 1 - \theta$, $p_\theta(1) = \theta$, and for $\Omega = [0, 1]$ we obtain $\hat{t}(x) = x$, attained at an endpoint of Ω where $\left|\dfrac{\partial p_\theta(x)}{\partial\theta}\right| = 1$.

EXAMPLE 7.10. Suppose $X_1, \ldots, X_n$ are independent and identically distributed with $\Omega = \{\theta : -\infty < \theta < \infty\}$ and

$$f_{\theta; X_1}(x_1) = \begin{cases} 1 & \text{if } \theta - 1/2 \leq x_1 \leq \theta + 1/2, \\ 0 & \text{otherwise.} \end{cases}$$

Then, for fixed $x = (x_1, \ldots, x_n)$,

$$f_{\theta; X}(x) = \begin{cases} 1 & \text{if } \max_i x_i - 1/2 \leq \theta \leq \min_i x_i + 1/2, \\ 0 & \text{otherwise.} \end{cases}$$

Thus, any value of θ between $\max_i x_i - 1/2$ and $\min_i x_i + 1/2$ maximizes $f_{\theta; X}(x)$, so that there is not a *unique* M.L. estimator; instead, any value between these limits is a possible value of an M.L. estimator. Now, an estimator is a well-defined function from S into $D = \{d : -\infty < d < \infty\}$, so that we must write down a specific choice of the value to be taken on by $\hat{t}(x)$ for each x, for any such estimator. Examples of M.L. estimators are

$$\hat{t}_1(x) = \max_i x_i - 1/2,$$

$$\hat{t}^2(x) = (\max_i x_i + \min_i x_i)/2,$$

$$\hat{t}_3(x) = \max_i x_i - 1/2 + (\sin^2 x_1)[1 - \max_i x_i + \min_i x_i].$$

To say only "$\hat{t}(x)$ is some value between $\max_i x_i - 1/2$ and $\min_i x_i + 1/2$" does not define a point estimator of θ.

We have seen that the M.L. estimator may not be unique. What is worse, it may not even exist, because the maximum of $f_{\theta;X}(x)$ with respect to θ may not be attained. A trivial and inconsequential example of this phenomenon is obtained by redefining $f_{\theta;X_1}(x_1)$ in Example 7.9 to be 0 at the endpoint θ of the interval where the probability is concentrated; then $f_{\theta;X}(x)$ is 0 for θ *up to and including* the value $\max_i x_i$ and is θ^{-n} for $\theta > \max_i x_i$; thus, the maximum of this function of θ is not actually attained. In the absolutely continuous case, where the pdf is not uniquely defined, this phenomenon is common. It is eliminated if one chooses the density to be "upper semicontinuous," i.e., so that it takes on the larger possible value at a discontinuity. Some examples of a more serious nature, which cannot be rectified merely by redefining a pdf in the absolutely continuous case on a set of zero probability as was just done in an example, are the following:

EXAMPLE 7.11(a). Let X be a single Poisson random variable with mean θ, where $\Omega = \{\theta : \theta > 0\}$. X may represent some physical phenomenon which, we know, has a positive probability of taking on a value >0; hence, the value $\theta = 0$ is excluded as a possible state of nature. If $X = x > 0$, the function $p_{\theta;X}(x) = e^{-\theta}\theta^x/x!$ is easily seen to attain its unique maximum at $\theta = x$; thus, $\hat{t}(x) = x$ is the M.L. estimator. If $x = 0$, we have $p_{\theta;X}(0) = e^{-\theta}$, which is strictly decreasing for $\theta > 0$, and this function thus does not attain its maximum on Ω. By adjoining the physically impossible value $\theta = 0$ to Ω, we could also have the maximum attained in this case, so that $\hat{t}(0) = 0$; but it hardly seems reasonable to enlarge Ω to include states of nature which are known to be impossible, merely in order to be able to use an intuitively appealing method which could not otherwise be used to define an estimator: an estimator is chosen to suit a Ω, not conversely. [Examples of this kind can be constructed easily in many problems for which Ω is an interval of values which does not contain one or both of its endpoints.]

Another example of this kind: a coin is flipped once, the probability θ of a head's coming up being known to satisfy $0 < \theta < 1$, the values $\theta = 0$ and 1 being excluded because the coin, having a head on one side and a tail on the other, is known not to be *that* biased; since $p_{\theta;X}(0) = 1 - \theta$ and $p_{\theta;X}(1) = \theta$, and since neither of these functions attains its maximum on the interval $0 < \theta < 1$, an M.L. estimator *never* exists in this example; if the endpoints

$\theta = 0$ and 1 are adjoined to make $\Omega = \{\theta : 0 \le \theta \le 1\}$, we obtain $\hat{t}(x) = x$, so that the M.L. estimator has now been made to exist, but only at the expense of its always taking on a value (0 or 1) which is known to be impossible. In Example 7.11(b) it makes even less sense to adjoin an additional point to Ω merely to make the M.L. estimator exist.]

EXAMPLE 7.11(b). An important probability law in many biological settings is the *truncated Poisson law*. As an illustration of its occurrence, suppose a certain disease infects Y individuals in a given population during a specified week, Y having a Poisson law with mean θ. Ordinarily, health officials are not concerned with the disease (and with estimating θ) unless at least one case of the disease is reported. Thus, they work not with Y, but with a random variable X which is defined only when $Y > 0$, in which case $X = Y$. The probability law of X is thus the conditional probability law of Y given that $Y > 0$: for $x = 1, 2, \ldots,$

$$p_{\theta; X}(x) = P_\theta\{Y = x \mid Y > 0\} = \frac{e^{-\theta}\theta^x}{(1 - e^{-\theta})x!},$$

and $p_{\theta; X}(x) = 0$ otherwise. Here $\theta > 0$; if $\theta = 0$, we have $P_0\{Y > 0\} = 0$, so X would never be defined (the disease would never be observed). Suppose, then, that $\Omega = \{\theta : \theta > 0\}$. If $X = 1$, we obtain

$$p_{\theta; X}(1) = \theta e^{-\theta}/(1 - e^{-\theta}) \text{ and } \partial p_{\theta; X}(1)/\partial\theta = [(1 - \theta)e^{-\theta} - e^{-2\theta}]/(1 - e^{-\theta})^2.$$

Since the graph of the function $g(\theta) = e^{-\theta}$ is tangent to that of the line $L(\theta) = 1 - \theta$ at $\theta = 0$ and $d^2e^{-\theta}/d\theta^2 \ge -1 = d^2L(\theta)/d\theta^2$, we see that $e^{-\theta} > 1 - \theta$ for all $\theta > 0$. Hence, the preceding derivative is <0 for all $\theta > 0$; that is, $p_{\theta; X}(1)$ *is strictly decreasing in* $\theta > 0$. Hence, the M.L. estimator $\hat{t}(x)$ is undefined when $x = 1$. Moreover, adjoining the value $\theta = 0$ to Ω and defining $p_{\theta; X}(1)$ as the limit ($= 1$) of $p_{\theta; X}(1)$ as $\theta \to 0$, simply in order to make the M.L. estimator exist, makes even less sense in this case than in the previous example; for, as we have remarked, the probability is zero that the disease would ever be observed if $\theta = 0$.

EXAMPLE 7.11(c). There are many examples in which, aside from the questionable practical meaning of the adjoining of a parameter value such as $\theta = 0$ in Example 7.11(b), it is not possible to do so while maintaining the desired continuity of the parametrization. Thus, for the geometric law $P_\theta(x_1) = (1 - \theta)\theta^{x_1}$, $\lim_{\theta \to 1} P_\theta(x_1) \to 0$ for each x_1, the probability mass "escaping to $+\infty$," so the limit P_1 defined by $P_1(x_1) \equiv 0$ is not a probability law. In this setting there is no difficulty of the M.L. estimator's failing to exist for $\Omega = [0, 1)$, but let us alter the example to make the density $P_\theta^*(x_1) = \left(1 - \frac{\theta}{2}\right)P_\theta(x_1) + \frac{\theta}{2}\tilde{P}(x_1)$, where P_θ is the geometric density and $\tilde{P}$ assigns all mass to $x_1 = 1$; since P_θ^* is a probability mixture of two probability laws, it is also a probability law for $0 \leq \theta < 1$, but $\lim_{\theta \to 1} P_\theta^*$ is not a probability law because of the geometric part, discussed previously. On the other hand, if $x_1 = 1$ (when $n = 1$) we obtain $P_\theta^*(1) = \frac{1}{2}(3\theta - 3\theta^2 + \theta^3)$, which is strictly increasing on $[0, 1)$ since its derivative is $3(1-\theta)^2/2$. Thus, $\hat{t}(1)$ does not exist, and $\{P_\theta^*, 0 \leq \theta < 1\}$ cannot be extended continuously to include $\theta = 1$.

Remarkably enough, there are cases in the literature in which statisticians wax mystical over the fact that, even in "nice" cases unlike those of Examples 7.9 and 7.10 (which are remarked upon just before the latter), the equation $\partial p_{\theta;X}(x)/\partial\theta = 0$ does not have a solution. The explanation is, of course, merely that the phenomenon of Examples 7.11 is present.

The equation $\partial \log p_{\theta;X}(x)/\partial\theta = 0$ is often referred to as the *likelihood equation.* (If θ has several real components θ_i, as in part (iii) of Example 7.8, this equation is replaced by the set of equations $\partial \log p_{\theta;X}(x)/\partial\theta_i = 0$, $i = 1, 2, \ldots$). In many common examples (like Examples 7.8), it is easy to solve; however, this is not always the case. For example, when $X = (X_1, \ldots, X_n)$ where the X_i's are iid with Cauchy density $f_{\theta;X_1}(x_1) = 1/\pi[1 + (x_1 - \theta)^2]$ and $\Omega = \{\theta: -\infty < \theta < \infty\}$, the likelihood equation is

$$\sum_{i=1}^{n} \frac{(x_i - \theta)}{1 + (x_i - \theta)^2} = 0,$$

which can be rewritten as a polynomial equation of degree $2n - 1$ in θ. When n is not very small, this would be tedious to solve without a computer; moreover, there may be many solutions, and the value of $f_{\theta;X}(x)$ must be computed for each solution θ of the likelihood equation, to find which solution yields a global maximum (that solution being $\hat{t}(x)$) and which ones are only local maxima. It is important in such problems not merely to obtain *a* solution to the likelihood equation, but to obtain the right one.

The computational aspects of the example of the previous paragraph indicate again the possibility that, even when the M.L. estimator (or some other estimator, as mentioned in connection with Example 7.5(c)) is good, a less accurate but more easily computable estimator may be preferred. We shall discuss this in more detail in Section 7.6.

One interesting aspect of M.L. estimation is that, in the setting of linear estimation in Section 5.2, if the X_i's are independent and normal with common variance σ^2, then it is easy to show that the Markoff (best linear unbiased) estimator of $\theta = (\phi_1, \ldots, \phi_k)$ coincides with the M.L. estimator; in fact, the logarithm of the likelihood is simply

$$-\frac{n}{2}\log(2\pi\sigma^2) - Q/2\sigma^2,$$

where Q is the sum of squares we minimized in Section 5.2 in order to obtain the least squares (best linear unbiased) estimators, which proves the result. Moreover, if σ^2 is unknown, an argument like that of Example 7.8(c) shows that n^{-1} times the residual sum of squares (see the end of Section 5.2) is the

M.L. estimator of σ^2; as in Example 7.8(c), this estimator is not unbiased. In some books the equivalence of M.L. and least squares estimation in the normal case is used as a justification for the method of least squares, without anything (other than intuition) being given as a justification of M.L. estimation. (Moreover, M.L. and least squares estimators need not coincide in other parametric cases in which the X_i's are not normal, as is already evident when $k = 1$ and the X_i's are independent and identically distributed with common mean $\theta = \phi_1$.)

You may have noticed the appearance of the minimal sufficient statistic in Examples 7.8, 7.9, and 7.10. These examples reflect a general phenomenon, often inaccurately stated as "the M.L. estimator is a function of the minimal sufficient statistic," the statement then being incorrectly interpreted (see Section 6.1) as a justification for the M.L. estimator. In fact, the estimator $\hat{t}_3$ of Example 7.10 illustrates the fact that an M.L. estimator may depend on X other than through the minimal sufficient statistic. What is true is that *if the M.L. estimator* $\hat{\mathrm{t}}$ *is unique, then for every sufficient statistic* s *(in particular, for a minimal one),* $\hat{\mathrm{t}}$ *depends on* X *only through* s(X); *in any event, if an M.L. estimator exists, then there exists an M.L. estimator which depends only on* s(X). The proof is quite simple: the Neyman decomposition theorem (see Section 6.2) states that we can write

$$p_{\theta;X}(x) = h(x)g(\theta, s(x)).$$

(We can assume that h and g are nonnegative functions since $0 \le p = hg = |h|\,|g|$.) Hence, for fixed x such that $p_{\theta;X}(x) > 0$ for some θ, $h(x)$ must be positive and thus if we choose θ to maximize $g(\theta, s(x))$, then we also maximize $h(x)g(\theta, s(x))$, and conversely; if for every x there is a unique maximizing value, it can only depend on $s(x)$, and in any event we can choose it to depend only on $s(x)$ (by letting $q(u)$ be such that the value $\theta = q(u)$ maximizes $g(\theta, u)$, and then letting $\hat{t}(x) = q[s(x)]$).

It is not hard to show that, if we replace $p_{\theta;X}$ by the probability function $p_{\theta;s}$ of a sufficient statistic s and compute an M.L. estimator based on the latter probability function, then the result could also have been obtained as an M.L. estimator based on the original probability function.

Similarly it can be shown that *if the M.L. estimator is unique, and if there are transformations* $(\mathrm{g_k}, \mathrm{h_k}, \mathrm{j_k})$, $\mathrm{k} \in \mathrm{K}$ *(see Section* 7.3*) which leave the specification of the problem unchanged, then the M.L. estimator will be an invariant function.* We show this in the case S is finite or countable and X has a probability function $p_{F;X}$. Our hypothesis of uniqueness is, then, for each $x \in S$ there is a unique $\hat{t}(x) \in \Omega$ such that

$$p_{\hat{t}(x);X}(x) = \max_{F \in \Omega} p_{F;X}(x).$$

Observe that

$$\begin{aligned} p_{F;X}(g_k(x)) &= P_F(X = g_k(x)) = P_F(g_k^{-1}(X) = x) \\ &= P_{h_k^{-1}(F)}(X = x) = p_{h_k^{-1}(F);X}(x). \end{aligned}$$

The function $p_{F;X}(g_k(x))$ is maximized by taking $F = \hat{t}(g_k(x))$. The function $p_{h_k^{-1}(F);X}(x)$ is maximized by taking $F = h_k(\hat{t}(x))$. These must be equal since both are the same function of F; that is,

$$\hat{t}(g_k(x)) = h_k(\hat{t}(x)).$$

In Section 7.6 we shall discuss the large-sample properties of M.L. estimators. We have mentioned earlier that M.L. estimators do not generally have any other good properties which are guaranteed by the M.L. theory itself, and it is easy to give examples to illustrate this fact. Two examples are now discussed.

EXAMPLE 4.4 (Continued). In Example 4.4 we had a single Bernoulli variable X with $\Omega = \{\theta : 1/3 \leq \theta \leq 2/3\}$ and $W(\theta, d) = (d - \theta)^2$. Since $p_{\theta;X}(1) = \theta$ and $p_{\theta;X}(0) = 1 - \theta$, the M.L. estimator is given by $\hat{t}(0) = 1/3$, $\hat{t}(1) = 2/3$. In Example 4.4 previously we have shown $t_{4/9}(0) = 4/9$, $t_{4/9}(1) = 5/9$ is Bayes and minimax.

Both of these estimators are of the form $t_a(0) = a$, $t_a(1) = 1 - a$ with $1/3 \leq a \leq 1/2$. We compute the risk function of t_a

$$\begin{aligned} r_{t_a}(\theta) &= (a - \theta)^2(1 - \theta) + ((1 - a) - \theta)^2\theta \\ &= (a - 1/2)^2 + (\theta - 1/2)^2 - 4(1/2 - a)(1/2 - \theta)^2. \end{aligned}$$

Therefore $r_{\hat{t}}(\theta) = 1/36 + (1/3)(\theta - 1/2)^2$ and $r_{t_{4/9}}(\theta) = 1/324 + (7/9)(\theta - 1/2)^2$. The extreme case is $t_{1/2}$, which always estimates $1/2$ for θ and has risk $r_{t_{1/2}}(\theta) = (\theta - 1/2)^2$. It is easy to show that

$$\begin{aligned} r_{\hat{t}}(\theta) - r_{t_a}(\theta) &= (1/6)^2 - (a - 1/2)^2 + 4(1/3 - a)(\theta - 1/2)^2 \\ &= (a - 1/3)[(2/3 - a) - (1 - 2\theta)^2]. \end{aligned}$$

If $1/3 \leq \theta \leq 2/3$, then $|1 - 2\theta| \leq 1/3$. Since $1/3 \leq a \leq 1/2$ it follows that

$$(2/3 - a) \geq 1/6 > 1/9 \geq (1 - 2\theta)^2.$$

Therefore if $1/2 \geq a > 1/3$ and if $1/3 \leq \theta \leq 2/3$,

$$r_{\hat{t}}(\theta) > r_{t_a}(\theta).$$

The M.L. estimator is therefore not admissible.

EXAMPLE 7.5(a) (Continued).

(vi) We have supposed that S is the set of real numbers, $\Omega = \{\theta : -\infty < \theta < \infty\}$ and X has a density function $f_{\theta;X}(x) = f^*(x - \theta)$ for all real x, θ. It is possible to construct functions f^* which are continuous, satisfying $f^*(x) \geq 0$ for all real x, $\int_{-\infty}^{\infty} f^*(x)\,dx = 1$ and yet to have there exist a sequence of numbers $x_1, x_2, x_3, \ldots$ such that $f^*(x_n) = n$. [You should try to construct for yourself an example of such a function.] It is clear that no M.L. estimator exists since $\sup_{-\infty<\theta<\infty} f^*(x-\theta) = \infty$ for every real x, but $f^*(x-\theta) < \infty$ for each x and θ.

If we suppose f^* is a bounded and continuous function with $f^*(x) \to 0$ as $x \to \pm\infty$, then for each value x there will be numbers $\hat{t}(x)$ such that

$$f^*(x - \hat{t}(x)) = \max_{-\infty<\theta<\infty} f^*(x-\theta).$$

It is clear that $\hat{t}(x)$ is uniquely determined exactly when f^* has a unique maximum.

Note that

$$f^*(x + a - \hat{t}(x+a)) = f^*(x - (\hat{t}(x+a) - a)).$$

Therefore if $\hat{t}(x)$ is unique, $\hat{t}(x+a) = a + \hat{t}(x)$ follows; that is, $\hat{t}$ will be an invariant function. As shown in Section 7.3, $\hat{t}(x) = x + b$ for some b follows. The constant b is determined only by f^* but is completely independent of the way loss is measured. But to find a best invariant procedure one must choose a constant a so as to minimize $\int_{-\infty}^{\infty} w(x+a) f^*(x)\,dx$. That is to say, whether the M.L. estimator is a best invariant estimator depends on the particular choice of a loss function.

7.6. Asymptotic Theory

The distributions and risk function of an estimator t are often difficult to compute. When the sample size is large, the central limit theorem gives a simple approximation to certain probability laws whose exact form would be hard to write out analytically (e.g., by repeated use of the convolution formula). Thus, we might expect that the use of this and similar theorems will enable us to simplify the computation of the risk function of certain procedures, when the sample size is large. Such approximations will be discussed in the subsection on limiting distributions of certain sequences of statistics.

At the same time, as mentioned in Sections 7.2 and 7.5, there is a sense in which the M.L. estimator is, for large sample sizes, "almost uniformly best" in certain problems, and this will be discussed in the subsection on asymptotic optimality properties, along with the often used asymptotic notions of "consistency" and "efficiency."

Finally, as we have mentioned before, it may be that a good procedure is difficult to use computationally, and that a less good one will be preferred in practice (or, more precisely, that when the cost or effort of computation is included, then the latter procedure becomes better than the former). This can be true for small sample sizes, but many important examples (which are easy to analyze because of the results of these subsections) occur in the asymptotic theory (large sample size) and will be discussed in the subsection on the use of inefficient procedures.

An important aspect of the asymptotic theory which should always be kept in mind is that it may not be applicable unless the sample size n is extremely large. Just as the central limit theorem can be seen to be a reasonable approximation when $n = 100$ in some cases but only when $n = 10{,}000$ in other cases (e.g., depending on the value of $p = P\{X_1 = 1\}$ in the Bernoulli case), so the results of the following two subsections have the same character. Moreover, although it is often easy in the case of the central limit theorem to give an explicit bound which tells how large n must be in order that the normal approximation have a given accuracy, no such simple explicit bound is generally available for the normal approximation (when it applies) to the distribution of the M.L. estimator (to which the central limit theorem may not apply directly). Thus, particular caution is needed in deciding that n is large enough in a given example so that the M.L. estimator *is* fairly good. (We saw in Section 7.5 that the M.L. estimator need not be good for "small" n.)

The results of this section, like those of Sections 7.4 and 7.5, have very similar counterparts in other statistical problems, such as hypothesis testing.

Limiting Distributions of Certain Sequences of Statistics

Suppose $X_1, X_2, \ldots$ are independently and identically distributed and that t_n is a function of $X_1, X_2, \ldots, X_n$. We shall discuss the limiting distribution of certain sequences $\{t_n\}$ as $n \to \infty$.

The Sample Moments. We have discussed for $j \geq 1$, the *raw sample moments*

$$m'_{j,n} = n^{-1} \sum_{i=1}^{n} X_i^j$$

and, for $j > 1$, the *central sample moments*

$$m_{j,n} = n^{-1} \sum_{i=1}^{n} (X_1 - \bar{X}_n)^j,$$

where of course $\bar{X}_n = m'_{1,n}$. Writing $\mu'_{Fj} = E_F X_1^j$ and $\mu_{Fj} = E_F(X_1 - \mu'_{F1})^j$ for the corresponding population moments, we saw that $Y_i = X_i^j$ has expectation μ'_{Fj} when the X_i have probability law F, and that the Y_i's are independently and identically distributed, so that, if μ'_{Fj} exists and is finite, the law of large numbers implies that the sequence $\{m'_{j,n}, n = 1, 2, \ldots\}$ converges stochastically to μ'_{Fj}; that is,

$$P_F(|m'_{j,n} - \mu'_{Fj}| < \varepsilon) \to 1$$

as $n \to \infty$, for each $\varepsilon > 0$. If, also, $\mu'_{F,2j}$ is finite, then Y_i has finite variance $\mu'_{F,2j} - (\mu'_{F,j})^2$, so that, by the central limit theorem, $n^{1/2}(m'_{j,n} - \mu'_{Fj})$ is asymptotically normal with mean 0 and variance $\mu'_{F,2j} - (\mu'_{F,j})^2$; that is,

$$P_F\left(\frac{n^{1/2}(m'_{j,n} - \mu'_{Fj})}{[\mu'_{F,2j} - (\mu'_{F,j})^2]^{1/2}} < z\right) \to \int_{-\infty}^{z} (2\pi)^{-1/2} e^{-x^2/2}\, dx$$

as $n \to \infty$, for each real z.

Although $m'_{j,n}$ has mean μ'_{Fj} and variance $n^{-1}[\mu'_{F,2j} - (\mu'_{F,j})^2]$, the exact expressions for the mean and variance of $m_{j,n}$ are not so simple. For example, we saw earlier that $E_F m_{2,n} = (n-1)\mu_{F,2}/n$, and the expressions for $j > 2$ are increasingly complex. (See, e.g., Parzen's book, p. 370, Ex. 4.6.) However, it can be shown that, if $\mu_{F,j}$ is finite, then $E_F m_{j,n}$ differs from $\mu_{F,j}$ by an expression which is no larger than some constant multiple of n^{-1}, which expression (the bias) hence approaches zero rapidly as $n \to \infty$. It can easily be proved, even though $m_{j,n}$ is not an average of *independent* random variables, that, since $\{\bar{X}_n\}$ converges stochastically to $\mu'_{F,1}$, the sequence $\{m_{j,n}\}$ converges stochastically to $\mu_{F,j}$ (if $\mu_{F,j}$ is finite); that is, $n^{-1}\sum_1^n (X_i - \bar{X}_n)^j$ converges stochastically to the same limit as the average $n^{-1}\sum_1^n (X_i - \mu_{F,1})^j$ of independent summands:

$$P_F(|m_{jn} - \mu_{F,j}| < \varepsilon) \to 1$$

as $n \to \infty$ for each $\varepsilon > 0$. If $\mu_{F,2j}$ is finite, we find similarly that $\mathrm{var}_F(n^{1/2} m_{j,n})$ differs from $\mathrm{var}_F(n^{-1/2}\sum_1^n (X_i - \mu'_{F1})^j) = \mu_{F,2j} - (\mu_{Fj})^2$ by a term of order n^{-1}, and we can obtain the same limiting distribution for the average m_{jn} of the slightly dependent random variables as we would for the average $n^{-1}\sum (X_i - \mu'_{F1})^j$ of independent summands (each with mean μ_{Fj} and variance $\mu_{F,2j} - (\mu_{F,j})^2$), namely,

$$P_F\left(\frac{n^{1/2}(m_{jn} - \mu_{Fj})}{[\mu_{F,2j} - (\mu_{F,j})^2]^{1/2}} < z\right) \to \int_{-\infty}^{z} (2\pi)^{-1/2} e^{-x^2/2}\, dx$$

as $n \to \infty$, for every z.

Sample Quantiles. We have the *order statistics* $Y_1 \leq Y_2 \leq \cdots \leq Y_n$ obtained by reordering $X_1, X_2, \ldots, X_n$, and for $0 < \alpha < 1$, the *sample quantiles,*

$$Z_{\alpha,n} = Y_{[\overline{\alpha n}]},$$

where $[\overline{\alpha n}]$ denotes the smallest integer $\geq \alpha n$. You may recall that $Y_{[\overline{\alpha n}]}$ is the α-tile (α^{th} quantile) of the *sample (or empiric) distribution function*

$$\bar{F}_n(x) = n^{-1} \text{ (number of } X_i \leq x, 1 \leq i \leq n).$$

(If αn is not an integer, $Y_{[\overline{\alpha n}]}$ is the unique α-tile; if $[\alpha n]$ is an integer, any value y with $Y_{\alpha n} \leq y \leq Y_{\alpha n+1}$ satisfies our definition of an α-tile of S_n. For definiteness, we have defined $Z_{\alpha,n}$ to be $Y_{\alpha n}$ in this case, but any other allowable choice would yield the same asymptotic results that follow.) Recall Section 4.8 for the intuition behind use of a sample quantile to estimate a corresponding population quantile.

The exact expressions for the probability laws of the order statistics are rather complicated in many cases, and simple formulas for probabilities, means, variances, etc., may not be available. When n is large, however, useful approximations are again available. If the probability law of X_1 has a *unique* α-tile $\gamma_{F,\alpha}$, it is easy to see that $Z_{\alpha,n}$ *converges stochastically to* $\gamma_{F,\alpha}$ *as* $n \to \infty$.

[PROOF. The uniqueness of the α-tile implies that, for every $\varepsilon > 0$, there is a $\delta > 0$ such that

$$P_F(X_1 \le \gamma_{F,\alpha} - \varepsilon) \le \alpha - \delta.$$

Let $U_i = 1$ or 0 according to whether or not $X_i \le \gamma_{F,\alpha} - \varepsilon$, so that $\sum_1^n U_i =$ [number of $X_i \le \gamma_{F,\alpha} - \varepsilon, 1 \le i \le n$]. We then have the following equivalent descriptions of the event $\{Z_{\alpha,n} \le \gamma_{F,\alpha} - \varepsilon\}$:

$$\begin{aligned}\{Z_{\alpha,n} \le \gamma_{F,\alpha} - \varepsilon\} &= \{Y_{[\overline{\alpha n}]} \le \gamma_{F,\alpha} - \varepsilon\} \\ &= \{\text{at least } [\overline{\alpha n}]\ X_i\text{'s are} \le \gamma_{F,\alpha} - \varepsilon\} = \left\{\sum_1^n U_i \ge [\overline{\alpha n}]\right\}.\end{aligned} \tag{7.57}$$

Since the U_i's are Bernoulli random variables with common means $\le \alpha - \delta$, the random variables $n^{-1}\sum_1^n U_i$ converge stochastically (by the law of large numbers) to $E_F U_1 \le \alpha - \delta$, so that ($[\overline{\alpha n}]$ being within 1 unit of αn) for the event (7.57) we have

$$P_F\left(\sum_1^n U_i \ge [\overline{\alpha n}]\right) \to 0$$

as $n \to \infty$. Similarly, $P_F(Z_{\alpha,n} \ge \gamma_{F,\alpha} + \varepsilon) \to 0$ as $n \to \infty$, so that we finally have, for each $\varepsilon > 0$,

$$P_F(|Z_{\alpha,n} - \gamma_{F,\alpha}| < \varepsilon) \to 1 \quad \text{as} \quad n \to \infty.]$$ □

We now investigate the asymptotic normality of $n^{1/2}(Z_{\alpha,n} - \gamma_{F,\alpha})$, under the assumption that X_1 has a density function $f_{F;X}$ which is positive and continuous in an interval about $\gamma_{F,\alpha}$.

Using (7.57) with $\varepsilon = -n^{-1/2}z$ where z is any real number, we have

$$\begin{aligned}P_F(n^{1/2}(Z_{\alpha,n} - \gamma_{F,\alpha}) \le z) &= P_F(Z_{\alpha,n} \le \gamma_{F,\alpha} + n^{-1/2}z) \\ &= P_F\left(\sum_1^n U_i \ge [\overline{\alpha n}]\right),\end{aligned} \tag{7.58}$$

where the U_i's are now Bernoulli random variables with common mean $P_F(X_1 \le \gamma_{F,\alpha} + n^{-1/2}z) = p_n$(say). Applying the central limit theorem to the U_i's, we have, as $n \to \infty$,

$$P_F\left(\frac{\sum_1^n U_i - np_n}{[np_n(1-p_n)]^{1/2}} \ge t\right) \to \int_t^\infty (2\pi)^{-1/2} e^{-x^2/2}\, dx. \tag{7.59}$$

The last expression of (7.58) can be put into the form of the left side of (7.59) if we let $t = ([\overline{\alpha n}] - np_n)/[np_n(1-p_n)]^{1/2}$. Since $n^{-1/2}z \to 0$ as $n \to \infty$, we have, as $n \to \infty$,

$$p_n = P_F(X_1 \le \gamma_{F,\alpha}) + P_F(\gamma_{F,\alpha} < X_1 < \gamma_{F,\alpha} + n^{-1/2}z) \doteq \alpha + n^{-1/2} z f_{F;X_1}(\gamma_{F,\alpha}),$$

in the sense that $(p_n - \alpha)/n^{-1/2} z f_{F;X_1}(\gamma_{F,\alpha}) \to 1$ as $n \to \infty$. Thus, we obtain that the previously stated value of t approaches $-zf_{F;X_1}(\gamma_{F,\alpha})/[\alpha(1-\alpha)]^{1/2} = q$ (say) as $n \to \infty$, so that (7.58) and (7.59) yield

$$P_F(n^{1/2}(Z_{\alpha,n} - \gamma_{F,\alpha}) \le z) \to \int_q^\infty (2\pi)^{-1/2} e^{-x^2/2}\, dx$$

as $n \to \infty$. Writing $z = \{[\alpha(1-\alpha)]^{1/2}/f_{F;X_1}(\gamma_{F,\alpha})\}v$, and using the symmetry of the normal density, we obtain

$$P_F\left(\frac{n^{1/2} f_{F;X_1}(\gamma_{F,\alpha})(Z_{\alpha,n} - \gamma_{F,\alpha})}{[\alpha(1-\alpha)]^{1/2}} \le v\right) \to \int_{-\infty}^{v} (2\pi)^{-1/2} e^{-x^2/2}\, dx \quad \text{as} \quad n \to \infty.] \tag{7.60}$$

In other words, $n^{1/2}(Z_{\alpha,n} - \gamma_{F,\alpha})$ is *asymptotically normal with mean* 0 *and variance* $\alpha(1-\alpha)/[f_{F;X_1}(\gamma_{F,\alpha})]^2$. [*Note*: In most practical cases the mean and variance tend to these limits, which are not necessarily the correct nonlimiting values; for example, $Z_{\alpha,n}$ is not necessarily an unbiased estimator of $\gamma_{F,\alpha}$, although the bias usually goes to 0 as $n \to \infty$.]

EXAMPLE 7.12. Suppose the X_i's are normal with unknown mean $\theta(-\infty < \theta < \infty)$ and known variance one. If the sample mean $\bar{X}_N$ based on N observations is used to estimate θ, this estimator is normal with mean equal to the true θ and variance $1/N$. If, on the basis of n observations, we instead use the sample median $Z_{1/2,n}$ to estimate θ (equal to $\gamma_{\theta,1/2}$), and if n is large, we have from our earlier results that $Z_{1/2,n}$ is distributed, approximately, according to the normal law about the true θ with variance

$$\frac{\alpha(1-\alpha)}{n[f_{F;X_1}(\gamma_{F,\alpha})]^2} = \frac{(1/2)^2}{n[(2\pi)^{-1/2}e^0]^2} = \frac{\pi}{2n} = \frac{1.57}{n}.$$

Thus, if $n = N$, the estimator $Z_{1/2,n}$ has a variance approximately 1.57 times that of $\bar{X}_n$, when n is large. Looking at it another way, in order for us to obtain the same precision from using the sample median $(Z_{1/2,n})$ as from using the sample mean $(\bar{X}_N)$, it is necessary in this example to base the former on 1.57 times as many observations as the latter $(n = 1.57N)$. Taking 157,000 observations and computing the sample median yields about the same accuracy as does taking 100,000 observations and computing $\bar{X}_{100,000}$.

Remark (Change of Variables). Suppose $\{U_n\}$ is any sequence of rv's converging stochastically to a value η in the interval (A, B), and that $n^{1/2}(U_n - \eta)$ is asymptotically $\mathcal{N}(0, \sigma^2)$, where $\sigma > 0$. Suppose h is monotone, with continuous derivative $h' > 0$ throughout (A, B). [The case $h' < 0$ throughout (A, B) yields the same conclusion.] By Taylor's theorem, given $\varepsilon > 0$ and η in (A, B), there is a $\delta > 0$ such that $(\eta - \delta, \eta + \delta) \subset (A, B)$ and such that

$1 - \varepsilon < \dfrac{h(u) - h(\eta)}{h'(\eta)(u - \eta)} < 1 + \varepsilon$ if $0 < |u - \eta| < \delta$; that is, h is approximately linear in a small interval about η. By assumption, $P\{|U_n - \eta| < \delta\} \to 1$ as $n \to \infty$. Hence, $P\left\{1 - \varepsilon < \dfrac{n^{1/2}[h(U_n) - h(\eta)]}{n^{1/2}h'(\eta)[U_n - \eta]} < 1 + \varepsilon\right\} \to 1$. (Note here that we need not worry about the factor $[U_n - \eta]$ in the denominator being 0, since $P\{n^{1/2}[U_n - \eta] \neq 0\} \to 1$ by the asymptotic normality.) Since $n^{1/2}[U_n - \eta]$ is asymptotically $\mathcal{N}(0, \sigma^2)$ and ε is arbitrarily small, *we conclude that* $n^{1/2}[h(U_n) - h(\eta)]$ *is asymptotically* $\mathcal{N}(0, [h'(\eta)\sigma].)$

[In Example 7.12, by taking $h(u) = u^r$, $h'(u) = ru^{r-1}$, we conclude that, when θ is the true parameter value (with $\theta \neq 0$ if $r < 0$), $n^{1/2}(Z_{1/2,n})^r - \theta^r$ is asymptotically $\mathcal{N}(0, \pi r^2|\theta|^{2r-2}/2)$.]

M.L. Estimators. Still in the setting where $X_1, \ldots, X_n$ are independent and identically distributed, it can be shown that, for most practical parametric Ω's (but not all Ω's!), if $\hat{t}_n$ is an M.L. estimator of θ (where Ω is parametrized in terms of θ) based on $X_1, \ldots, X_n$, then the sequence $\{\hat{t}_n\}$ converges stochastically to the true θ as $n \to \infty$. This is more difficult to show than the previous results and will not be proved here. It can also be shown that, if θ is real and $f_{\theta; X_1}(x)$ is sufficiently well behaved (this is called the *regular case*, the precise conditions being similar to those invoked in Section 7.2), then $n^{1/2}(\hat{t}_n - \theta)$ *is asymptotically normal with mean* 0 and variance $1/I(\theta)$, where $I(\theta) = E_\theta[\partial \log f_{\theta; X_1}(x_1)/\partial \theta]^2$ is the "information of X_1 about θ," as defined in Section 7.2, so that $nI(\theta)$ is the "information of $(X_1, \ldots, X_n)$." [A proof of the asymptotic normality can be found in Cramér's book.] There is a corresponding result when θ is a vector. In examples which are not "regular," for example, if $f_{\theta; X_1}$ is rectangular from 0 to θ, $\{\hat{t}_n\}$ will still converge stochastically to θ, but the limiting distribution need not be normal; thus, in this rectangular case, for which $\hat{t}_n = \max(X_1, \ldots, X_n)$, we have, for $z > 0$,

$$P_\theta(n(\hat{t}_n - \theta) < -z) = P_\theta\left(\hat{t}_n < \theta - \frac{z}{n}\right) = \left[\left(\theta - \frac{z}{n}\right)\Big/\theta\right]^n \to e^{-z/\theta}$$

as $n \to \infty$; here $\hat{t}_n - \theta$ must be multiplied by n rather than by $n^{1/2}$ to obtain a nondegenerate limiting law, and that limiting law is not normal.

There are sequences of estimators other than sample moments, sample quantiles, and M.L. estimators, whose asymptotic behavior is also useful to know (e.g., see the discussion of B.A.N. estimators); we have taken up these three because they are of greatest practical importance.

Asymptotic Optimality Properties

Let $\{t_n\}$ be a sequence of estimators of a parameter $\phi(F)$, t_n being based on n observations $X_1, \ldots, X_n$. Such a sequence is said to be a *consistent sequence of estimators of* ϕ *if* $\{\mathrm{t_n}\}$ *converges stochastically to* $\phi(\mathrm{F})$, *whatever* F *is true, as* n $\to \infty$; that is, if

$$P_F(|t_n - \phi(F)| < \varepsilon) \to 1$$

as $n \to \infty$, for every $\varepsilon > 0$ and every F in Ω. (Note: Do not confuse this "consistency" with that of Section 7.3.)

If we were to give a customer a list of estimators $\{t_n\}$, one for each possible sample size which he or she might take for a given n and function ϕ, then it seems reasonable to ask that this sequence be consistent; that is, that by taking a large enough sample size, the customer can attain as large a probability as he or she desires that the estimator will be within a specified amount ε of the true value $\phi(F)$. However, consistency, like any asymptotic property, does not by itself say *how large* n *must be in order for* t_n *to be fairly accurate*. For, consistency states only what happens to t_n as $n \to \infty$, not what happens for any fixed n. For example, in Example 7.12 (normal) with $\phi(\theta) = \theta$, the sequence

$$t_n(X_1, \ldots, X_n) = \begin{cases} e^{X_1} & \text{if} \quad n < 10^{70} \\ \bar{X}_n & \text{if} \quad n \geq 10^{70} \end{cases}$$

is just as "consistent" as is $\{\bar{X}_n\}$, although not very useful (unless the customer is likely to take more than 10^{70} observations). Similarly, in general, if any consistent sequence of estimators in a problem is altered by changing any *finite number* of the estimators in the sequence, the resulting sequence is also consistent.

In Example 7.12, $\{Z_{1/2,n}\}$ and $\{\bar{X}_n\}$ are both consistent, but the former gives a worse estimator than the latter, for any particular n. Thus, consistency is not a fine enough property to distinguish such a difference. In general, for any fixed n, we could compare two estimators t_n and t_n' by considering the ratio

$$r_{t_n}(\theta)/r_{t_n'}(\theta).$$

However, as we have mentioned, the computation of r_{t_n} for many procedures is quite difficult, *although the approximate behavior of* r_{t_n} as n $\to \infty$ *may not be so difficult to obtain* (using results like those of the subsection on limiting distributions of certain sequences of statistics). This suggests that we may sometimes be able to compute the *limit* of the preceding ratio as $n \to \infty$. Thus, with squared error in mind as the loss function, we may define

$$e_{\{t_n'\};\{t_n\}}(\theta) = \lim_{n\to\infty} [E_\theta(t_n - \phi(\theta))^2 / E_\theta(t_n' - \phi(\theta))^2],$$

when this limit exists, as the *asymptotic relative efficiency of the sequence* $\{t_n'\}$ to the sequence $\{t_n\}$ *when* θ *is true*; if this ratio is <1 for all θ, it would seem that t_n might be preferred to t_n' for large n. (We shall see later that care is needed in making such an interpretation.)

The notion of "efficiency" of a sequence of estimators in the regular case that was adopted by R. A. Fisher was that, if $n^{1/2}(t_n - \theta)$ is asymptotically normal (otherwise Fisher does not define efficiency) with mean 0 and variance $v(\theta)$ as $n \to \infty$ when θ is true, then $1/I(\theta)v(\theta)$ is defined as the *asymptotic efficiency of the sequence* $\{t_n\}$. The rationale for this definition can be found in the information inequality of Section 7.2: loosely speaking, under Fisher's assumption of asymptotic normality of $t_n - \theta$ with mean asymptotically 0, t_n would seem to be "almost unbiased," so that $n^{1/2}(t_n - \theta)$ could not have variance much smaller than $1/I(\theta)$, and $(1/I(\theta))/v(\theta)$ thus would tell us how bad $\{t_n\}$ actually is compared with the lower bound $1/I(\theta)$, for large n. A sequence of procedures is called efficient by Fisher if $1/I(\theta)v(\theta) = 1$. As with the definition of the previous paragraph, some care is needed in interpreting this notion, as we shall see in Example 7.13.

We note that the same criticism we applied to "consistency" also applies to all asymptotic "efficiency" definitions: there is no guarantee that any given n is large enough for t_n to have the behavior indicated by its asymptotic efficiency (or relative efficiency compared to t_n'), and by altering any finite number of members of the sequence $\{t_n\}$ we do not alter the asymptotic efficiency.

A sequence of M.L. estimators in the regular case, according to the subsection on M.L. estimators, satisfies Fisher's definition of efficiency. Thus, M.L. estimators were often described by Fisherians as being "consistent, efficient, and sufficient" (see the discussion in Section 7.5 on the inaccuracy of this last), and the choice of adjectives and a lack of thorough understanding of their meaning has made M.L. estimators widely accepted as the answer to all the statistician's prayers (see, e.g., Mood's book, p. 161, last sentence before references). We have seen in Section 7.5 that M.L. estimators are not really such panaceas.

Fisher's definition of asymptotic efficiency in the regular case should be contrasted with the definition of efficiency discussed in Section 7.2, which was criticized there because even a *best* unbiased estimator might not have "efficiency one" by that definition; in the Fisher definition, $1/I(\theta)$ is asymptotically attained in the regular case as $n\ \mathrm{var}_\theta(\hat{t}_n - \theta)$ or as $nE_\theta(t_n - \theta)^2$, so the same criticism does not arise. However, other criticisms (aside from that of the previous paragraph) do arise. For example, why restrict attention to asymptotically *normal* estimators, and what definition should we use in the nonregular case? These two questions are not addressed by a definition of relative efficiency like that given previously, but there is another difficulty which we shall discuss next and which is present in essentially any attempt to define asymptotic efficiency.

The difficulty is that, in almost any estimation problem, for any sequence $\{t_n\}$ there is an asymptotically "more efficient sequence" $\{t_n'\}$; i.e., one for which $e_{\{t_n\};\{t_n'\}}(\theta) \leq 1$ with strict inequality for some θ. Thus, there are *no* "asymptotically admissible sequences $\{t_n\}$," and in the regular case there are sequences of procedures $\{t_n\}$ for which $\{t_n\}$ is asymptotically normal with $v(\theta) \leq 1/I(\theta)$ for all θ, *with strict inequality for some* θ. (Such a sequence $\{t_n\}$ is called "superefficient.") We now illustrate this phenomenon by an example.

EXAMPLE 7.13. With the same normal setup as in Example 7.12, let t_n be defined by

$$t_n(x) = \begin{cases} 0 & \text{if} \quad |\bar{x}_n| < n^{-1/4}, \\ \bar{x}_n & \text{if} \quad |\bar{x}_n| \geq n^{-1/4}. \end{cases}$$

If the true θ is a value *other* than zero, we know that, since $\bar{X}_n$ approaches θ stochastically, $P_\theta(|\bar{X}_n| < n^{-1/4}) \to 0$ as $n \to \infty$; hence, $P_\theta(t_n(X) = \bar{X}_n) \to 1$ as $n \to \infty$, and thus t_n and $\bar{X}_n$ have the same limiting distribution: $n^{1/2}(t_n - \theta)$ is asymptotically normal with mean 0 and variance 1, and $nE_\theta(t_n - \theta)^2 \to 1$ as $n \to \infty$. On the other hand, if $\theta = 0$, we have

$$P_0(n^{1/2}(t_n - 0) = 0) = P_0(|\bar{X}_n| < n^{-1/4}) = P_0(n^{1/2}|\bar{X}_n - 0| < n^{1/4}) \to 1$$

as $n \to \infty$, by the central limit theorem; thus, when $\theta = 0$, $n^{1/2}(t_n - 0)$ is "asymptotically normal with mean 0 *and variance* 0," and $nE_0(t_n - 0)^2 \to 0$ as $n \to \infty$. Thus, $I(\theta)$ being 1 in the normal case, for this sequence $\{t_n\}$ we have, as $n \to \infty$,

$$\frac{E_\theta(t_n - \theta)^2}{E_\theta(\bar{X}_n - \theta)^2} \to 1/I(\theta)v(\theta) = \begin{cases} 0 & \text{if} \quad \theta = 0 \\ 1 & \text{if} \quad \theta \neq 0. \end{cases} \tag{7.61}$$

At first glance, this may seem to contradict the information inequality (7.10) and also the fact that $\bar{X}_n$ is admissible for squared error loss (see Section 7.4). The explanation is that t_n is not actually unbiased (so that (7.9) and not (7.10) applies), and that for large n the quantity $nE_\theta(t_n - \theta)^2$ is slightly greater than 1 for all θ outside a small neighborhood of the point $\theta = 0$, and dips down close to 0 for θ very close to 0. Thus, t_n is not better than $\bar{X}_n$ for any n, but in the limit we still have (7.61).

It is easy to see that a similar construction would produce a sequence $\{t_n'\}$ which asymptotically betters the preceding sequence $\{t_n\}$ at one point (or more), e.g., at $\theta = 1$. Similarly, in general any sequence $\{t_n^*\}$ can be "improved asymptotically" in the sense (7.61) that $\bar{X}_n$ was "improved to $\{t_n\}$," namely, in terms of the limiting risk on a small set of θ-values.

In what sense, then, can we meaningfully talk about asymptotic efficiency? The answer is that, in the regular case, for any sequence $\{t_n\}$ for which $nE_\theta(t_n - \theta)^2$ approaches a limit $q(\theta)$ as $n \to \infty$, it must be that

$$q(\theta) \geq 1/I(\theta),$$

except possibly on a small set of θ-values, namely, on a set of θ-values which has "measure zero," i.e., a set of θ-values which has probability zero according to any a priori density $f_\xi(\theta)$ on Ω (for example, the set may be finite, as in the preceding example, but could also be an infinite set of zero measure, e.g., a countable set). In particular, the M.L. estimators $\{\hat{t}_n\}$, for which $q(\theta) = 1/I(\theta)$, are asymptotically efficient in the sense that any other sequence of estimators can have a limiting risk function which is better on at most such a small set of θ-values.

This is the sense we alluded to earlier, in which the M.L. estimator could be described loosely as being an "almost uniformly best" procedure for very large sample sizes.

There is another way of describing the asymptotic behavior of M.L. estimators which is quite illuminating. Suppose we are in the regular case with Ω an interval, and that f_ξ is any a priori density which is positive and continuous throughout Ω. Then it can be shown that, as $n \to \infty$, the a posteriori law of θ will, with probability approaching one (under the true parameter value θ_0), turn out to be such that θ is asymptotically normal about the M.L. estimator $\hat{t}_n$, $n^{1/2}(\theta - \hat{t}_n)$ being approximately normal with mean 0 and variance $1/I(\theta_0)$. Then if $W(\theta, d)$ is an increasing function of $|\theta - d|$, it follows that the decision $d = \hat{t}_n$ will approximately minimize the integral of $W(\theta, d)$ times the a posteriori density. Thus, *for all a priori laws of the preceding form, the same procedure* $\hat{t}_n$ *is approximately Bayes for large* n. This also reflects the fact noted two paragraphs back, that, for any other sequence $\{t_n\}$, the quantity $h(\theta) = \lim_{n\to\infty} E_\theta(t_n - \theta)^2/E_\theta(\hat{t}_n - \theta)^2$ can be <1 on at most a "small set" of θ-values; for, if $h(\theta) < 1$ on (for example) an interval, then by taking f_ξ to concentrate most a priori probability on that interval we would obtain $\lim_{n\to\infty} R_{t_n}(\xi)/R_{\hat{t}_n}(\xi) < 1$, contradicting the asymptotic Bayes character of $\hat{t}_n$ which we have just discussed.

You are reminded again that, despite the appealing asymptotic properties of M.L. estimators, there is in general no result telling us "how large n must be" in order that a M.L. estimator have approximately its asymptotic behavior. In some simple cases, for example, in the K-D case in which an M.L. estimator is a sum of the form $\sum_1^n k(X_i)$ so that the central limit theorem can be applied directly, one can use known results on the error of the normal approximation; in other cases, it is more difficult to obtain such results.

One aspect of asymptotic theory which has received some discussion in the literature is this: the fact that a sequence $\{U_n\}$ (where U_n may be $n^{1/2}(t_n - \theta)$ in our examples) has a given limiting law, for example,

$$P(U_n < z) \to \int_{-\infty}^{z} (2\pi)^{-1/2} e^{-x^2/2}\, dx,$$

does *not* mean that EU_n or var U_n approaches the corresponding quantities for the limiting law. For example, if U_n has density $(1 - n^{-1})(2\pi)^{-1/2}e^{-x^2/2} + n^{-1}/\pi[1 + x^2]$, it is easy to see that $\{U_n\}$ is asymptotically normal, the limiting law having mean 0 and variance one. But EU_n is not even defined, and $EU_n^2 = \infty$ for each n. The difficulty is, of course, that the introduction of a little probability into the tails of a distribution can affect the moments greatly without affecting the probability of any event by much. In most practical examples this phenomenon will not occur, and we will have $\lim_{n\to\infty} \text{var}(U_n) =$ var(limiting law of $\{U_n\}$). Where this is not the case, as in the preceding example, it will usually be the case that the variance of the limiting law is the more relevant quantity; for it is this quantity, and not $\lim_{n\to\infty} \text{var}(U_n)$, which will reflect the dispersion of the distribution of U_n (or $t_n - \theta$) for large n. Furthermore, the consideration of var U_n (or $E_\theta(t_n - \theta)^2$) in practice probably does not usually mean that squared error is the relevant loss function, but rather that it gives a computationally simple indication of the spread of U_n (accuracy of t_n); when a little probability far out in the tails affects the moments greatly, this is no longer the case. Thus, in our example involving U_n, the fact that var(limiting law of $\{U_n\}$) $= 1$ much better indicates to us the approximate value of a quantity like $P(|U_n| > 2)$, which may be related to the actual risk, than would the value $EU_n^2 = \infty$.

B.A.N. Estimators. Neyman noted that the only proved "optimum" properties of M.L. estimators are the asymptotic ones we have discussed. He therefore suggested in certain settings (especially where M.L. estimators are tedious to compute) the use of estimators which have the same asymptotic properties as the M.L. estimators *but which are much easier to compute.* These are called *best asymptotically normal* (B.A.N.) estimators. It would take too long to discuss them here, but you should refer to the literature on them if you have occasion to work on a problem in which M.L. estimators are difficult to obtain, especially multinomial problems for which the unknown probabilities are specified functions of more primitive parameters. (Such problems arise in genetics, for example.) Some of the B.A.N. estimators other than M.L. that are available in such settings are (i) Bayes procedures relative to smooth f_ξ's; (ii) estimators (originally suggested by Neyman) obtained by expanding $\partial \log P_{\theta;x}(x)/\partial\theta$ in a Taylor's series in θ about a suitably accurate (inefficient but easy-to-compute) preliminary estimator t_n' of θ, retaining only 0th and 1st degree terms in $(\theta - t_n')$, setting the resulting expression equal to 0 (in place of using the likelihood equation), and using the solution for θ as estimator; (iii) the "maximum probability estimator" of Wolfowitz (also efficient in various nonregular settings). We now illustrate (ii).

Suppose $\{t_n'\}$ is consistent for θ, with deviations of order smaller than $n^{-1/4}$, in probability; that is, $P_\theta\{|t_n' - \theta|n^{1/4} > \varepsilon\} \to 0$ as $n \to \infty$ for each $\varepsilon > 0$. (Such a $\{t_n'\}$, easily computable, will often be simple to find, since estimators based on sample moments or sample quantiles can often be used, with deviations of the smaller order $n^{-1/2}$, as is evident from the asymptotic normality results for such estimators.) For likelihood function $L_n(\theta)$, assuming appropriate regularity, the likelihood equation for the M.L. estimator $\hat{t}_n$ (assumed unique), expanded in a series in $\hat{t}_n$ about t_n', is

$$0 = \left.\frac{\partial \log L_n(\theta)}{\partial\theta}\right|_{\theta=\hat{t}_n} = \left.\frac{\partial \log L_n(\theta)}{\partial\theta}\right|_{\theta=t_n'} + \left.\frac{\partial^2 \log L_n(\theta)}{\partial\theta^2}\right|_{\theta=t_n'}(\hat{t}_n - t_n') + \cdots .$$

The terms not exhibited can be shown to be negligible in the step that follows, because of the assumption on $\{t_n'\}$. Assuming this, we obtain

$$\hat{t}_n \approx t_n' - \frac{\partial \log L_n(\theta)/\partial\theta|_{\theta=t_n'}}{\partial^2 \log L_n(\theta)/\partial\theta^2|_{\theta=t_n'}} \stackrel{\text{def}}{=} \tilde{t}_n. \tag{7.62}$$

The estimator $\tilde{t}$ defined by (7.62) can be shown to have the same asymptotic efficiency as $\hat{t}_n$.

EXAMPLE. In the Cauchy case, where M.L. computations are intractable, it will be seen in Example 7.14, that $t_n' = Z_{1/2,n}$ is already 81 percent efficient asymptotically. Using it in (7.62), we obtain the 100 percent asymptotically efficient $\tilde{t}_n$, upon substituting using the fact that the first two derivatives with respect to θ of $\log \dfrac{1}{\pi[1+(x-\theta)^2]}$ are $\dfrac{-2(\theta-x)}{1+(\theta-x)^2}$ and $\dfrac{-2[1-(\theta-x)^2]}{[1+(\theta-x)^2]^2}$:

$$\tilde{t}_n = Z_{1/2,n} - \frac{\sum_1^n \{(X_i - Z_{1/2,n})/[1+(X_i - Z_{1/2,n})^2]\}}{\sum_1^n \{[(X_i - Z_{1/2,n})^2 - 1]/[1+(X_i - Z_{1/2,n})^2]^2\}}.$$

This is much easier to compute than Bayes estimators, too, in this example.

A Nonparametric Result. Suppose $X_1, X_2, \ldots$ are independent and identically distributed with unknown distribution function F, about which nothing is known. The object may be to estimate not merely some parameter like $E_F X_1$, but rather the entire law F. An estimator is then a function on S *whose value is a distribution function.* For example, with $n = 3$, an estimator t may be defined by stating that $t(x_1, x_2, x_3)$ is that df G_{y_1,y_2,y_3} which is given for each fixed x_1, x_2, x_3 in terms of the order statistics, y_1, y_2, y_3, as

$$G_{y_1,y_2,y_3}(x) = \begin{cases} 0 & \text{if } x < y_1, \\ 1/5 & \text{if } y_1 \le x < y_2, \\ 3/4 & \text{if } y_2 \le x < y_3, \\ 1 & \text{if } x \ge y_3. \end{cases}$$

You can check that the *sample* df $\bar{F}_n$ defined in Section 4.8 is just such an estimator, with "1/5 and 3/4" replaced by "1/3 and 2/3." There are many other such estimators, and they need not be of this form where the estimate is a step function. We can measure the loss incurred when a given df G is the decision and F is true by something like the maximum vertical deviation $\max_{-\infty<x<\infty} |F(x) - G(x)|$ between F and G. A good procedure, e.g., a minimax procedure, is very hard to compute for each n. However, it can be shown that the sample df $\bar{F}_n$ is "asymptotically minimax" for essentially *any reasonable loss function* (such as mentioned previously), in the sense that

$$\lim_{n\to\infty} \left(\frac{\max_F r_{\bar{F}_n}(F)}{\min_{t_n} \max_F r_{t_n}(F)} \right) = 1.$$

The Use of Inefficient Procedures

We shall now illustrate the use of easily computed inefficient procedures whose employment is in some instances preferable to that of efficient procedures which are harder to compute. As we have mentioned, a complete discussion of this topic would include a precise consideration of the computing costs associated with using various procedures.

One of the most commonly used simple estimators is a sample quantile, such as the sample median for an estimator of the population median (as in Example 7.12). Instead of using a single sample quantile, one may instead use a weighted average of several sample quantiles. In fact, it can be shown that, in the regular case, one can obtain an estimator whose efficiency is any specified number <1 (e.g., .99) by using the proper weighted average of enough different sample quantiles. In obtaining the limiting distribution of such an estimator, we use the following extension of the results of the subsection on sample quantiles: If $0 < \alpha_1 < \alpha_2 < \cdots < \alpha_k < 1$, *then the limiting joint distribution of* $n^{1/2}(Z_{\alpha_1,n} - \gamma_{F,\alpha_1}), \ldots, n^{1/2}(Z_{\alpha_k} - \gamma_{F,\alpha_k})$ *is the* k-*variate normal distribution with means* 0, *variances* $\alpha_i(1-\alpha_i)/[f_{F;X_1}(\gamma_{F,\alpha_i})]^2$ *and covariances* $\alpha_i(1-\alpha_j)/f_{F;X_1}(\gamma_{F,\alpha_i})f_{F;X_1}(\gamma_{F,\alpha_j})$ *for* $\alpha_i < \alpha_j$. The proof is similar to that of the subsection on sample quantiles. We now illustrate the use of a weighted average of quantiles.

EXAMPLE 7.12 (Continued). In our normal example, suppose we try to estimate θ by the average $t_{\alpha,n} = (Z_{\alpha,n} + Z_{1-\alpha,n})/2$ of two symmetric sample quantiles (that is, the α_1- and α_2- sample quantiles with $\alpha_1 = 1 - \alpha_2$). Because of the symmetry of the normal density, it is clear that $t_{\alpha,n}$ is symmetrically distributed around the true value θ. Writing α for the α_i which is $<1/2$ and $1-\alpha$ for the other α_i, we thus have, by the result of the previous paragraph that $n^{1/2}(t_{\alpha,n} - \theta)$ is asymptotically normal with mean 0 and variance equal to that of $(U_\alpha + U_{1-\alpha})/2$, where $\text{var}\, U_\alpha = \text{var}\, U_{1-\alpha} = \alpha(1-\alpha)/[(2\pi)^{-1/2}e^{-\gamma_{0,\alpha}^2/2}]^2$ and $\text{cov}(U_\alpha, U_{1-\alpha}) = \alpha^2/[(2\pi)^{-1/2}e^{-\gamma_{0,\alpha}^2/2}]^2$. This limiting variance is thus

$$\frac{1}{4}[\text{var}(U_\alpha) + \text{var}(U_{1-\alpha}) + 2\,\text{cov}(U_\alpha, U_{1-\alpha})] = \frac{\alpha}{2[f_{0;X_1}(\gamma_{0,\alpha})]^2}.$$

One could naturally ask, what choice of α $(0 < \alpha < 1/2)$ gives the smallest value of this limiting variance? Writing $f(x) = (2\pi)^{-1/2}e^{-x^2/2}$ and $F(x)$ for the corresponding distribution function, and $\gamma = \gamma_{0,\alpha}$ (so that $F(\gamma) = \alpha$), we must minimize

$$g(\gamma) = \frac{F(\gamma)}{2[f(\gamma)]^2}$$

with respect to γ, $-\infty < \gamma < 0$. Differentiating $g(\gamma)$ with respect to γ and setting the result equal to zero, we obtain the equation $2\gamma F(\gamma) + f(\gamma) = 0$, which can be solved by trial and error using the normal tables, to yield $\gamma = -.61$, and it is easy to check that this extremum is a maximum (alternatively, we could merely compute $g(\gamma)$ directly and find, by trial and error, the maximizing value). For $\gamma = -.61$, we obtain $\alpha = F(-.61) = .27$ and $g(\gamma) = 1.19$. Thus, instead of computing $\bar{X}_{100000}$ or $Z_{1/2,157000}$, we could compute $(Z_{.27,119000} + Z_{.73,119000})/2$ and achieve about the same accuracy. With only a small amount of "arithmetic" (namely, averaging two sample quantiles) we reduce the 157,000 observations needed with the sample median by 24 percent, to 119,000 needed to obtain about the same accuracy using $t_{.27,n}$.

As we have seen in Section 7.2, in other examples in the regular case it may be that $1/nI(\theta)$ is not attained as $E_\theta(t_n - \theta)^2$ for any unbiased (or biased) t_n in the way that it was by $\bar{X}_n$ in the normal example, and that a good (e.g., minimax) procedure is hard to find or computationally unwieldy (as even in Example 7.5(b)). For small n we can still compare $E_\theta(t_n - \theta)^2$ for a computationally simple t_n with $1/nI(\theta)$, but, as we have discussed, this comparison may make t_n seem less good than it really is. For large n, the value $1/nI(\theta)$ is almost attained as $E_\theta(\hat{t}_n - \theta)^2$ for an M.L. estimator $\hat{t}_n$, but the latter may be unwieldy to use even with considerable computing facilities; for example, this is so in the Cauchy case (see the discussion following Example 7.11(c) in Section 7.5).

EXAMPLE 7.14. In the Cauchy case, in which $f_{\theta;X_1}(x_1) = 1/\pi[1 + (x_1 - \theta)^2]$, the quantity $I(\theta)$ does not depend on θ (this being a location parameter problem). Since

$$\frac{\partial \log f_{\theta;X_1}(x_1)}{\partial \theta} = \frac{2(x - \theta)}{[1 + (x - \theta)^2]^2},$$

we thus have

$$I(\theta) = I(0) = \int_{-\infty}^{\infty} \frac{4x^2}{\pi[1 + x^2]^3}\,dx.$$

This integral can be evaluated, for example, by making the trigonometric substitution $x = \tan y$. We obtain $I(\theta) = 1/2$. By the result of the subsection on M.L. estimates, $n^{1/2}(\hat{t}_n - \theta)$ is thus asymptotically normal with mean 0 and variance $1/I(\theta) = 2$. However, as discussed in Section 7.5, $\hat{t}_n$ is difficult to compute. On the other hand, since $\gamma_{\theta,1/2} = \theta$, we have by the subsection on sample quantiles that $n^{1/2}(Z_{1/2,n} - \theta)$ is asymptotically normal with mean 0 and variance $(1/2)^2/[f_{\theta;X_1}(\theta)]^2 = \pi^2/4$. The asymptotic efficiency of the sample median is thus $8/\pi^2 = .81$, which is quite high. (Of course, $\bar{X}_n$ is a very poor estimator in this example, since it has the same law as X_1; thus, $\{\bar{X}_n\}$ is not even consistent.)

An additional reason for using estimators based on one or a few sample quantiles is that they retain properties of accuracy when the true law is slightly different from the form assumed in Ω, better than do estimators like $\hat{t}_n$ or $\bar{X}_n$. This ability is referred to as *robustness*.

PROBLEMS

The problems for this chapter are divided into sets corresponding to sections of the chapter.

Problems on Section 7.1. Unbiasedness, completeness, Blackwell-Rao method.
Suggested: At least one of 7.1.1–7.1.4 and at least one of 7.1.7–7.1.11.

7.1.1. Suppose $X_1, X_2, \ldots, X_n$ have common unknown mean μ_F and common unknown variance σ_F^2, and *known* correlation ρ between each pair X_i, X_j $(i \neq j)$. Find a number A (depending on ρ, n) such that $A \sum_{i=1}^{n} (X_i - \bar{X}_n)^2$ is an unbiased estimator of σ_F^2.

[*Note*: This problem can be solved by direct computation of the expectation of any term in the sum. An alternative approach (which is needlessly complex in this simple example) uses the covariance-changing method of Section 5.4 to reduce the problem to the case $\rho = 0$.]

7.1.2. A common "time series" model is the following: random variables U_i are independent with common unknown variance σ_F^2 but are not observed directly. One observes $X_1 = U_1$, $X_2 = U_2 + \frac{1}{2}U_1$, $X_3 = U_3 + \frac{1}{2}U_2 + \frac{1}{4}U_1$, etc.; in other words, the phenomenon X_i observed at time i is a sum of effects from the present (U_i) and past, but the contribution of the effect of U_{i-t} from t time units ago is reduced by multiplication by 2^{-t}. (See Problem 5.5(b) for a practical setting of such a model.) If the X_i have common unknown mean μ_F, find a value c such that $c\sum_1^3 (X_i - \bar{X}_3)^2$ is an unbiased estimator of σ_F^2. [*Hint*: $\sum_1^3 (X_i - \bar{X}_3)^2 = \sum_1^3 (X_i - \mu_F)^2 - 3(\bar{X}_3 - \mu_F)^2$; compute its expectation by showing that $\operatorname{Cov}\begin{pmatrix} X_1 \\ X_2 \\ X_3 \end{pmatrix} = \sigma_F^2 \begin{pmatrix} 1 & 1/2 & 1/4 \\ 1/2 & 5/4 & 5/8 \\ 1/4 & 5/8 & 21/16 \end{pmatrix}$ and using this. Alternatively, write $\sum_1^3 (X_i - \bar{X}_3)^2$ in terms of the U_i's.]

7.1.3. (a) [Go directly to (b) if you prefer.] Suppose U_1, U_2, U_3, U_4 are uncorrelated with common variance σ_F^2, that $X_i = \sum_{j=1}^{i} U_j = U_1 + \cdots + U_i$, and that $E_F X_i = \mu_F$, independent of i. [This may arise in a time series similar to one studied in Problem 5.4(b).] Find a constant c such that $c\sum_1^4 (X_i - \bar{X}_4)^2$ is an unbiased estimator of σ_F^2. [*Hint:* Write $X_i - \mu_F = Y_i$. Start by showing that $\sum_1^4 (X_i - \bar{X}_4)^2 = \sum_1^4 (Y_i - \bar{Y}_4)^2 = \frac{3}{4}\sum_1^4 Y_i^2 - \frac{1}{2}\sum_{i<j} Y_i Y_j$. Find the expectation of this last expression by showing

$$\operatorname{Cov}\begin{pmatrix} X_1 \\ X_2 \\ X_3 \\ X_4 \end{pmatrix} = \sigma_F^2 \begin{pmatrix} 1 & 1 & 1 & 1 \\ 1 & 2 & 2 & 2 \\ 1 & 2 & 3 & 3 \\ 1 & 2 & 3 & 4 \end{pmatrix}.]$$

(b) Work problem 7.1.3(a), but for the estimator $c_n \sum_1^n (X_i - \bar{X}_n)^2$ where again the U_j are uncorrelated with common variance σ_F^2 and $X_i = \sum_{j=1}^{i} U_j$. [*Hint*: Show that

$$\sum_1^n (X_i - \bar{X}_n)^2 = \left(\frac{n-1}{n}\right)\sum_1^n X_i^2 - \frac{2}{n}\sum_{i<j} X_i X_j,$$

and obtain an expression for this in terms of $\sum_1^n j$ and $\sum_1^n j^2$ by showing and using the fact that

$$\text{Cov}(X) = \sigma_F^2 \begin{pmatrix} 1 & 1 & \cdots & 1 \\ 1 & 2 & \cdots & 2 \\ \vdots & \vdots & & \vdots \\ 1 & 2 & \cdots & n \end{pmatrix}.$$

Use the not-to-be-proved facts that $\sum_1^n j = n(n+1)/2$ and $\sum_1^n j^2 = n(n+1)(2n+1)/6$ to compute the sum of the n^2 elements of $\text{Cov}(X)$ and thus to obtain $E_F \sum_1^n (X_i - \bar{X}_n)^2$. (Check the result against that obtained in part 7.1.3(a).)]

7.1.4. The result that $E_F m_{4,n} = \mu_{4F}(n-1)(n^2-3n+3)/n^2 + \mu_{2F}^2(n-1)(2n-3)/n^3$ and that $E_F m_{2,n}^2 = \mu_{4F} n^{-3}(n-1)^2 + \mu_{2F}^2 n^{-3}(n-1)(n^2-2n+3)$, to find constants c and d (depending on n but not on F) such that $cm_{4,n} + d(m_{2,n})^2$ is an unbiased estimator of $\mu_{4,F}$, where $\Omega = \{F : \mu_{4,F} < \infty\}$, the X_i's being independent and identically distributed. [*Optional:* Prove the two formulas by methods like those at the end of Section 7.1.]

7.1.5. Suppose $\Omega = \{\theta : \theta \text{ is an integer}\}$ and

$$P_\theta\{X = \theta\} = P_\theta\{X = \theta - 1\} = P_\theta\{X = \theta + 1\} = 1/3.$$

The random variable X is easily seen to be minimal sufficient (not to be proved). Let $t^*(x)$ = closest integer to x that is divisible by 3 (e.g., $t^*(-3) = -3$, $t^*(1) = 0$, etc.) (a) Show that t^* is an unbiased estimator of θ, and hence that $E_\theta[t^*(X) - X] \equiv 0$; conclude that X is not complete. (b) Show that $\text{var}_\theta(t^*(X)) = 0$ if θ is divisible by 3. (c) For each of $i = 1, 2$, find an unbiased estimator t_i of θ whose variance is 0 provided $\theta + i$ is divisible by 3. (d) Use the conclusion of (b) and (c) to show that no uniformly best unbiased estimator exists in this case, since its variance would have to be 0 for all θ. [This is a discrete analogue of Ex. 6.6 in Section 7.1.]

7.1.6. Suppose $\Omega = \{\theta : 0 \le \theta < \infty\}$ and that X has discrete law $P_0\{X = 0\} = 1$ and $P_\theta\{X = -1\} = \dfrac{\theta}{1+2\theta} = P_\theta\{X = \theta\}$, $P_\theta\{X = 0\} = \dfrac{1}{1+2\theta}$, $\theta > 0$. The rv X is minimal sufficient (not to be proved). For an arbitrary estimator t of θ, taking values in $D = \{d : 0 \le d < \infty\}$, write $A = t(-1)$ and $B = t(0)$. (a) If t is an unbiased estimator of θ, show that $t(x) = 2x + 1 - A - Bx^{-1}$ for $x > 0$, and consequently that $B = 0$ and $0 \le A \le 1$ for t to take on values only in D. (b) Show that among unbiased estimators with $B = 0$ and $0 \le A \le 1$, the unbiased estimator that has (locally) best variance at $\theta = \theta_0$ is

$$t_{\theta_0}(x) = \begin{cases} A_{\theta_0} & \text{if } x = -1 \\ 0 & \text{if } x = 0 \\ 2x + 1 - A_{\theta_0} & \text{if } x > 0, \end{cases}$$

where

$$A_{\theta_0} = \begin{cases} \theta_0 + \frac{1}{2} & \text{if} \quad 0 \le \theta_0 \le \frac{1}{2}, \\ 1 \quad \text{if} & \theta_0 > \frac{1}{2}. \end{cases}$$

Hence, there is no uniformly best unbiased estimator. Graph $r_{t_{\theta_0}}$ for $\theta_0 = 1/2$, 0, and 1. (c) Show that the biased estimator t^* defined by $t^*(0) = t^*(-1) = 0$, $t^*(x) = x$ for $x > 0$, is better than every t_{θ_0}.

7.1.7. $X_1, X_2, \ldots, X_n$ are iid with common geometric probability function

$$f_{\theta;X_1}(x) = \begin{cases} (1-\theta)\theta^x & \text{if} \quad x = 0, 1, 2, \ldots, \\ 0 \quad \text{otherwise}, \end{cases} \tag{7.63}$$

where $\Omega = \{\theta : 0 \le \theta < 1\}$. (When $\theta = 0$, we have $P_0\{X_1 = 0\} = 1$.) This arises from a coin-flipping experiment: X_i is the *number of heads* before the first tail in a sequence of iid coin flips with probability θ of a head on a single flip. It is desired to estimate $\phi(\theta) = (1 - \theta)$ (from which θ can also be estimated). From the results of Chapter 6 and Problem 6.1 we know that (7.63) is K-D so that $T = \sum_1^n X_i$ is minimal sufficient. [It is also complete.] Moreover, since $T + n$ is, for example, the number of flips required to get exactly n tails, it is easily derived that

$$P_\theta\{T = u\} = \binom{u+n-1}{u}(1-\theta)^n\theta^u \quad \text{for} \quad u = 0, 1, 2, \ldots. \tag{7.64}$$

Although $1 \notin \Omega$, we include 0 in $D = [0, 1]$, to help construct a simple example.

(a) In order to appreciate the method which follows, spend a few minutes trying to *guess* a function of T which is an *unbiased* estimator of this $\phi(\theta)$. Then proceed. (You need not succeed!)

(b) A commonly used "try" for some *simple* unbiased estimator (not a function of T) in such problems is a 0–1 "indicator random variable" I_A of some event A, for which A is a subset of S (as in the discussion of Ex. 7.2). We showed there that $E_\theta\{I_A\} = P_\theta\{A\}$ in general. In the present example try various A's of the form $A_B = \{(X_1, \ldots, X_n) : X_1 \in B\}$, and show that $B = \{0\}$ is of the desired form $P_\theta\{A_B\} = 1 - \theta$. Thus, the simple dependence of A_B *on* X_1 *alone* makes it easy for us to verify unbiasedness.

(c) Show that, because of the simple 0–1 nature of I_A, one always has $E_\theta\{I_A | T = s_0\} = P_\theta\{A | T = s_0\}$. Use the fact that $\sum_2^n X_i$ has the law (7.64) *with* n *replaced by* n − 1, along with method (ii) of Section 7.1 to find an unbiased estimator t^* of θ which is uniformly best among unbiased estimators for all convex (in d) W. [The point of the preceding example is that it was difficult to *guess* which function of T was unbiased, but that we obtain the function t^* mechanically, starting with any old easy-to-guess unbiased t, no matter how bad (in variance) t is!]

7.1.8. Work Problem 7.1.7 with $\phi(\theta) = \theta(1 - \theta)$ instead. (This is the variance of the Bernoulli law that gave rise to (7.63), not of (7.63) itself, but is often of interest.) [*Hint*: The answer is that, for $n > 2$, the estimator $t^*(X) = (n-1)T/(T+n-1)(T+n-2)$ is an unbiased estimator of $\theta(1-\theta)$ which, *among all unbiased* *estimators, has uniformly smallest risk function, for every* W(θ, d) *which is convex in* d. For $n = 1$, $t^*(X) = I_{A_B}$: for $n = 2$, $t^*(X) = \begin{cases} (T+1)^{-1} & \text{if} \quad T > 0 \\ 0 \quad \text{if} \quad T = 0 \end{cases}$.]

7.1.9. Work 7.1.7 with $\phi(\theta) = P_\theta\{X_i \le 1\} = 1 - \theta^2$. [*Hint*: In part (c), $A_B \cap \{T = s_0\} = \left\{X_1 = 1, \sum_2^n X_i = s_0 - 1\right\} \cup \left\{X_1 = 0, \sum_2^n X_i = s_0\right\}$. *Answer*: For $n > 2$, $t^*(X) = (n-1)(2T+n-2)/(T+n-1)(T+n-2)$ is an unbiased estimator of $1 - \theta^2$, which, *among all unbiased estimators,has uniformly smallest risk function, for every* W(θ, d) *which is convex in* d. For $n = 1$, one can show that $t^*(X) = I_{A_B}$; for $n = 2$,

$$t^*(X) = \begin{cases} (n-1)(2T+n-2)/(T+n-1)(T+n-2) & \text{if} \quad T > 0 \\ 1 \quad \text{if} \quad T = 0 \end{cases}.]$$

7.1.10. Suppose $X = (X_1, \ldots, X_n)$, the X_i being discrete "uniform" iid random variables from 1 to $\theta : \Omega = \{\theta : \theta = \text{positive integer}\}$ and $P_\theta\{X_1 = j\} = \theta^{-1}$ for $j = 1, 2, \ldots, \theta$. (This is the discrete case analogue of Example 6.5.)
(a) Show that $s(X) = \max(X_1, \ldots, X_n)$ is sufficient.
(b) Find the pdf of $s(X)$. [*Hint:* Compute $P_\theta\{s(X) \le j\}$, first.]
(c) Show that $s(X)$ is complete by considering $E_\theta h(s(X))$ successively for $\theta = 1, 2, 3, \ldots$.
(d) Use the method indicated in 7.1.7(b) (after trying 7.1.7(a), with s for T), to show that the best unbiased estimator of $\phi(\theta) = \theta^{-1}$ for all convex (in d) W is $t^*(s) = \dfrac{1}{2s-1}$ if $n > 1$, and $= I_{A_B}$ if $n = 1$. [Here D can be taken as $[0, 1]$ to make it an interval, but in fact t^* takes on only values in $\{d : d = 1/(\text{positive integer})\}$ for $n > 1$. In other examples in which $\{\phi(\theta) : \theta \in \Omega\}$ is not an interval, t^* obtained in this way may turn out to take on values other than possible values of $\phi(\theta)$.]

7.1.11. Let $X = (X_1, X_2, \ldots, X_n)$ where the X_i and iid, each with the absolutely continuous case uniform density from 0 to θ (Example 6.5); $\Omega = \{\theta : 0 < \theta < \infty\}$. Thus, $s(X) = \max(X_1, \ldots, X_n)$ is a complete minimal sufficient statistic. It is desired to find an unbiased estimator of $\phi(\theta) = P_\theta\{X_1 < 1\} = \min(\theta^{-1}, 1)$.
(a) Try 7.17(a) (with s for T).
(b) Find an event B of the form prescribed in 7.1.7(b).
(c) Show that $P_\theta\{s(X) = X_1\} = n^{-1}$ and that, for $n > 1$, the joint probability law of X_1 and $s(X)$ is $P_\theta\{X_1 < t, s(X) < u\} = \theta^{-n} t u^{n-1}$ for $0 < t < u < \theta$, from which the density part of this law on $0 < t < u < \theta$ can be computed. From that density, conclude that the conditional law of X_1, given that $s(X) = u$, is

$$\begin{cases} \text{discrete probability } n^{-1}, \quad \text{at} \quad x_1 = u, \\ \text{conditional density } g(x_1 | u) = (n-1)/nu, \quad 0 < x_1 < u; \end{cases}$$

this law is "mixed", neither entirely discrete nor absolutely continuous.
(d) Use "Method (ii)" with (c) to show that the uniformly best unbiased estimator of $\phi(\theta)$ for all convex (in d) W is

$$t^*(s(X)) = \begin{cases} 1 & \text{if} \quad s(X) < 1, \\ (n-1)/ns(X) & \text{if} \quad s(X) \geq 1. \end{cases}$$

(For $n = 1$, $d = 0$ is thus possible.)

(e) Using the fact that $s(X)$ has density $f_\theta^*(s) = n\theta^{-n}s^{n-1}$ on $0 < s < \theta$, check that t^* does indeed have expectation $\phi(\theta)$.

Problems on Section 7.2. Information inequality.

7.2.1. Suppose $X_1, X_2, \ldots, X_n$ are independent and identically distributed, each with $\mathcal{N}(\theta, 1)$ density:

$$f_{\theta; X_1}(x_1) = (2\pi)^{-1/2} e^{-(x_1-\theta)^2/2}, \quad -\infty < x_1 < \infty.$$

Here $X = (X_1, \ldots, X_n)$, $\Omega = \{\theta : \theta \text{ real}\}$, $D = \{d : d \text{ real}\}$, $W(\theta, d) = [\phi(\theta) - d]^2$.

(a) Compute $I_n(\theta)$ and use the information inequality to show that $\bar{X}_n$ is uniformly best among unbiased estimators of $\phi(\theta) = \theta$. [Part of answer: $I_1(\theta) \equiv 1$.]

(b) Suppose it is desired to estimate $\phi(\theta) = \theta^2$.

(i) Show that the information inequality for any unbiased estimator of $\phi(\theta)$ is $\text{var}_\theta(t_{\text{unbiased}}) \geq 4\theta^2/n$.

(ii) Show that $t_n^* = \bar{X}_n^2 - n^{-1}$ is an unbiased estimator of $\phi(\theta)$. (Once more, we must include impossible values of ϕ in D to get *any* unbiased estimator!) Show that $\text{var}_\theta(t_n^*) = 4n^{-1}\theta^2 + 2n^{-2}$, by checking that the first four moments of the $\mathcal{N}(0, n^{-1})$ *rv* $(\bar{X}_n - \theta)$ are $0, n^{-1}, 0, 3n^{-2}$.

(iii) Recalling Section 7.1, show that t_n^* is a best unbiased estimator of θ^2. Using (i) and (ii), show that this conclusion that t_n^* is best unbiased of θ^2 could *not* have been deduced from the information inequality. [*Fact*: $\sum_1^n X_i$ is a complete sufficient statistic.]

(iv) Show that the ratio $H_n(\theta)/\text{var}_\theta(t_n^*)$ approaches 1 as $n \to \infty$. Hence (although (iii) demonstrates that the information inequality is not always adequate to verify that a best unbiased estimator t_n^* is indeed best), show that this inequality suffices to establish that t_n^* is *approximately* best unbiased for estimating θ^2 for large n, in the sense that, for all values θ,

$$\text{var}_\theta(t_n^*)/\min_{t_n' \in L_n} \text{var}_\theta(t_n') \to 1 \quad \text{as} \quad n \to \infty,$$

where L_n is the class of all unbiased estimators of θ^2 based on $(X_1, \ldots, X_n)$. [*Comment*: A corresponding result is valid for other functions ϕ which one might want to estimate.]

7.2.2. $X_1, X_2, \ldots, X_n$ are independent and identically distributed, each with density

$$f_{\theta; X_1}(x) = \begin{cases} \theta^{-1}e^{-x/\theta} & \text{if} \quad x > 0, \\ 0 & \text{otherwise.} \end{cases}$$

Here $X = (X_1, \ldots, X_n)$, $\Omega = \{\theta : \theta > 0\}$, $D = \{d : d > 0\}$, $W(\theta, d) = [\phi(\theta) - d]^2$.

(a) Compute $I(\theta)$ and use the information inequality to show that $\bar{X}_n$ is a best unbiased estimator of $\phi(\theta) = \theta$. [Part of answer: $I_1(\theta) = \theta^{-2}$.]

(b) Suppose it is desired to estimate the variance of X_1, so that $\phi(\theta) = \text{var}_\theta(X_1) = \theta^2$.

(i) Compute the information inequality lower bound $H(\theta)$ on any unbiased estimator t of $\phi(\theta)$ in this case.

(ii) Show that $t_n^* = [n(n+1)]^{-1}\left(\sum_1^n X_i\right)^2$ is an unbiased estimator of θ^2. [*Hint*: See Appendix A regarding the Γ – distribution, for the fact that the density of $Y = \sum_1^n X_i \big/ \theta$ is $y^{n-1}e^{-y}/(n-1)!$ for $y > 0$.]

(iii) Show [Using the same hint] that

$$\mathrm{var}_\theta(t_n^*) = (4n+6)\theta^4/n(n+1).$$

(iv) Recalling Section 7.1, show that t_n^* is a best unbiased estimator of θ^2. Recalling (i) and (iii), show that this conclusion that t_n^* is best unbiased for θ^2 could *not* have been deduced from the information inequality. [*Fact*: $\sum_1^n X_i$ is a complete sufficient statistic.]

(v) Show that the ratio $H(\theta)/\mathrm{var}_\theta(t_n^*)$ approaches 1 as $n \to \infty$. Hence (although (iv) illustrates that the information inequality is not always adequate to prove that a best unbiased estimator t_n^* is indeed best), show that the inequality suffices to show that t_n^* is *approximately* best unbiased for θ^2 for large n. (This means, show that $\mathrm{var}_\theta(t_n^*)$ divided by $\min_{\{t_n' \text{ unbiased for } \phi\}} \mathrm{var}_\theta(t_n')$ is close to 1 for large n.)

[*Comment*: A corresponding result is valid for other functions ϕ which one might want to estimate.]

7.2.3. Same setting and same part (a) as in Problem 7.2.2.

(b) Suppose it is desired to estimate $\phi(\theta) = 1/\theta$. [As you can check in a probability book, this is the expected number of counts per unit time of the "Poisson process" for which the X_i are the lengths of time between counts. It turns out that there is no unbiased estimator of $1/\theta$ if $n = 1$; we will assume $n > 1$ later.]

(i) Show that the information inequality lower bound $H(\theta)$ on any unbiased estimator t of $\phi(\theta)$ in this case is $\mathrm{var}_\theta(t_{\text{unbiased}}) \geq 1/\theta^2$.

(ii) If $n \geq 2$, show that $t_n^* = (n-1)\Big/\sum_1^n X_i$ is an unbiased estimator of $1/\theta$. [*Hint*: See Appendix A on the Γ-law, for the fact that the density of $Y = \sum_1^n X_i$ is $\theta^{-n}y^{n-1}e^{-y/\theta}/(n-1)!$ for $y > 0$.]

(iii) If $n \geq 3$, show [using the same hint] that $\mathrm{var}_\theta(t_n^*) = 1/(n-2)\theta^2$. [If $n = 2$, t_n^* is unbiased but has infinite variance.]

(iv), (v) Same as in 7.2.2, but with θ^{-1} for θ^2.

7.2.4. Suppose $X = (X_1, X_2, \ldots, X_n)$ where the X_i are iid Bernoulli random variables, so that

$$f_{\theta;X_1}(x_1) = \begin{cases} \theta^{x_1}(1-\theta)^{1-x_1} & \text{if } x_1 = 0 \text{ or } 1, \\ 0 & \text{otherwise.} \end{cases}$$

Here $\Omega = \{\theta : 0 \leq \theta \leq 1\}$, $D = \{d : 0 \leq d \leq 1\}$, $W(\theta, d) = [\phi(\theta) - d]^2$.

(a) Show that $I_1(\theta) = 1/[\theta(1-\theta)]$. Use this with the information inequality to show that $\bar{X}_n$ is uniformly best among all unbiased estimators of $\phi(\theta) = \theta$.

(b) Suppose it is desired to estimate *not* the mean θ of the underlying Bernoulli law, but rather its variance; that is, $\phi(\theta) = \theta(1-\theta)$. Since this has no unbiased estimator if $n = 1$ (Section 4.6), we hereafter assume $n \geq 2$.

(i) Show that the information inequality lower bound $H_n(\theta)$ for any unbiased estimator of ϕ in this case is

$$\operatorname{var}_\theta(t_{\text{unbiased}}) \geq n^{-1}(1-2\theta)^2\theta(1-\theta) \stackrel{\text{def}}{=} H_n(\theta).$$

(ii) It can be shown, but you need not do so, that the first four moments of $S_n = \sum_1^n X_i$ (binomial random variable) are

$$\begin{aligned} E_\theta S_n &= n\theta, \\ E_\theta S_n^2 &= n\theta + n(n-1)\theta^2, \\ E_\theta S_n^3 &= n\theta + 3n(n-1)\theta^2 + n(n-1)(n-2)\theta^3, \\ E_\theta S_n^4 &= n\theta + 7n(n-1)\theta^2 + 6n(n-1)(n-2)\theta^3 \\ &\quad + n(n-1)(n-2)(n-3)\theta^4. \end{aligned} \tag{7.65}$$

From these, verify that $t_n^* \stackrel{\text{def}}{=} [nS_n - S_n^2]/[n(n-1)]$ is an unbiased estimator of $\phi(\theta)$. Use (7.65) to compute $E_\theta(t_n^*)^2$ and thus show that

$$\operatorname{var}_\theta(t_n^*) = \frac{1}{n(n-1)}\{(n-1)\theta + (7-5n)\theta^2 + (8n-12)\theta^3 + (6-4n)\theta^4\}.$$

(iii) From (i) and (ii), show that

$$\operatorname{var}_\theta(t_n^*) - H_n(\theta) = \frac{2\theta^2(1-\theta)^2}{n(n-1)}, \tag{7.66}$$

which is ≥ 0 (and >0 unless $\theta = 0$ or 1). From this conclude that *the information inequality method could not have been used to show* t_n^* *is a* minimum variance unbiased estimator (M.V.U.E.). [Comparing (a), this illustrates that *the success of this method depends on whether the* $\phi(\theta)$ *under consideration* is the "right one", namely, a linear function of $E_\theta k(X_1)$ in the K-D representation.]

(iv) Given the knowledge that S_n is a complete sufficient statistic, what could you have concluded about t_n^* from the method of Section 7.1?

(v) To see that, despite the conclusion of (iii), the method of the information inequality shows that t_n^* is at least *approximately* best unbiased for estimating $\theta(1-\theta)$ when n is large, show from (7.66) that

$$\lim_{n\to\infty} nH_n(\theta) = \lim_{n\to\infty} n \operatorname{var}_\theta(t_n^*) = \theta(1-\theta)(1-2\theta)^2,$$

which can be thought of as the limiting (normalized) risk of the sequence of procedures $\{t_n^*\}$. [A corresponding result holds for other functions ϕ.]

Problems on Section 7.3. Invariance. [Most of these problems are in the setting of the "scale-parameter estimation problem" described in Problem 7.3.1. Although this setting is sometimes arithmetically more complex than that of typical location-parameter problems, it has the advantage of greater intuitive naturalness (change of units) and

also gives the opportunity to exhibit some natural misgivings regarding the use of certain much-used loss functions. Examples of scale-parameter problems were given earlier, too; e.g., see the note on Problems 4.25–4.28.]
Suggested: At least one of 7.3.1–7.3.3.

7.3.1. *Scale-parameter estimation.*

Suppose $S = \{x : x > 0\}$ and that X is a real random variable with continuous case density function

$$f_{\theta;X}(x) = \begin{cases} \theta^{-1}g(x/\theta) & \text{if} \quad x > 0, \\ 0 & \text{otherwise,} \end{cases}$$

where $g(x) = f_{1;X}(x)$ for $x > 0$. The density g is given. Here $\Omega = \{\theta : \theta > 0\}$ and $D = \{d : d > 0\}$. The object is to estimate the unknown "*scale-parameter*" θ on the basis of a single observation X. The "family" of possible distributions $\{f_{\theta;X}, 0 < \theta < \infty\}$ is known, being specified in the form of the known function g; what is unknown is which member of the family is "true". We assume W depends only on the ratio d/θ, so that $W(\theta, d) = w(d/\theta)$ for an appropriate function w. This means that the loss depends only on the *relative* error $(d/\theta) - 1$, and not on the units of measurement; this form of dependence on (d, θ) is an aspect of the invariance of this problem. You can check that Problem 10101 of Section 7.3 falls into this framework with $g(x) = e^{-x}$. Moreover, the transformations considered there also leave the present problem invariant, so that the general form of an invariant estimator is $t_b(x) = bx$ where $b > 0$, and each invariant procedure has a constant risk function.

(a) How would b be chosen in general so as to yield the best invariant estimator? [Answer as "choose b to minimize ...," since you *cannot* hope to give the answer as "$b = \ldots$" unless W is more explicit.]

(b) Suppose $W(\theta, d) = [1 - (d/\theta)]^2$, "squared relative error." Show that $r_{t_b}(\theta) = b^2\mu_2 - 2b\mu_1 + 1$ for all θ, where $\mu_i = E_1 X^i = \int_0^\infty x^i g(x)\,dx$ (i^{th} moment of X when $\theta = 1$). Hence, show that the best invariant estimator is t_{b^*}, where $b^* = \mu_1/\mu_2$. Compute $r_{t_{b^*}}$. [Assume that μ_1 and μ_2 are finite in this problem.]

(c) Show that $\bar{t}(x) = x/\mu_1$ is an unbiased estimator of θ (which is best linear unbiased and, if X is complete for Ω, best unbiased; $\bar{t}$ is also invariant.) Compute $r_{\bar{t}}$ and $r_{t_{b^*}}$ and show that, for all θ,

$$\frac{r_{\bar{t}}(\theta)}{r_{t_{b^*}}(\theta)} = \frac{\mu_2}{\mu_1^2} \geq 1,$$

and that $\mu_2/\mu_1^2 = 1$ could hold for a positive random variable X only in the trivial case when $P_\theta(X = c\theta) = 1$ for some constant c (which must be known, since the law of X is assumed known except for the unknown scale parameter θ), and here $r_{\bar{t}} = r_{t_{b^*}} \equiv 0$ (no errors of observations, hence not our case of a density function). [As in the special models of Problems 4.25–4.28, this illustrates the questionable rationale of the "unbiasedness" criterion.]

(d) Suppose the problem is the same as previously, except that $W(\theta, d) = (\theta - d)^2$. Show that this form of W makes the problem no longer invariant under the same family of transformations. Show that any t has θ^2 times the risk function it has in (b), so that the ratio of $r_{\bar{t}}$ to $r_{t_{b^*}}$ for this W is the same as in (c). What is this ratio if (i) $g(x) = e^{-x}$ (as in the problem of Section 7.3); (ii) $g(x) = 1$ or 0 according to whether or not $0 < x < 1$ (problem of rectangular density from 0 to θ)? [This is analogous to the computations of Problems 4.25–4.28.]

(e) If $X = (X_1, \ldots, X_n)$ where the X_i's are independently and identically distributed with common density $f_{\theta;X}(x)$ of the form considered previously, show that the minimal sufficient statistic $\bar{X}_n$ of part (d), case (i), and $U_n = \max x_i$ of part (d), case (ii), both have a density of the form $\theta^{-1}f^*(s/\theta)$. Find the analogues of t_{b*} and $\bar{t}$ (the role of X in (b) and (c) being taken by the sufficient statistic). [As mentioned in Example 7.5(c), the t_{b*} based on the sufficient statistic coincides with the best invariant estimator which would have been obtained directly from $(X_1, \ldots, X_n)$ using the method of Example 7.5(b); do *not* prove this.] Find the ratio $r_{\bar{t}}(\theta)/r_{t_{b*}}(\theta)$ of (c) (d) in each of these two cases, noting what happens to this ratio as $n \to \infty$.

(f) In the general scale-parameter setup considered at the outset, let $Y = \log X$ and $\psi = \log \theta$, and write $\overline{W}(\psi, d') = W(e^{\psi}, e^{d'})$. Show that the problem of estimating ψ on the basis of Y with sample space $S' = \{y : -\infty < y < \infty\}$, decision space $D' = \{d' : -\infty < d' < \infty\}$, loss function $\overline{W}$, and $\Omega' =$ the set of all possible values of ψ corresponding to θ in Ω, is a location-parameter problem in the sense of Example 7.5(a). Show that, if t is an invariant estimator for the scale-parameter problem, and if t' is defined for the location-parameter problem by $t'(y) = \log t(e^y)$, then t' is invariant for the latter problem, and conversely; show also that, whether or not t is invariant, if t and t' are related in this fashion, we have, for each θ,

$$\underset{\text{(scale problem)}}{r_t(\theta)} = \underset{\text{(location problem)}}{r_{t'}(\log \theta)}.$$

[*Conclusion*: the comparison of procedures, and, in particular, the characterization of the best invariant procedure, for any location-parameter problem can be obtained by a simple transformation from the analogues for a corresponding scale-parameter problem, and conversely.]

7.3.2. The setting, and part (a), are as in 7.3.1.

(b) Suppose that $\int_0^\infty (\log x)^i g(x) = c_i$ (say) where c_i is finite for $i = 1, 2$. Suppose that $W(\theta, d) = (\log(d/\theta))^2$. [This loss function can be seen to be qualitatively of the right form to express the relative importance of various losses, being 0 when $d/\theta = 1$ and going to ∞ when $d/\theta \to 0$ or ∞; this W may be preferred to simple "squared relative error" $(d/\theta - 1)^2$, since the latter does not adequately express the seriousness of errors when d/θ is close to zero.] Show that

$$r_{t_b}(\theta) = (\log b)^2 + 2c_1 \log b + c_2 \quad \text{for all } \theta, \tag{7.67}$$

and hence that a uniformly best estimator among invariant estimators is obtained by choosing t_{b*} where b^* minimizes (7.67). Minimizing (7.67) as a function of $\log b$, show that the best choice of b is $b^* = e^{-c_1}$, for which

$$r_{t_{b*}}(\theta) = c_2 - c_1^2 \quad \text{for all } \theta.$$

(c) W is as in (b). From the fact that $(-\log x)$ is strictly convex for $x > 0$, and from Jensen's inequality, it follows in the present continuous case that $(-c_1) = E_1(-\log X) > -\log(E_1 X)$, so that $e^{-c_1} > (1/E_1 X)$. *Do not prove the previous sentence*, but use the conclusion of it with (7.67) of (b) to determine whether or not the estimator $\bar{t}(x) = x/\mu_1$, where $\mu_1 = \int_0^\infty xg(x)\,dx$, is admissible. Show that $\bar{t}$ is an unbiased estimator of θ. [And is, moreover, (i) the only unbiased (nonrandomized) invariant estimator of θ, (ii) the B.L.U.E. of θ, and (iii) if X complete for Ω, $\bar{t}$ is best among *all* unbiased estimators (including nonlinear and noninvariant ones), for every loss function that is convex in d. This is another illustration of the questionable rationale of using the unbiasedness criterion. It should be noted that $(\log u)^2$ is not convex for large u, so that the argument of Sections 6.5 and 7.1 does not imply that $\bar{t}$ is better than every *randomized* unbiased estimator, for the W of (b).]

(d) We now illustrate the preceding result. Suppose that

$$g(x) = \begin{cases} 1 & \text{if} \quad 0 < x < 1, \\ 0 & \text{otherwise,} \end{cases}$$

so that $f_{\theta;X}$ is uniform from 0 to θ. Verify (or believe) that $\mu_1 = \frac{1}{2}$, $c_1 = -1$, and $c_2 = 2$ in this case, and compute b^* and $r_{t_{b^*}}$ from (b). Now show that only the value $b = 2$ makes t_b an *unbiased* estimator of θ. [In fact, from Ex. 6.5 in Section 7.1 we know that X is complete in this case, so t_2 is a U.M.V.U. estimator of θ; it is the $\bar{t}$ discussed in (c).] Use (7.67) to compute r_{t_2}, and conclude that r_{t_2} is about 9 percent greater than $r_{t_{b^*}}$.

(e) Work case (ii) (the present g) of 7.3.1(e) for the W of the present problem ($[\log(d/\theta)]^2$), using the change of variable $u = e^{-x}$ and properties of the Γ-function (Appendix A) to compute that

$$\int_0^1 (\log x)^i x^{r-1}\,dx = (-1)^i i!/r^{i+1} \quad \text{for} \quad r > 0.$$

(f) Work 7.3.1(f), and show that the W of (b) of the present problem yields "squared error loss" for the corresponding location-parameter problem.

7.3.3. The setting, and part (a), are as in 7.3.1.

(b) Suppose that $\int_0^\infty x^i g(x)\,dx = c_i$ (say) where c_i is finite for $i = 1, 2, -1, -2$. Suppose also that $W(\theta, d) = (1 - d/\theta)^2 + (1 - \theta/d)^2$. [This loss function can be seen to be qualitatively of the right form to express the relative importance of various losses; it may be preferred to simple "squared relative error" $(d/\theta - 1)^2$, since the latter does not adequately express the seriousness of errors when d/θ is close to zero.] Show that

$$r_{t_b}(\theta) = c_2 b^2 - 2c_1 b + 2 - 2c_{-1} b^{-1} + c_{-2} b^{-2} \quad \text{for all } \theta, \qquad (7.68)$$

and hence that a uniformly best estimator among invariant estimators is obtained by choosing t_{b^*} where b^* minimizes (7.68). [Don't try to minimize (7.68) explicitly until the example of (c).]

(c) Now suppose $g(x) = x^2 e^{-x}/2$. (This $\theta^{-1} g(x/\theta)$ could arise, for example, as the density of the minimal sufficient statistic $X_1 + X_2 + X_3$ for iid observations with $n = 3$ in the "problem 10101" of Section 7.3.) Show that $c_2 = 12$, $c_1 = 3$, $c_{-1} = c_{-2} = 1/2$, and verify numerically (e.g., by solving, approximately, $\frac{\partial}{\partial b} r_{t_b} = 0$) that (7.68) is minimized by $b^* \doteq .47$, yielding $r_{t_{b^*}} \doteq 1.86$.

(d) Verify that the only invariant *unbiased* estimator is $t_{1/3}$. [The completeness of X follows from Example 7.3 in Section 7.1. Hence, $t_{1/3}$ is best unbiased for all W convex in d. *Optional*: Show that the W of (b) is convex in d.] Show that $r_{t_{1/3}}$ is more than 50 percent larger than r_{t_b*}. [Another illustration of what can happen from using the "unbiasedness" criterion.]

(e) Work 7.3.1(e) for the g of the present part (c), with the present $W(\theta, d) = (\theta^{-1}d - 1)^2 + (d^{-1}\theta - 1)^2$, using the fact (Appendix A) that the sum of n iid Γ-random variables with index parameter 2 (of g) has a Γ-law with index $2n$. That sum is minimal sufficient, by Problem 6.1.

(f) Work 7.3.1(f).

Problem on Section 7.4.

7.4.1. Suppose $X_1, \ldots, X_n$ are iid, with $\Omega = \{F : F \text{ is concentrated on } [0, 1]\} = \{F : P_F\{0 \le X_i \le 1\} = 1\}$. It is desired to estimate $\phi(F) = E_F X_1$ with squared error loss, in this nonparametric setting. The following result from Problems 2.1 and 4.19 should be used: there are constants A_n, B_n such that the estimator $t_n^* = A_n \bar{X}_n + B_n$ is minimax for the *subproblem* of the preceding, where Ω is replaced by $\Omega' = \{\text{all Bernoulli laws for } X_1\}$.

(a) Use the inequality between x and x^2 for $0 \le x \le 1$ to conclude that, for F in Ω, we have $E_F X_1^2 \le E_F X_1$, and consequently $\text{var}_F(X_1) \le \phi(F)[1 - \phi(F)]$, with equality if $F \in \Omega'$.

(b) Use the conclusion of (a) to verify that $r_{t_n^*}(F) \le A_n^2 n^{-1} \phi(F)[1 - \phi(F)] + [B_n + (A_n - 1)\phi(F)]^2$, with equality if $F \in \Omega'$.

(c) Use the result in the first of the remarks on minimax procedures at the end of Section 7.4 to conclude that t_n^* is minimax for the problem concerning Ω.

Problems on Section 5. Maximum likelihood.
Suggested: *One or two from* 7.5.1–7.5.6 or 7.5.9; 7.5.7 or 7.5.8.

7.5.1. In Problem 6.1 on sufficiency (notation of (6.12)) with $n = 1$, suppose that $q(\theta)$ is differentiable and strictly increasing in θ (this last merely ensures that different values of θ yield different probability laws) and that Ω is an interval of θ-values such that the M.L. estimator can be obtained by solving the likelihood equation $\partial \log f_{\theta; X_1}(x_1)/\partial\theta = 0$. (a) Show that $k(X_1)$ is the unique M.L. estimator of $\phi(\theta) = E_\theta k(X_1)$ so that $\phi^{-1}[k(X_1)]$ is the M.L. estimator of θ, where ϕ^{-1} is the inverse function of ϕ, i.e., $\phi^{-1}(\phi(\theta)) = \theta$. [*Hint*: Use the fact that

$$1 = c(\theta) \sum_{x_1} b(x_1) e^{k(x_1)q(\theta)}$$

to evaluate $c'(\theta)/c(\theta)$ directly or by implicit differentiation.] (b) See (iv) of Ex. 7.8 regarding the use of this result for $n > 1$, and by substituting $C(\theta)$ for $c(\theta)$, etc., show that (again, if the likelihood equation yields the M.L. estimator) the M.L. estimator of θ when $X = (X_1, \ldots, X_n)$ with iid X_i's having law (6.12), is $\phi^{-1}\left(\sum_1^n k(X_i)/n\right)$, where ϕ^{-1} is as defined in (a) for $n = 1$. [*Remark*: The assumption that the M.L. estimator can be found by solving the likelihood equation is not alway satisfied, even in K-D settings, as can be seen in the remark just before Ex. 7.10. If the true θ_0 lies in the interior of the interval Ω (always true if Ω is open), it can be shown under reasonable regularity conditions (even outside the K-D case) that, as $n \to \infty$, $P_{\theta_0}\{\hat{t}(X_1, \ldots, X_n) \text{ is a solution of the likelihood equation}\} \to 1$.

7.5.2. Find the M.L. estimator $\hat{t}_n$ of ϕ in each of the following continuous cases where $X = (X_1, \ldots, X_n)$ with the X_i's iid. After finding the formula for $\hat{t}_n$ in each case, compute the value taken on by this estimator if $n = 4$ and the X_i's take on values 5, 2, 9, 8.

(a) The Ω is that of Problem 6.3 and (two problems)

(i) $\phi(\theta) = \theta$,

(ii) $\phi(\theta) = \theta^2$.

(b) The Ω is $\{\theta : -\infty < \theta < \infty\}$ with $f_{\theta; X_1}(x_1) = e^{-(x_1 - \theta)}$ if $x_1 \geq \theta$ (and 0 otherwise) and $\phi(\theta) = P_\theta\{X_1 \geq 3\}$.

(c) The X_i are iid normal with mean and standard deviation both equal to θ. Here $\Omega = \{\theta : 0 < \theta < \infty\}$ and $\phi(\theta) = \theta$. [This is the family of Problem 6.3(v), but with θ restricted to be >0.] [*Answer*: $\hat{t}(x_1, \ldots, x_n) = -\frac{1}{2}\bar{x}_n + \frac{1}{2}\left[\bar{x}_n^2 + 4n^{-1}\sum_1^n x_i^2\right]^{1/2}$.]

7.5.3. If $X_1, X_2, \ldots, X_n$ are iid, each with density [motivated from a "practical point of view" at the end of this problem]

$$f_{(\theta_1, \theta_2); X_1}(x_1) = \begin{cases} \theta_2^{-\theta_1} \theta_1 x_1^{\theta_1 - 1} & \text{if } 0 < x_1 \leq \theta_2, \\ 0 & \text{otherwise}, \end{cases}$$

(where $\theta_1 > 0$, $\theta_2 > 0$, and Ω will be completely specified later), show that

(i) if θ_1 is known (positive) and $\Omega = \{\theta_2 : \theta_2 > 0\}$, the M.L. estimator of θ_2 is $\max(X_1, X_2, \ldots, X_n)$;

(ii) if θ_2 is known (positive) and $\Omega = \{\theta, 0 < \theta_1 < \infty\}$, the M.L. estimator of θ_1 is $n \Big/ \sum_1^n \log(\theta_2 / X_i)$;

(iii) if $\Omega = \{(\theta_1, \theta_2) : 0 < \theta_1 < \infty, 0 < \theta_2 < \infty\}$, the M.L. estimator of (θ_1, θ_2) is $\left(n \Big/ \sum_1^n \log[\max(X_1, \ldots, X_n)/X_i], \max(X_1, \ldots, X_n)\right)$;

(iv) the estimator of (i) is biased, but in case (ii) the M.L. estimator of $1/\theta_1$ is unbiased [*Hint*: $-\int_0^1 x^{\alpha - 1} \log x \, dx = \alpha^{-2}$; incidentally, the M.L. estimator of θ_1 itself *is* biased];

(v) in case (iii) the M.L. estimator of $\phi(\theta_1, \theta_2) = P_{(\theta_1, \theta_2)}\{X_1 > 1\}$ is ________?

[Model that could yield such a problem: There are iid random variables Y_j, uniformly distributed from 0 to θ_2. You send an observer out on each of n successive days to observe some Y_j's. He doesn't record the Y_j's. Instead, knowing that "the maximum of the Y_j's is sufficient and an M.L. estimator," he decides to observe a certain number, θ_1, of the Y_j's each day and compute the maximum of these θ_1 observations. X_i is the maximum he computes on the ith day. Unfortunately, he forgets to tell you the θ_1 he used, so it is also unknown. Then X_i has the density function stated for this problem, where we have simplified matters by allowing θ_1 to be any positive value instead of restricting it to integers.]

7.5.4. If $X_1, X_2, \ldots, X_n$ are iid, each with density (of Problem 6.3(ix))

$$f_{(\theta_1, \theta_2); X_1}(x_1) = \begin{cases} \theta_2^{-1} e^{-(x_1 - \theta_1)/\theta_2} & \text{if } x_1 \geq \theta_1, \\ 0 & \text{otherwise}, \end{cases}$$

show that

(i) if θ_1 is known and $\Omega = \{\theta_2 : \theta_2 > 0\}$, the M.L. estimator of θ_2 is $\sum_1^n (X_i - \theta_1)/n$;

(ii) if θ_2 is known and $\Omega = \{\theta_1 : -\infty < \theta_1 < \infty\}$, the M.L. estimator of θ_1 is $\min(X_1, \ldots, X_n)$;

(iii) if $\Omega = \{(\theta_1, \theta_2) : -\infty < \theta_1 < \infty, \theta_2 > 0\}$, the M.L. estimator of (θ_1, θ_2) is $\left(\min(X_1, \ldots, X_n), n^{-1}\sum_1^n [X_i - \min(X_1, \ldots, X_n)]\right)$;

(iv) the estimator of (i) is unbiased but that of (ii) is not [*optional*: what about (iii)?];

(v) the M.L. estimator of $\phi(\theta_1, \theta_2) = P_{(\theta_1,\theta_2)}\{X_1 > 0\}$ is ________?

7.5.5. Find the M.L. estimator $\hat{t}_N$ of $\phi(\theta)$ in each of the following cases. After computing the general formula, find the value taken on by $\hat{t}_N$ when $N = 5$ and the X_i's take on values 1.3, 1.7, 2.1, 1.4, 2.5.

(a) In the setup of Problem 6.9 with $\phi(\theta) = \theta$, find three values of p such that, assuming the known value of p is one of these three, $\hat{t}_N$ can be obtained by solving an equation of degree at most 2; obtain $\hat{t}_N$ in these three cases. How would you obtain $\hat{t}_N$ for other values of p?

(b) The setup is that of Problem 6.11 with $\phi(\theta) = \theta$, assuming $g(x)$ positive for $0 < x < \infty$. [*Hint*: Show that $c(\theta)$ is strictly decreasing and that the likelihood is 0 for $\theta < \max(x_1, \ldots, x_N)$.]

7.5.6. Suppose $X = (X_1, \ldots, X_n)$ where the iid X_i's have density given by Problem 6.6(x) with $\Omega = \{\theta : 0 < \theta < \infty\}$.

(a) If $n = 1$, show (i) that an M.L. estimator of θ is $\hat{t}_1(x_1) = 2x_1$. Show also that

(ii) $\hat{t}_1$ is not an unbiased estimator of θ $\left[\text{verify or believe: } \int_0^\infty \frac{x\,dx}{(x+1)^3} = \frac{1}{2}\right]$;

(iii) if $W(\theta, d) = (\theta - d)^2$, then $\hat{t}_1$ is much worse than $t^*(x_1) \equiv 17$ [*Hint*: $\int_0^\infty \frac{x^2}{(1 + x^3)}\,dx = +\infty$.]

(b) If $n = 2$, show that an M.L. estimator of θ is $\hat{t}_2(x_1, x_2) = \frac{1}{4}[x_1 + x_2 + (x_1^2 + 34x_1x_2 + x_2^2)^{1/2}]$. For general n, describe the computation of the M.L. estimator $\hat{t}_n$ in terms of solving a polynomial equation of some degree, checking whether a local maximum is a global maximum, etc.

7.5.7. Suppose $X = (X_1, \ldots, X_n)$ where the iid X_i's have density given by Problem 6.3, part (vi), where again $\Omega = \{\theta : -1 \leq \theta \leq 1\}$.

(a) [*Easy*] If $n = 1$, show that an M.L. estimator of θ is

$$\hat{t}_1(x_1) = \begin{cases} 1 & \text{if } x_1 \geq 0 \\ -1 & \text{if } x_1 < 0 \end{cases},$$

the only arbitrariness in this definition occurring when $x_1 = 0$. Show also that

(i) $\hat{t}_1$ is not an unbiased estimator of θ.

(ii) If $W(\theta, d) = (\theta - d)^2$, then $r_{\hat{t}}(\theta) = 1$ for all θ, and the estimator $t_0(x_1) \equiv 0$ has a better risk function.

(b) [*Harder*] If $n = 2$, show that an M.L. estimator

$$\hat{t}_2(x_1, x_2) = \begin{cases} -\frac{1}{2}(x_1^{-1} + x_2^{-1}) \text{ if } x_1 x_2 < 0 \text{ and } -2 \le (x_1^{-1} + x_2^{-1}) \le 2, \\ 1 \quad \text{if} \begin{cases} x_1 x_2 < 0 \quad \text{and} \quad (x_1^{-1} + x_2^{-1}) < -2 \\ \qquad \text{or} \\ x_1 x_2 \ge 0 \quad \text{and} \quad x_1 + x_2 \ge 0, \end{cases} \\ -1 \quad \text{if} \begin{cases} x_1 x_2 < 0 \quad \text{and} \quad (x_1^{-1} + x_2^{-1}) > 2 \\ \qquad \text{or} \\ x_1 x_2 \ge 0 \quad \text{and} \quad x_1 + x_2 < 0, \end{cases} \end{cases}$$

and that the only arbitrariness in this definition occurs when $x_1 + x_2 = 0$. In the region $S = \{(x_1, x_2): -1 \le x_1, x_2 \le 1\}$, show where $\hat{t}_2 = 1, -1, 0, 1/2$.

(c) [For what follows, guess, or try when $n = 3$, etc.] Describe the computation of the M.L. estimator $\hat{t}_n$ in terms of solving a polynomial equation of some degree, checking whether a local maximum is a global maximum, etc.

7.5.8. The probability law is that of Problem 6.22.

(a) (i) Show that, if Ω is taken to be $\{\theta: 0 < \theta < \infty\}$, and if $n = 1$, then $\hat{t}_1$ does not exist unless $X_1 = (K + 1)/2$, in which event it is not uniquely defined.

(ii) Show that, in this example, $p_{\theta; X_1}(x_1)$ approaches a limit *which is a discrete pdf*, as $\theta \to 0$, and also as $\theta \to +\infty$. Denoting the limits as $p_{0; X_1}(x_1)$ and $p_{\infty; X_1}(x_1)$, we thus also have a continuous parametrization on $\Omega^* = \{\theta: 0 \le \theta \le \infty\}$. Show that, if Ω is replaced by Ω^*, the M.L. estimator always exists. [*Remarks*: (1) The "extension" (compactification) of Ω to Ω^*, as performed here, is not always possible, as is illustrated in Ex. 7.11(c). (2) If one does not like to include "$+\infty$" as the name of a parameter value, one can reparametrize the problem, e.g., by writing $\tau = \theta/(1 + \theta)$ if $0 \le \tau < 1$ and $\tau = 1$ if $\theta = +\infty$. The density $p_\tau^* = p_{\theta_\tau}$, defined by this correspondence ($\theta_\tau = \tau/(1 - \tau)$ if $0 \le \tau < 1$, $\theta_1 = +\infty$) is then continuous in τ on $[0, 1]$.]

(b) Find $\hat{t}_2$ when $n = 2$, using the Ω^* of (a) in place of Ω, to make the considerations simpler. [If $\min(X_1, X_2) > (K + 1)/2$ or $\max(X_1, X_2) < (K + 1)/2$, the consideration is similar to that of (i) of (a). Other cases are conveniently divided in terms of the sign of $\bar{X}_2 - (K + 1)/2$.]

(c) Same as 7.5.7(c). $\left[\partial \log \prod_1^n p_{\theta; X_1}(x_i)\right] / \partial\theta$ is a rational function of θ whose numerator is of degree $K - 1$; thus, for $K = 2$ or 3, the M.L. estimator is still manageable.]

7.5.9. *A non-iid model*: Three iid flips of a coin are performed with probability of a head $= \theta$ on each flip, $0 \le \theta \le 1$. However, the experimenter records only X_1 and $Y = X_2 X_3$ (the latter being the indicator of the event that the second and third flips are *both* heads). Find the M.L. estimator of θ based on X_1 and Y, writing out the four numerical values of $\hat{t}(x_1, y)$ for $x_1 = 0, 1$ and $y = 0, 1$.

7.5.10. Suppose $\Omega = \{\theta: A \le \theta \le B\}$ where $-\infty < A < B < \infty$, and that, for each x in S, $p_{\theta; X}(x)$ is continuous in θ. (a) Show that every possible value of $\hat{t}(x)$ is a *mode* of the posterior law of θ given that $X = x$, when the prior law is *uniform* density on Ω. (b) If $W(\theta, d) = \begin{Bmatrix} 0 & \text{if} \quad |\theta - d| \leq \varepsilon, \\ 1 & \text{otherwise} \end{Bmatrix}$ with $D = [A, B]$, show that, whenever $A + \varepsilon \leq \hat{t}(x) \leq B - \varepsilon$, the estimator $\hat{t}(x)$ is *approximately* a Bayes estimator relative to the uniform prior law on Ω, provided ε is small enough. (How small ε has to be depends on the form of $p_{\theta;X}(x)$ for θ near $\hat{t}(x)$.)

Problems on Section 7.6. *Asymptotic theory*. These problems are divided into two parts. Problems 7.6.1–7.6.5 cover some of the asymptotic considerations in "nice" settings where the computations are simple and one can illustrate the ideas by *explicit* comparison of such asymptotically efficient estimators as M.L., Bayes, best invariant, and best unbiased estimators. The remaining problems are devoted to more complex cases in which, typically, the M.L. and other asymptotically efficient estimators are intractable, and an estimator such as the $\tilde{t}_N$ of (7.62) may be used (or else, a less efficient one is used). *Suggested*: At least one of 7.6.1–7.6.2 and at least one of 7.6.6–7.6.9.

7.6.1. [Parts (a)–(f) are easy.] This problem illustrates, in a "nice" (K-D) setting, the fact that several different methods of constructing estimators lead to approximately the same risk functions for large N in the setting $X = (X_1, \ldots, X_N)$ with the X_i's iid, and that all of these ((b)–(e)) are "asymptotically efficient" in the sense described in Section 7.6, at least in this example. The setting is that of Problem 7.2.3, estimating $1/\theta$ for the exponential density of Problem 7.2.2. The loss function, squared error, is $(d - \theta^{-1})^2$. Write $T_N = N\bar{X}_N = \sum_1^N X_i$.

(a) (i) [*Optional*: Computations used later.] From Problem 6.1, T_N is sufficient (and it is complete). We recall that T_N has, in the notation of Appendix A, the gamma-density $\gamma(\theta, N)$. Hence $E_\theta T_N^k = \theta^k(N + k - 1)!/(N - 1)!$ for k an integer ($+\infty$ if $N + k - 1 < 0$).

(ii) Use (i) or Problem 7.2.3(b) (ii), (iii) to show that, for $N \geq 3$, any estimator of $1/\theta$ of the form $c_N T_N^{-1}$ (with c_N constant) has risk function

$$E_\theta(c_N T_N^{-1} - \theta^{-1})^2 = \theta^{-2}\{c_N^2/(N - 1)(N - 2) - 2c_N/(N - 1) + 1\}. \tag{7.69}$$

(b) Use (7.69) and the result of Problem 7.5.1 or 7.5.4(i) to show that the M.L. estimator of θ^{-1} corresponds to $c_N = N$ and yields risk function

$$\theta^{-2}(N + 2)/(N - 1)(N - 2).$$

(c) From (a) (i) or from 7.3.2(b) (ii), recall that, for $N \geq 2$, $c_N = N - 1$ gives an unbiased estimator of θ^{-1} (which is in fact "uniformly best unbiased" from our completeness-sufficiency theory) and yields risk function

$$\theta^{-2}(N - 1)/(N - 1)(N - 2)$$

[written this way for comparison with (b)].

(d) The form $c_N T_N^{-1}$ of an estimator is motivated by "invariance" considerations. Show that the "best" choice of c_N in (7.69) is $c_N = N - 2$, and that the resulting estimator has risk function

$$\theta^{-2}(N - 2)/(N - 1)(N - 2).$$

(e) Recall from Problem 7.2.3(b) (i) that the information inequality lower bound for the risk function $E_\theta(\tilde{t}_N - \theta^{-1})^2$ of an *unbiased* estimator $\tilde{t}_N$ of θ^{-1} is θ^{-2}/N.

Consequently, assuming that biased estimators also cannot have risk much smaller than this bound on a large set of θ-values when N is large, we define a *sequence* of estimators $\{t_N\}$ (one for each sample size) to be *asymptotically efficient* if $E_\theta(t_N - \theta^{-1})^2/[\theta^{-2}/N] \to 1$ as $N \to \infty$, for each θ in Ω. Show that each of the sequences of (b), (c), (d) is asymptotically efficient. [This is what was computed in Problem 7.2.3(b), part (v), under the restriction to *unbiased* estimators.]

(f) As an indication of the care one must exercise in using "asymptotic theory," compare the risk functions of (b), (c), (d) when $N = 3$, $N = 6$, $N = 25$. When would you judge that N is large enough that each of the "asymptotically equivalent" estimators of (b) and (c) behaves roughly like that of (d)? (A vague, but practical question; your answer may differ in the two cases.)

(g) [*Harder*] Another type of estimator t_N, *not of the form* $c_N T_N^{-1}$, can be obtained as a Bayes procedure relative to a "smooth" prior density f_ξ (not depending on N). The density is chosen here to make the integration as simple as possible, and many other prior densities would work. One does not have to be a Bayesian to use this method, the point of which is that, *for any suitably regular prior law* ξ, *the Bayes procedures* $\{t_{\xi,N}\}$ *(one for each sample size), will be asymptotically efficient* (as discussed in Section 7.6). A disadvantage of this approach for unbounded parameter sets Ω (as in the present example) is that no ξ can assign much probability to very large θ values, so that the risk function will require very large N to become "efficient" for such values. (Values near 0 are also troublesome in our example, because of the analytic form of f_ξ we have chosen in order to make the necessary integration simple.)

$$\text{Let } f_\xi(\theta) = \begin{cases} \theta^{-2}e^{-1/\theta}, & \theta > 0, \\ 0 & \text{otherwise.} \end{cases}$$

First show that this is indeed a probability density function by integrating it, using the transformation $\rho = \theta^{-1}$. Next, note that the result of Problem 4.6 (on Bayes estimation with quadratic loss, valid here although $\psi = \theta^{-1}$ and Ω are not bounded) implies that the Bayes procedure relative to ξ is the *a posteriori expected value of* ψ *given* X, which by sufficiency we can alter to "given T_N"; when $T_N = \tau$, this estimator takes on the value (in terms of integration with respect to the variable θ rather than ψ)

$$t_{\xi,N}(\tau) = \frac{\int_0^\infty \theta^{-1} f_{\theta,T_N}(\tau) f_\xi(\theta)\, d\theta}{\int_0^\infty f_{\theta;T_N}(\tau) f_\xi(\theta)\, d\theta}, \tag{7.70}$$

where f_{θ,T_N} is the density of T_N. Perform the integration (by changing variables) to obtain

$$t_{\xi,N}(X) = (N+1)/(T_N+1).$$

Note that, for fixed "true" $\theta > 0$, T_N/N is very likely to be close to θ when N is large, so that $(T_N + 1)/T_N$ is likely to be close to 1, and thus $t_{\xi,N}$ is likely to be close in value to the estimators of (b), (c), (d). [*Optional*: Make this asymptotic efficiency of $\{t_{\xi,N}\}$ precise by finding the risk function of $t_{\xi,N}$ and comparing it with θ^{-2}/N.]

(h) [*Not too hard*] The estimators of (b), (c), (d), (g) are easy enough to compute from the data, so that one would rarely use alternative estimators based on a few sample quantiles in this setting. Nevertheless, we illustrate the arithmetic of such estimators for one of the simplest form, the *sample median* $t_N^\#$ (say). [More efficient estimators could be based, e.g., on a weighted average of $t_N^\#$ and the upper and lower sample quantiles.] Show that (still in the setting of Problem 7.3.2 with $f_{\theta;X_1}$ being abbreviated f_θ),

(i) f_θ has median $\theta \log 2$;

(ii) $t_N^\#$ is for large N approximately normal with mean $\theta \log 2$ and variance $.25N^{-1}/[f_\theta(\theta \log 2)]^2 = \theta^2 N^{-1}$. (See the result preceding Ex. 7.12.)

(iii) Since $t_N^\#$ is close to $\theta \log 2$ with high probability when N is large, an appropriate estimator of θ may be $t_N^\#/\log 2$, which is approximately $\mathcal{N}(\theta, \theta^2/N(\log 2)^2)$.

(iv) For estimating θ^{-1}, we may hence use $(t_N^\#/\log 2)^{-1} = t'_N$ (say). Its risk function is hard to compute exactly, but one can use the result given at the end of the subsection on sample quantiles. Putting $h(u) = u^{-1}$ there (for $u > 0$), conclude from the result there and from (iii) that $N^{1/2}[(t_N^\#/\log 2)^{-1} - \theta^{-1}]$ is asymptotically normal, the limiting normal law having mean zero and variance

$$1/\theta^2(\log 2)^2.$$

Hence, comparing this with (e), we see that the risk function of t'_N is asymptotically 2.1 (i.e. $1/(\log 2)^2$) times that of the estimator of (b), (c), or (d). By using a linear combination of enough sample quantiles, we can reduce this "2.1" to as near 1 as desired.

7.6.2. The setting is the same exponential Ω as in Problem 7.6.1, but now it is desired to estimate θ^2, again with squared error loss, as in Problem 7.2.2. Part (i) of (a) is as in Problem 7.6.1.

(a) (ii) Show that any estimator of θ^2 of the form $c_N T_N^2$ (with c_N constant) has risk function

$$\begin{aligned} &E_\theta(c_N T_N^2 - \theta^2)^2 \\ &\quad = \theta^4\{c_N^2 N(N+1)(N+2)(N+3) - 2c_N N(N+1) + 1\}. \end{aligned} \tag{7.71}$$

(b) Use (7.71) and the result of Problem 7.5.1 or 7.5.4(i) to show that the M.L. estimator of θ^2 corresponds to $c_N = N^{-2}$ and yields risk function

$$\theta^4\left\{\frac{4N^2 + 11N + 6}{N^3}\right\}.$$

(c) From (a) (ii) show that $c_N = 1/N(N+1)$ gives an unbiased estimator of θ^2 (which is in fact "uniformly best unbiased" from our completeness-sufficiency theory) and yields risk function

$$\theta^4\left\{\frac{4N+6}{N(N+1)}\right\}.$$

(d) The form $c_N T_N^2$ of an estimator is motivated by "invariance" considerations. Show that the "best" choice of c_N in (7.71) is $c_N = 1/(N+2)(N+3)$, and that the resulting estimator has risk function

$$\theta^4\left\{\frac{4N+6}{(N+2)(N+3)}\right\}.$$

(e) Use the result of Problem 7.2.2(b) (i) (or compute it) to show that the information inequality lower bound for the risk function $E_\theta(t_N^* - \theta^2)^2$ of an *unbiased* estimator t_N^* of θ^2 is $4\theta^4/N$. Consequently, assuming that biased estimators also cannot have risk much smaller than this bound on a large set of θ-values when N is large, we define a *sequence* of estimators $\{t_N\}$ (one for each sample size) to be *asymptotically efficient* if $E_\theta(t_N - \theta^2)^2/[4\theta^4/N] \to 1$ as $N \to \infty$, for each θ in Ω. Show that each of the sequences of (b), (c), (d) is asymptotically efficient.

(f) Same statements as in Problem 7.6.1(f).

(g) Same statement down to (7.70) as Problem 7.6.1(g), replacing $\psi = \theta^{-1}$ by $\psi = \theta^2$ and θ^{-1} in the numerator of (7.70) by θ^2. Thereafter, proceed as follows: Assuming $N \geq 2$, perform the integration to obtain, for $N \geq 2$,

$$t_{\xi,N}(X) = \frac{(T_N+1)^2}{N(N-1)}.$$

Note that, for fixed "true" $\theta > 0$, T_N/N is very likely to be close to θ when N is large, so that $t_{\xi,N}$ is likely to be close to $\bar{X}_N^2$, and hence to the estimators of (b), (c), (d). [*Optional*: Make this asymptotic efficiency of $\{t_{\xi,N}\}$ precise by finding the risk function of $t_{\xi,N}$ and comparing it with $4\theta^2/N$.]

(h) Same as Problem 7.6.1(h) through (iii). Then proceed as follows:

(iv) For estimating θ^2, we may hence use $(t_N^\#/\log 2)^2 = t_N'$ (say). Its risk function is hard to compute exactly, but one can use the result given at the end of the subsection on sample quantiles as follows: put $h(u) = u^2$, $h'(u) = 2u$ on $0 < u < \infty$, and from that result and (iii) conclude that $N^{1/2}[t_N' - \theta^2]$ is asymptotically normal, the limiting normal law having mean 0 and variance

$$4\theta^4/(\log 2)^4.$$

Hence, comparing this with (e), we see that the risk function of t_N' is asymptotically $(\log 2)^{-4}$ (approximately 4) times that of the estimator of (b), (c), or (d). By using enough sample quantiles, we could reduce this "4" to as near 1 as desired.

7.6.3. The setting is that of Problem 7.2.1(a), of estimating the mean θ of the exponential law, with squared error loss. It is decided to use an inefficient estimator based on a single sample α-tile $Z_{\alpha,N}$, and it is desired to find the asymptotically best choice of α, where $0 < \alpha < 1$.

(a) Show that the population α-tile is $-\theta\log(1-\alpha)$, and use the result stated just before Ex. 7.12 to show that $Z_{\alpha,N}$ is approximately $\mathcal{N}(-\theta\log(1-\alpha), \theta^2N^{-1}\alpha/(1-\alpha))$.

(b) From (a), it seems appropriate to estimate θ by $t_{\alpha,N} = -Z_{\alpha,N}/\log(1-\alpha)$. Show that $t_{\alpha,N}$ is approximately $\mathcal{N}(\theta, \theta^2N^{-1}\alpha/(1-\alpha)[\log(1-\alpha)]^2)$.

(c) From (b), show that the variance of the limiting normal law of $N^{1/2}(t_{\alpha,N} - \theta)$ is minimized by the choice $\alpha = .80$ (or, more accurately, .797), for which the asymptotic efficiency compared with the M.L. estimator $\bar{X}_N$ is 65 percent. [The best α may be found numerically, after differentiation.]

(d) Show that the corresponding estimator $t_{1/2,N}$ based on the sample median has approximately 35 percent larger variance than $t_{.80,N}$.

7.6.4. For estimating the mean θ of the $\mathscr{N}(\theta, 1)$ law with $\Omega = \{\theta: -\infty < \theta < \infty\}$, suppose it is decided to use an inefficient estimator $t_{\alpha;N} = \frac{1}{2}(Z_{\alpha,N} + Z_{(1-\alpha),N})$, the average of two symmetrically related sample quantiles. Here $0 < \alpha < \frac{1}{2}$; the estimator $t_{1/2,N}$ is the sample median.

(a) Use the result stated (in the subsection on inefficient procedures) for the joint limit law of two sample quantiles, to show that $t_{\alpha,N}$ is approximately $\mathscr{N}(\theta, \alpha/2N[\phi(t_\alpha)]^2)$, where ϕ is the $\mathscr{N}(0,1)$ density and t_α is the α-tile of the $\mathscr{N}(0,1)$df Φ.

(b) By differentiation and numerical solution, show that $\Phi(t)/[\phi(t)]^2$ on $-\infty < t < 0$ is minimized at $t = .612$. Conclude that the best choice of α, for minimizing the variance of the limiting normal law of $N^{1/2}(t_{\alpha,N} - \theta)$, is $\alpha = .27$, and that $t_{.27,N}$ is asymptotically 81 percent efficient compared with $\bar{X}_N$.

(c) Referring to Ex. 7.12, show that the sample median $Z_{1/2,N}$ has approximately 27 percent larger variance than $t_{.27,N}$.

7.6.5. *Robustness consideration*: Referring to the setting of Problem 7.6.2, of estimating θ^2 where θ is the mean of the exponential law, suppose that, unknown to the statistician, the exponential density is slightly (1 percent) "contaminated" by the density $\theta(x_1 + \theta)^{-2}$, $\theta > 0$; thus, the X_i are iid with density

$$\tilde{f}_\theta(x_1) = .99\theta^{-1}e^{-x_1/\theta} + .01\theta(x_1 + \theta)^{-2}$$

on $\{x_1 : x_1 > 0\}$, and once more $\Omega = (0, \infty)$. One may think of this "mixture" as arising because, for each observation, there is a 99 percent chance that it has an exponential density and a 1 percent chance that something goes wrong and it has density $\theta(x_1 + \theta)^{-2}$ instead.

(a) If the statistician uses one of the estimators of Problem 7.6.2(b), (c), (d), or (g), of the form $c_N T_N^2$, show that the expected loss is infinite, no matter what the true value of θ may be.

(b) It may be objected that the conclusion of (a) is not very upsetting, since it depends on the unboundedness of W, which, after all, was chosen largely for computational convenience as long as it had the right "shape." Show that any of the estimators $c_N T_N^2$ mentioned previously in fact behaves catastrophically in any reasonable nonmasochistic sense, by showing that, under $\tilde{f}_\theta$, $\lim_{N\to\infty} P_\theta\{\bar{x}_N > B\} = 1$ *for every* $B > 0$, so that "$\bar{X}_N \to +\infty$ stochastically." [One method, if one does not know a "law of large numbers" for such a density: Fix θ. Since $\int_0^\infty x_1 \tilde{f}_\theta(x_1)\,dx_1 = +\infty$, you can show that for each B there is a finite number k (depending on θ and B) such that $\int_0^k x_1 \tilde{f}_\theta(x_1)\,dx_1 > B + 1$. Let $X_i' = \begin{Bmatrix} X_i & \text{if} \quad X_i \le k \\ 0 & \text{otherwise} \end{Bmatrix}$. Then the X_i' are iid *bounded* random variables with $E_\theta X_i' > B + 1$. Use the law of large numbers to show $\lim_{N\to\infty} P_\theta\{\bar{X}_N' > B\} = 1$, and show that $\bar{X}_N \ge \bar{X}_N'$, from which the desired result follows.]

(c) Show that $\tilde{f}_\theta$ has median approximately $\theta(\log 2 + .003)$. From this and the result just before Ex. 7.12, show that, when the X_i have law $\tilde{f}_\theta$, the random variable $Z_{1/2,N}/\log 2$ (considered in Problems 7.6.1(h) and 7.6.2(h)) converges stochastically to a value 1.004θ as $N \to \infty$ (so that there is a persistent bias as $N \to \infty$), with a corresponding convergence to $1.01\theta^2$ for the estimator t'_N of θ^2, of Problem 7.6.2(h)(iv); but that the variance of the limiting law of $N^{1/2}(t'_N - 1.004\theta)$ is close to what it was under the exponential law in Problem 7.6.2. [Thus, one says that the behavior of t'_N is *robust* under such slight contamination, but that of the estimators of Problem 7.6.2(b), (c), (d), (g) is not. Since the exact law of the contamination one might be faced with would not be known, it is pointless to try to correct for the asymptotic bias $.01\theta^2$ of t'_N under $\tilde{f}_\theta$, since it will be another value—positive or negative—for a different type of contamination.]

7.6.6. Throughout this problem, $X_1, X_2, \ldots$ are independent and identically distributed with common density

$$f_{\theta;X_1}(x_1) = \begin{cases} 3\theta^3/(x_1+\theta)^4 & \text{if } x_1 > 0; \\ 0 & \text{otherwise.} \end{cases}$$

Here $\Omega = \{\theta : \theta > 0\}$. {*Note:* Most of the computations in this problem can be simplified by noting that Ω is a scale-parameter family; thus, $E_\theta X_1^j = \theta^j E_1 X_1^j$, and most of the computations of (b), (c), (d) later need be carried out only for $\theta = 1$; moreover, for this density it is easiest to compute $E_1(X+1)^j$ and use $E_1 X^2 = E_1\{(X+1)^2 - 2(X+1) + 1\}$, etc., rather than to compute $E_1 X^2$ directly.]

(a) Discuss briefly the complexity of a minimal sufficient statistic (some guessing is allowed), the difficulty of computing an M.L. procedure, and the complexity of form of a best invariant procedure for even an analytically simple loss function. [For this last, make a guess from looking at Examples 7.5(b) and 7.5(c), which give corresponding results for a location-parameter problem.] Also note the difficulty of computing a Bayes procedure relative to any density f_ξ on Ω.

(b) Show that, for a single observation, $I_1(\theta) = 3/5\theta^2$. What is the asymptotic behavior of an M.L. estimator as $n \to \infty$?

(c) Show that $2\bar{X}_n$ is an unbiased estimator of θ and that the sequence $\{2\bar{X}_n\}$ has asymptotic (Fisher) efficiency 5/9 for all θ. [If you can't compute the moments of $\bar{X}_n$, use $E_\theta X_1 = \theta/2$ and $E_\theta X_1^2 = \theta^2$; use the *conclusion* of (b) in any case.]

(d) Show that the "population median" is $[2^{1/3} - 1]\theta$, and hence that $t = Z_{1/2,n}/[2^{1/3} - 1]$ (where $Z_{1/2,n}$ is the population median) is an estimator of θ for which $n^{1/2}(t_n - \theta)$ is asymptotically normal with mean 0 and variance $\theta^2/9[1 - (\frac{1}{2})^{1/3}]^2$. Hence, use (b) to show that the asymptotic efficiency of $\{t_n\}$ is $15[1 - (\frac{1}{2})^{1/3}]^2 \doteq .64$. [See the result just before Ex. 7.12 for the limiting distribution of the sample median.]

[*Note*: For even fairly small n, an "optimum" estimator, in any reasonable sense, is difficult to compute. For large n, the method of moments produces the estimator of (c), which is not very bad. The estimator of (d) is still better (which depends on the example), and is more robust; see Problem 7.6.5. Sometimes a linear combination of two or three sample quantiles is used to increase the efficiency to 80 or 90 percent in such settings. See also (f), for an asymptotically efficient sequence of estimators.]

(e) *Choice of an asymptotically best sample quantile.* Show that the "population" α-tile is given by $\gamma_\alpha \theta$, where $\gamma_\alpha = [(1-\alpha)^{-1/3} - 2]$, and hence (using the result just before Ex. 7.12) that

$$t_{\alpha,n} = Z_{\alpha,n}/\gamma_\alpha$$

is an estimator of θ for which $n^{1/2}(t_{\alpha,n} - \theta)$ is asymptotically normal with mean 0 and variance

$$\theta^2 \frac{(1+\gamma)^2[(1+\gamma)^3 - 1]}{9\gamma^2},$$

where we have written γ for γ_α.

Show that the asymptotic efficiency of the estimators $\{t_{\alpha,n}\}$ is $15\gamma^2/(1+\gamma)^2\{(1+\gamma)^3 - 1\}$. Show that this is approximately maximized by $\gamma = .446$ (that is, $\alpha = .67$), so that the asymptotically most efficient sequence of estimators of the form $\{t_{\alpha,n}\}$ is that based on the sample .67-tile, for which the asymptotic efficiency is about .7.

(f) Use the development in the subsection about B.A.N. estimators to show that, for t'_N chosen to be the estimator of (c), (d), or (e), the sequence $\{\tilde{t}_N\}$ defined by

$$\tilde{t}_N = t'_N + \frac{(3/t'_N) - 4N^{-1}\sum_{i=1}^{N}(X_i + t'_N)^{-1}}{3/(t'_N)^2 - 4N^{-1}\sum_{i=1}^{N}(X_i + t'_N)^{-2}}$$

is asymptotically efficient (i.e., is as efficient asymptotically as $\{\tilde{t}_N\}$).

(g) If $n = 5$ and the X_i's are 2.8, 1.5, 3.0, 16.2, 8.1, compute each of the estimators of (c), (d), and (f) (using the t'_5 of (d) in (f)).

7.6.7. In place of the density of Problem 7.6.6, suppose the X_i are iid with density

$$f_{\theta;X_1}(x_1) = \begin{cases} h\theta^h/(x_i+\theta)^{h+1} & \text{if } x_1 > 0, \\ 0 & \text{otherwise,} \end{cases}$$

where h is a *known positive constant*. Again $\Omega = \{\theta : \theta > 0\}$.

(a) Same statement as Problem 7.6.6(a).

(b) Show that $I_1(\theta) = h/(h+2)\theta^2$.

(c) Show that, if $h > 2$, $(h-1)\bar{X}_n$ is an unbiased estimator of θ, whose asymptotic efficiency relative to $\hat{t}_n$ is $1 - 4h^{-2}$. How would an estimator $c_n\bar{X}_n$ work if $h \le 2$? [*Very optional*: can you find an estimator $c'_n\left(n^{-1}\sum_i X_i^\varepsilon\right)^{1/\varepsilon}$ based on the "fractional ε^{th} sample moment" that works in this case?]

(d) Show that $f_{1;X_1}$ has p-tile $\gamma_p = (1-p)^{-1/h} - 1$, that $n^{1/2}(\gamma_p^{-1}Z_{p,n} - \theta)$ is asymptotically $\mathcal{N}(0, \theta^2 p/\gamma_p^2 h^2(1-p)^{2h^{-1}+1})$, and that $\gamma_p^{-1}Z_{p,n}$ has asymptotic efficiency $h(h+2)p^{-1}(1-p)[1-(1-p)^{h^{-1}}]^2$ relative to $\hat{t}_n$.

(e) In particular, the estimator $\gamma_{1/2}^{-1}Z_{1/2,n}$ based on the sample median has asymptotic efficiency $h(h+2)[1-2^{-1/h}]^2$. (i) Show that, when h is large, this efficiency may be satisfactory, since it is approximately $(\log 2)^2 \doteq .5$, but that it is quite unsatisfactory when h is small. (ii) On the other hand, if p is approximately bh (b fixed and positive) when h is near 0, show that the asymptotic relative efficiency of $\gamma_p^{-1}Z_{p,n}$ is approximately $2b^{-1}(1 - e^{-b})^2$, which attains its maximum of .81 at $b = 1.256$. (iii) Show that, when h is large, the asymptotic relative efficiency of $\gamma_p^{-1}Z_{p,n}$ is approximately $(p^{-1} - 1)[\log(1 - p)]^2$, and that this attains its maximum of .65 at $p = .80$. [This last turns out to be the same maximization problem as that solved in Problem 7.6.3(c).]

(f) Find the form of an asymptotically efficient sequence $\{\tilde{t}_n\}$ of the type developed in the discussion of B.A.N. estimators in Section 7.6.

7.6.8. Suppose the model is that considered in Problems 6.22 and 7.5.8. In the latter the difficulty of computing the M.L. estimator was seen. Note also the difficulty of computing a Bayes procedure in this case unless K and n are small. [In all that follows we take $\Omega = (0, \infty)$, although the endpoints $\theta = 0$ and $+\infty$ can also be included, as discussed in Problem 7.5.8.]

(a) Show that there is no unbiased estimator of θ. [Write out $E_\theta\{t(X_1, \ldots, X_n)\}$ and see that it is a rational function which cannot simplify to θ.]

(b) Use the formulas $\sum_{j=1}^{K} j = K(K+1)/2$ and $\sum_{j=1}^{K} j^2 = K(K+1)(2K+1)/6$ to show that $E_\theta \bar{X}_n = g_K(\theta)$, where $g_K(\theta) = \frac{K+1}{2}\left\{1 + \frac{K-1}{6(\theta + \frac{K+1}{2})}\right\}$, and also to show that $\operatorname{var}_\theta(\bar{X}_n)$ is of order n^{-1} as $n \to \infty$. You do not have to compute an explicit formula for this variance (unless you work part (e)).

(c) Note that the function g_K of (b) is strictly decreasing on the domain $(0, \infty)$, which it maps into the interval $\left(\frac{K+1}{2}, \frac{2K+1}{3}\right)$. If g_K^{-1} is the inverse function of g_K, conclude that, whenever $\bar{x}_n$ is close to $g_K(\theta)$, it follows that θ is closed to $g_K^{-1}(\bar{x}_n)$. Using the law of large numbers for $\bar{X}_n$ (whatever be the true θ in $(0, \infty)$), conclude that the estimators $\{\bar{t}_n\}$ defined by

$$\bar{t}_n = \begin{cases} \dfrac{K-1}{6\left(\dfrac{2}{K+1}\bar{X}_n - 1\right)} - \dfrac{K+1}{2} & \text{if } \dfrac{K+1}{2} < \bar{X}_n < \dfrac{2K+1}{3}, \\ 1 & \text{otherwise,} \end{cases} \tag{7.72}$$

give a consistent sequence (see the subsection on asymptotic optimality) of estimators of θ. (The top line of (7.72) is $g_K^{-1}(\bar{X}_n)$; the bottom line could be taken to be anything definite instead of 1, the important fact being that it will not arise often for large n, as you can show by proving that $\lim_{n\to\infty}\left\{\frac{K+1}{2} < \bar{X}_n < \frac{2K+1}{3}\right\} = 1$ for $0 < \theta < \infty$.)

(d) Write an explicit formula for the asymptotically efficient sequence $\{\tilde{t}_n\}$ of (7.62). It will be shown in (e) that $\bar{t}_n$ of (c) is accurate enough to be taken as the t_n' of (7.62). When $K = 5$, $n = 50$, and the $s^*(X)$ of Problem 6.22 is (7, 6, 10, 13, 14), compute $\bar{t}_{50} = t_{50}'$ and also $\tilde{t}_{50}$. [*Optional*: Compute also $\hat{t}_{50}$, by numerically solving the likelihood equation.]

[*Note*: The estimator $\tilde{t}_n$, although slightly complicated looking, is in explicit closed form, suitable for simple calculation—unlike $\hat{t}_n$ or a Bayes estimator.]

(e) [*Very optional*] (i) Using the fact that $\sum_{j=1}^{K} j^3 = K^2(K+1)^2/4$, compute $\text{var}_\theta(X_1)$ and thus $\text{var}_\theta(\bar{X}_n)$. (ii) Using the result of (i) and the result on change of variables just after Ex. 7.12, with $h = g_K^{-1}$, find the variance of the limiting normal law of $n^{1/2}(\bar{t}_n - \theta)$. (iii) Compute $I(\theta)$ in terms of the function $L_K(\theta) \stackrel{\text{def}}{=} \sum_{j=1}^{K} (j+\theta)^{-1}$. At least for $K = 2$ or 3, compare the result of (ii) with $1/I(\theta)$, e.g., at $\theta = \frac{1}{2}$, 1, 2, and as $\theta \to 0$ or $+\infty$. [This would be of interest to see whether it is worthwhile computing $\hat{t}_n$ or the estimator $\tilde{t}_n$ of (d) rather than simply using $\bar{t}_n$.]

7.6.9. In the setting of Problem 7.5.7 and Problem 6.3(vi), we already know that the minimal sufficient statistic is n-dimensional and that the M.L. estimator of θ is intractable. Invariance considerations yield little, the only possible invariance being found in the symmetry about $\theta = 0$ (analogous to that in the Bernoulli case about $\theta = \frac{1}{2}$). We might try estimation based on one or more sample quantiles, or on sample moments ($3\bar{X}_n$ being unbiased but not very efficient). Since the computations are then similar to those in other problems of this section, we consider here only the B.A.N. estimators obtained from (7.62) or the Bayes method.

(a) With $t_n' = 3\bar{X}_n$, exhibit the asymptotically efficient estimator $\tilde{t}_n$ of (7.62).

(b) For simplicity in the computations, we take $f_\xi(\theta) = 1/2$ for $-1 \le \theta \le 1$ (uniform prior density). Then (Problem 4.5(b) again) the Bayes procedure for estimating θ with squared error loss is

$$t_{\xi,n}(X_1,\ldots,X_n) = \frac{\int_{-1}^{1} \theta \prod_1^n (1+\theta X_i)\,d\theta}{\int_{-1}^{1} \prod_1^n (1+\theta X_i)\,d\theta}.$$

Compute this, at least when $n = 1$ and 2; and, if possible, show that, for n even,

$$t_{\xi,n} = \frac{2^{-1}\sum X_i + 4^{-1}\sum_{i<j<k} X_iX_jX_k + \cdots + n^{-1}\sum_{i_1<\cdots<i_{n-1}} X_{i_1}X_{i_2}\cdots X_{i_{n-1}}}{1 + 3^{-1}\sum_{i<j} X_iX_j + \cdots + (n+1)^{-1}\prod_i X_i},$$

with a corresponding expression for n odd. This $t_{\xi,n}$ looks messy compared with that of (a), and a more tractable *highly* efficient estimator could also be based on a few sample quantiles, but $t_{\xi,n}$ *is at least explicit, unlike* $\hat{t}_n$, and for large n it is nearly efficient.

CHAPTER 8

Hypothesis Testing

8.1. Introductory Notions

Recall Chapters 3 and 4 for brief mention of the setup of hypothesis testing and the way it fits into our general picture. (Also, review the parts of Section 4.4 and Problems 4.3, 4.12–4.17 on the geometry of risk points in the case where Ω contains two elements.)

The space Ω is the union of disjoint sets Ω_0 and Ω_1, and $D = \{d_0, d_1\}$ where d_i means "I guess the true F is in Ω_i." Thus, d_i is regarded as "correct" if $F \in \Omega_i$, and in the spirit of the earlier discussion of W this is usually reflected by having

$$W(F, d_i) = 0 \quad \text{if} \quad F \in \Omega_i.$$

When $F \notin \Omega_i$, the form of $W(F, d_i)$ may reflect the seriousness of the incorrect decision; for example, if we flip a coin with probability θ of coming up H and if $\Omega_0 = \{\theta : 0 \leq \theta \leq 1/2\}$ and $\Omega_1 = \{\theta : 1/2 < \theta \leq 1\}$, then one may have

$$W(\theta, d_i) = |\theta - 1/2| \quad \text{if} \quad \theta \notin \Omega_i$$

to reflect increase of loss as the true θ moves away from the set where d_i guesses it to lie. To simplify our discussions, we will consider, except for some examples in Section 8.7, and some comments in Section 8.3, only the *simple loss function*

$$W(F, d_i) = \begin{cases} 1 & \text{if} \quad F \notin \Omega_i, \\ 0 & \text{if} \quad F \in \Omega_i, \end{cases} \tag{8.1}$$

for which we saw earlier that

$$r_\delta(F) = P_F\{\delta \text{ makes the wrong decision}\}. \tag{8.2}$$

[In the literature you will sometimes see a hypothesis testing problem formulated with Ω the disjoint union of Ω_0, Ω_1, and $\Omega_{\text{INDIFFERENCE}}$, the latter being the set of Fs for which we are "indifferent" as to which decision we make, and both decisions are regarded as correct. Thus, we add "$W(F, d_i) = 0$ if $F \in \Omega_{\text{INDIFF}}$" to (8.1) and note that, as a consequence, part of (8.2) now says that $r_\delta(F) = 0$ for all F in Ω_{INDIFF} and for all δ. Theoretical developments may therefore proceed by ignoring Ω_{INDIFF} and considering the problem with $\Omega_0 \cup \Omega_1 \cup \Omega_{\text{INDIFF}}$ replaced by $\Omega_0 \cup \Omega_1$, in the manner that one considers problems for which no Ω_{INDIFF} was mentioned.]

Each Ω_i is sometimes called a *hypothesis*. The preceding Bernoulli example is sometimes written in the literature

"Test between the hypotheses

$$H_0 : \theta \leq 1/2$$

and

$$H_1 : \theta > 1/2."$$

Often in the classical treatment of this subject one hypothesis, labeled H_0 or Ω_0 or *null hypothesis*, is singled out for different treatment from the other, H_1 or Ω_1, called *the alternative* (i.e., the alternative, for an F in Ω, to F being a member of Ω_0). The risk function is not considered in the same way in Ω_0 and Ω_1 in this sometimes questionable treatment, which we shall outline because it is still common in the literature. In the early development of the subject H_0 often represented an old theory or the equality of two treatments or varieties, and H_1 stood for a departure from this; the lack of symmetry in the roles of the Ω_i proceeds from this background in classical applications.

The reader should recall the "initial" and "postexperimental" modes of randomization described in Chapter 4. The former was useful in illustrating the role of randomization in the geometry of risk points (Chapter 4); the latter is useful in constructing tests with particular properties (e.g., "Neyman-Pearson test of level α").

A *test* (a statistical procedure in this setting), allowed to be randomized, is, in terms of the "postexperimental" mode of randomization (the "general randomization of Section 4.3), a nonnegative function δ on the product space of S and D such that

$$\delta(x, d_0) + \delta(x, d_1) = 1; \tag{8.3}$$

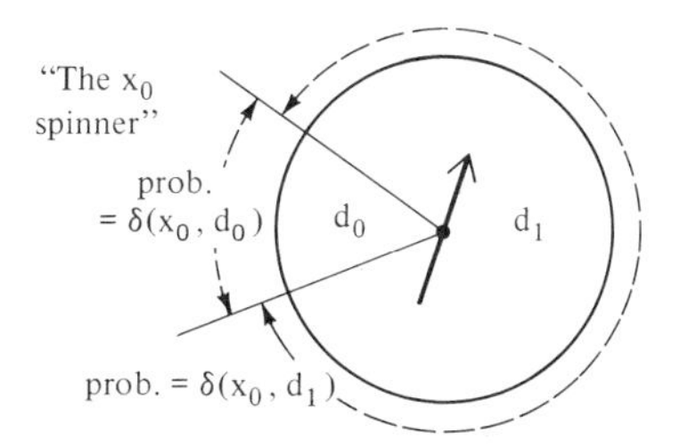

if δ is used and $X = x_0$, we use the x_0-spinner as described earlier. Since there are only two d_i's, knowledge of $\delta(x, d_1)$ determines $\delta(x, d_0)$ by (8.3), and hence it is customary in the literature to use "test" or "test function" for the function $\delta(x, d_1)$ and to shorten the notation by writing

$$\delta(x) \quad \text{for} \quad \delta(x, d_1).$$

We hereafter do this. Thus, a test δ is a function from S into the interval $[0, 1]$. The performance of a test δ is often described in terms of its *power function*

$$\beta_\delta(F) \stackrel{\text{def}}{=} P_F\{\delta \text{ makes decision } d_1\} = E_F\delta(X), \tag{8.4}$$

the last equality of (8.4) following from summing (on x) the results of multiplying by $P_F\{X = x\}$ on each side of the equation $\delta(x) = P_F\{\delta \text{ makes decision } d_1 | X = x\}$. By (8.2), we see that either of β_δ and r_δ determines the other, since

$$r_\delta(F) = \begin{cases} \beta_\delta(F) & \text{if} \quad F \in \Omega_0, \\ 1 - \beta_\delta(F) & \text{if} \quad F \in \Omega_1. \end{cases} \tag{8.5}$$

Typically, if Ω is parametrized in terms of some real parameter θ, for example with $\Omega_0 = \{\theta : a \le \theta \le b\}$ and $\Omega_1 = \{\theta : b < \theta \le c\}$, and if $p_\theta(x)$ is continuous in θ, then for every δ it is easy to see that $\beta_\delta(\theta) = \sum_x \delta(x) p_\theta(x)$ is continuous in θ. Thus, $r_\delta(\theta)$ will be continuous except at the point $\theta = b$, where it will have a jump (unless $\beta_\delta(b) = 1/2$): An ideal (and unattainable) test δ would have $r_\delta(F) = 0$, or

$$\beta_\delta(F) = \begin{cases} 0 & \text{for} \quad F \in \Omega_0, \\ 1 & \text{for} \quad F \in \Omega_1. \end{cases}$$

Thus, roughly, one seeks a test whose power function is close to 0 in Ω_0 and close to 1 in Ω_1. The fact that β_δ can usually be written as a more succinct expression than the discontinuous r_δ leads us usually to compare tests in terms of the former rather than the latter; any such discussion can be given, equivalently, in terms of r_δ instead.

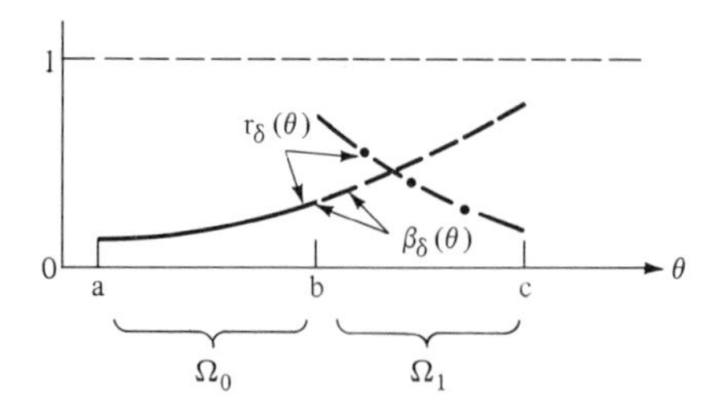

If a test is *nonrandomized* (δ takes on only the values 0 and 1), we may represent it as a function from S into D, as follows, (see Chapter 4): if $B_\delta = \{x : \delta(x) = 1\}$ and δ is nonrandomized, then

$$t(x) = \begin{cases} d_1 & \text{if} \quad x \in B_\delta, \\ d_0 & \text{if} \quad x \notin B_\delta \end{cases} \tag{8.6}$$

is the representation of δ as such a *function* (our original definition of a *procedure*), as discussed earlier. B_δ is called the *critical region* or *rejection* (of H_0) *region* of the test δ (or t), and $S - B_\delta$ is the *acceptance region.* The nonrandomized tests correspond to the [measurable] subsets of S, since any such subset B can be regarded as the B_δ of the test δ (or t) defined by (8.6) by putting B for B_δ. Thus, instead of comparing tests one sometimes writes the comparison in terms of corresponding critical regions. Note that, for a non-randomized t or B, the power function of (8.4) becomes

$$\beta_t(F) = P_F\{t(X) = d_1\}, \quad \text{or} \quad \beta_B(F) = P_F\{X \in B\}. \tag{8.4'}$$

Making decision d_1 when $F \in \Omega_0$ is often called an *error of type* I in the literature. Similarly, making decision d_0 when $F \in \Omega_1$ is called an *error of type* II.

The *level* α_δ of the test δ (sometimes called the *significance level* or *size*) is defined by

$$\alpha_\delta = \max_{F \in \Omega_0} \beta_\delta(F).$$

The special role of Ω_0 in the classical development, mentioned earlier, is seen in that statisticians often looked at $r_\delta(F)$ (or $\beta_\delta(F)$) on Ω_0 only through the *single number* α_δ, but looked at the entire behavior of the function $r_\delta(F)$ (or $\beta_\delta(F)$) on Ω_1.

If Ω_i consists of a single element, it is said to be *simple.* Otherwise it is said to be *composite.* One can have a problem with either of the Ω_i simple or composite—four possibilities. When Ω_1 is simple, the power function's value evaluated at the one F in Ω_1 is simply called the *power* of a test.

In the Neyman-Pearson approach, one specified a value α, $0 \leq \alpha \leq 1$. (Often $\alpha = .01$ or $.05$ by tradition and because of what is tabled, but for no rational reason; the choice of α is the most arbitrary and questionable part of the Neyman-Pearson development). Then, letting $\mathscr{C}_\alpha$ be the set of all tests δ with $\alpha_\delta \leq \alpha$, one restricted consideration to tests in $\mathscr{C}_\alpha$ and examined their power (or risk) functions on Ω_1. If some test δ^* in $\mathscr{C}_\alpha$ has uniformly highest power function (hence, by (8.5), uniformly smallest risk function) on Ω_1 among all tests in $\mathscr{C}_\alpha$, it is called a "*uniformly most powerful* (UMP) *test of level* $\leq \alpha$," and would be selected for use. If no such test exists, additional considerations are needed. This is in the spirit of previous discussions of unbiased or invariant estimation, etc., where, since no procedure had uniformly smallest risk in the class of *all* procedures, we restricted attention to some smaller class and hoped (often with success) that there existed a uniformly best procedure in the restricted class; the meaningfulness of this approach, of course, depends on the reasonableness of the restricting criterion. [That there is no uniformly best procedure in the class of *all* procedures was seen in Section 4.4 and Problems 4.3 and 4.12–4.17 on the geometry of the risk points for testing between two simple hypotheses; in particular, writing δ_i for the test given by $\delta(x, t_i(x)) = 1$ (which amounts to using nonrandomized t_i), we see that the nonrandomized tests $t_0 \equiv d_0$ and $t_1 \equiv d_1$, corresponding to

$$\text{critical regions} \quad \begin{Bmatrix} B_{\delta_0} = \phi \\ B_{\delta_1} = S \end{Bmatrix} \quad \text{have} \quad \begin{Bmatrix} r_{\delta_0}(F) = 0 & \text{for} & F \in \Omega_0 \\ r_{\delta_1}(F) = 0 & \text{for} & F \in \Omega_1 \end{Bmatrix}, \quad \text{so}$$

any uniformly best procedure δ^* would have to have $r_{\delta^*}(F) = 0$ for all F, which is impossible except in trivial examples in which there exists a set B in S of probability one (respectively, zero) for all F in Ω_1 (respectively, Ω_0). This and the Problems in Chapter 4 on risk points thus also indicate why one cannot hope to have a uniformly best procedure if one looks at $r_\delta(F)$ in both Ω_0 and Ω_1 among tests of *different* levels; the Neyman-Pearson approach considers only tests in $\mathscr{C}_\alpha$ and then compares their goodness by looking at $r_\delta(F)$ only in Ω_1. If $B_1 \subset B_2$ and $P_F\{B_2 - B_1\} > 0$ for all F and B_i is UMP of level α_i, you should convince yourself that $\beta_{B_1}(F) < \beta_{B_2}(F)$ for all F, but that r_{B_1} and r_{B_2} "cross".]

[The Neyman-Pearson approach mentioned in the previous paragraph is often stated with $\mathscr{C}_\alpha$ replaced by $\mathscr{C}'_\alpha = \{\delta : \alpha_\delta = \alpha\}$. The minor difference created by this is described in Problems 4.15 and 4.16 on the geometry of risk points. It usually will not arise in practice.]

8.2. Testing Between Simple Hypotheses

In Section 4.4 and Problems 4.3 and 4.12–4.17, we discussed the geometry of risk points in the case of *testing between two simple hypotheses.* Briefly, if X has density or probability function f_0 or f_1, where f_i is the single member of Ω_i, the set R of risk points of all tests is a convex closed set as shown in the figure on the next page. The risk points $(0, 1)$ and $(1, 0)$ correspond to the trivial tests δ_0 and δ_1 described previously. The curve from $(0, q')$ to $(q, 0)$ consists of risk points of admissible tests, all of which are Bayes; the line segments from $(0, 1)$ to $(0, q')$ and from $(1, 0)$ to $(q, 0)$, excluding the last endpoint $(0, q')$ or $(q, 0)$, correspond to Bayes tests which are inadmissible. [If the sets on which the two functions f_i are positive are the same, these segments are absent.] The tests in $\mathscr{C}_\alpha$ have risk points in the shaded area. The test δ_α whose risk point is labeled has smallest $r_\delta(F_1)$ value—and hence largest $\beta_\delta(F_1)$ value—among tests in $\mathscr{C}_\alpha$ or $\mathscr{C}'_\alpha$, and is hence called the *most powerful test of level* α or the *Neyman-Pearson* (*N-P*) *test of level* α. [If $q < \alpha \le 1$, there are many risk points of tests in $\mathscr{C}_\alpha$ which are "most powerful," having power 1: the risk points on the segment from $(q, 0)$ to $(\alpha, 0)$. Of these only $(q, 0)$ corresponds to an admissible test, which is why it is more sensible to consider $\mathscr{C}_\alpha$, allowing this risk point, rather than to insist on $\mathscr{C}'_\alpha$ and thus an inadmissible procedure with risk point $(\alpha, 0)$, if $q < \alpha$.]

Suppose $\xi_0 > 0$ and $\xi_1 > 0$ (where $\xi_0 + \xi_1 = 1$) are the a priori probabilities

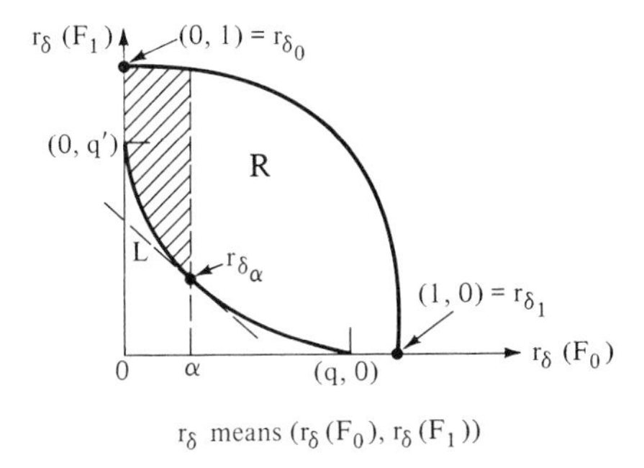

of f_0 and f_1 being true. Then according to the results of Section 4.1, δ is Bayes relative to $\xi = (\xi_0, \xi_1)$ if and only if it always chooses d_0 or d_1 so as to minimize $h_\xi(x, d)$, which equals $\xi_1 f_1(x)$ if $d = d_0$ and $\xi_0 f_0(x)$ if $d = d_1$. Since δ is determined by $\delta(x)$, the probability of choosing d_1 when x is observed, this prescription amounts to

$$\delta(x) = \begin{cases} 1 & \text{if} \quad \xi_1 f_1(x) > \xi_0 f_0(x), \\ \text{arbitrary} & \text{if} \quad \xi_1 f_1(x) = \xi_0 f_0(x), \\ 0 & \text{if} \quad \xi_1 f_1(x) < \xi_0 f_0(x). \end{cases} \tag{8.7}$$

If r_{δ_α} is as shown in the preceding figure, we saw in Chapter 4 that through any such point on the lower boundary of R there is at least one line L not passing through the interior of R. If such a line L has slope $-s$ $(0 \geq -s \geq -\infty)$, then δ_α must be Bayes relative to $(\xi_0, \xi_1) = \left(\dfrac{s}{s+1}, \dfrac{1}{s+1}\right)$ (interpreted as $(1, 0)$ if $s = \infty$), since this (ξ_0, ξ_1) gives $-\xi_0/\xi_1 = -s$. Thus, a *level α N-P test* is a Bayes test relative to (ξ_0, ξ_1), and hence by (8.7) is of the form

$$\delta(x) = \begin{Bmatrix} 1 \\ \text{arb.} \\ 0 \end{Bmatrix} \quad \text{if} \quad f_1(x) \begin{Bmatrix} > \\ = \\ < \end{Bmatrix} c f_0(x) \tag{8.8}$$

where we have written $c = \xi_0/\xi_1$. Here c is a finite nonnegative constant (if $\alpha > 0$, since then s is finite). We have still not related c (or, which is the same thing, (ξ_0, ξ_1), or L) to α. But as soon as we find a c such that δ defined by (8.8) (including specification of the "arbitrary" part when $f_1(x) = cf_0(x)$) has level α, the preceding considerations show that this test must be a *most powerful (MP) level α test*. Moreover, for every α, $0 < \alpha < 1$, the geometric picture shows that such a test can be found. This is the

Neyman-Pearson Lemma. *For testing between two simple hypotheses at level α, $0 < \alpha < 1$, every MP test of level α has the form*

$$\delta^*(x) = \begin{cases} 1 & \text{if} \quad f_1(x) > cf_0(x) \\ 0 & \text{if} \quad f_1(x) < cf_0(x) \end{cases}$$

(and $\delta^*(x)$ appropriately specified when $f_1(x) = cf_0(x)$ so as to make δ^* have level α) *where* c *is an appropriate constant (depending on* α*). Such a* c *and* δ^* *exist for each value of* α, $0 < \alpha < 1$.

This is often proved (not inappropriately, since Neyman-Pearsonites are usually non-Bayesians) without mention of Bayes procedures. Since we discussed the geometry of the latter in Chapter 4, it is simpler to use that material as we have done rather than to start from scratch. Moreover, it is instructive to realize the relationship between the class of all Bayes tests corresponding to all prior distributions ξ and the class of all N-P tests for all α (even though a Bayesian with prior distribution ξ may not know or care what value of α of the Neyman-Pearsonite yields the same test as the Bayes test relative to ξ, and vice versa).

EXAMPLE 8.1. $S = \{x : x > 0\}$, $f_1(x) = 2e^{-2x}$, $f_0(x) = e^{-x}$. Here the form of (8.8) can be arrived at by noting that

$$\frac{f_1(x)}{f_0(x)} > c \Leftrightarrow 2e^{-x} > c \Leftrightarrow x < -\log\left(\frac{c}{2}\right) = c'(\text{say}). \tag{8.9}$$

Since $f_1(x)/f_0(x) = c$ only on a set (one point) of probability 0 under each f_i, we will not have to worry about the definition of δ on that set, since changes of δ on that set will affect no probabilities. Thus, an N-P test here can be described simply as a *critical region*, since we can take δ to be 0 or 1 on this one-point set just discussed, and the resulting test will be nonrandomized. We thus seek a critical region of the form $B = \{x : x < c'\}$ such that $P_0\{B\} = \alpha$; that is,

$$\alpha = \int_B f_0(x)\,dx = \int_0^{c'} e^{-x}\,dx = 1 - e^{-c'}, \quad \text{or} \quad c' = \log\frac{1}{1-\alpha}. \tag{8.10}$$

We conclude that the test "reject H_0 if $X < \log\dfrac{1}{1-\alpha}$, accept H_0 otherwise" is the MP (N-P) test of level α; it is the "essentially unique" such test, in that any N-P test of level α coincides with it except on a subset of S which has probability 0 under both f_i.

An important point to realize in this derivation is that, having gone through the calculation (8.9) to simplify the way in which "$f_1(x)/f_0(x) > c$" was expressed in terms of x, we made subsequent calculations in (8.10) in terms of c' and *never found it necessary to rewrite the test in terms of the equivalent first form of* (8.9) (which would say $2e^{-2x}/e^{-x} > c$ where $c = 2e^{-c'} = 2(1-\alpha)$). This is standard procedure in all problems but is sometimes misunderstood: the form of B does not have to be written out as the first form of (8.9) with the f_1 and f_0 exhibited, but can be written as any equivalent expression which is simpler; one goes through a reduction like (8.9) to obtain such a simpler form in terms of one or more constants (here c') related to the original c, in order to be able to adjust them as in (8.10) to give a test of level α, the computation $P_0\{x : f_1(x)/f_0(x) > c\} = \alpha$ being less clear without such a reduction. The test having been found, we know we could compute c from c' and, reversing the calculation of (8.9), could obtain the test in the more complicated first form, but there is no reason to do so (and every reason not to), since we already have it written in simpler form.

EXAMPLE 8.1(a). If in the preceding example X were replaced by $(X_1, \ldots, X_n) = X$ where the X_i are iid with one of the two densities e^{-x_i} (under H_0) or $2e^{-2x_i}$ (under H_1) for $x_i > 0$, we would have $f_1(x_1, \ldots, x_n)/f_0(x_1, \ldots, x_n) = \prod_1^n (2e^{-2x_i}/e^{-x_i})$, and the last line of (8.9) would become $\sum_1^n x_i < c''$ (it being unnecessary to keep track of the relation between c'' and c, for the reason given in the previous paragraph). The density of $Y = \sum_1^n X_i$ under H_0 is $y^{n-1}e^{-y}/(n-1)!$ for $y > 0$. (See Appendix A on the Γ-distribution as it arises from adding exponentially distributed random variables.) Thus, in place of (8.10) we obtain

$$\alpha = \frac{1}{(n-1)!}\int_0^{c''} y^{n-1}e^{-y}\,dy.$$

For $n = 2$ or 3 you can write out the integral in terms of polynomials and an exponential and for given α (e.g., $\alpha = .05$) can solve for c'' by trial and error. For larger n you would perhaps use instead "tables of the incomplete Γ-function," found in the more extensive books of statistical tables, where $\frac{1}{(n-1)!}\int_0^y t^{n-1}e^{-t}\,dt$ is tabled, or use numerical integration on a computer. For most α encountered in practice and $n > 10$ or 20, one would use the fact that, by the central limit theorem, $\sum_1^n X_i$ is approximately normal with mean n and variance n under H_0, so that

$$P_0\left\{\sum_1^n X_i < c''\right\} = P_0\left\{\left(\sum_1^n X_i - n\right)\Big/\sqrt{n} < \frac{c''-n}{\sqrt{n}}\right\} \approx \Phi\left(\frac{c''-n}{\sqrt{n}}\right),$$

where Φ is the standard normal df. Given α, one determines k_α such that $\Phi(k_\alpha) = \alpha$, and then sets $(c''-n)/\sqrt{n} = k_\alpha$.

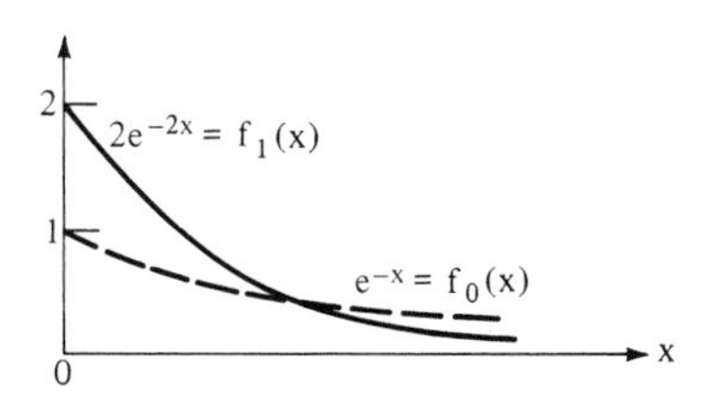

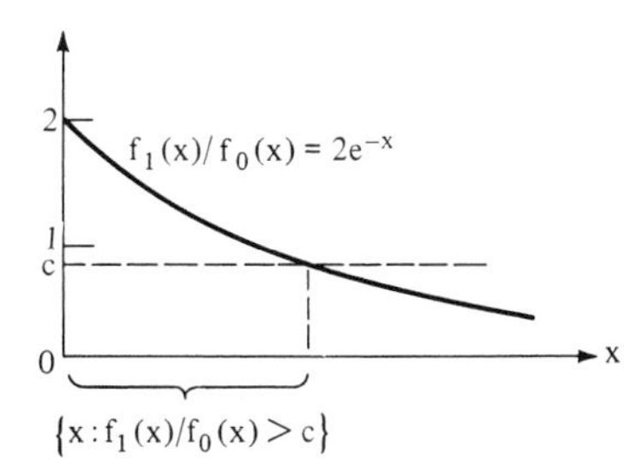

In Example 8.1, one can think intuitively that "f_1 is concentrated more on small values than f_0 so that we should vote for f_1 when x is small," and this turned out to be the case. More precisely, we can think of graphing f_1/f_0 and seeing where this ratio is large. This last is the way in which one must proceed, and the next example shows that intuitive arguments about f_1 and f_0, rather than a precise look at their ratio, can lead one astray.

EXAMPLE 8.2. (Cauchy distribution.) Suppose $S =$ reals, $f_0(x) = \dfrac{1}{\pi[1 + (x + 1)^2]}$, $f_1(x) = \dfrac{1}{\pi[1 + (x - 1)^2]}$. Here $f_1(x)/f_0(x) > c \Leftrightarrow \dfrac{1 + (x + 1)^2}{1 + (x - 1)^2} > c$. The equation $f_1(x) = cf_0(x)$ is a quadratic which is easily solved. One finds five different possible forms of critical region, depending on the value of c. Two of these, for c very large or very small, are the trivial tests δ_0 and δ_1 ($\alpha = 0$ or 1) discussed earlier. If $c = 1$ ($\alpha = 1/2$ then, it turns out), the critical region is $\{x : x > 0\}$. If c is slightly less than 1 or slightly larger than 1, the result is not so simple; for example, by graphing f_1/f_0, one sees in the latter case that the critical region is of the form $\{x : a < x < b\}$ where the finite constants a and b are related to c (hence, to each other); thus, the N-P test says to vote for f_0 when $X \leq a$ or when $X \geq b$, and the latter part would never have been obtained from the fuzzy reasoning "since f_0 is to the left of f_1 (being obtained from it by shifting f_1 two units to the left), we should vote for f_0 when X is small (or negative) and for f_1 when X is large and positive." Problems 8.1 and 8.2 are related to this.

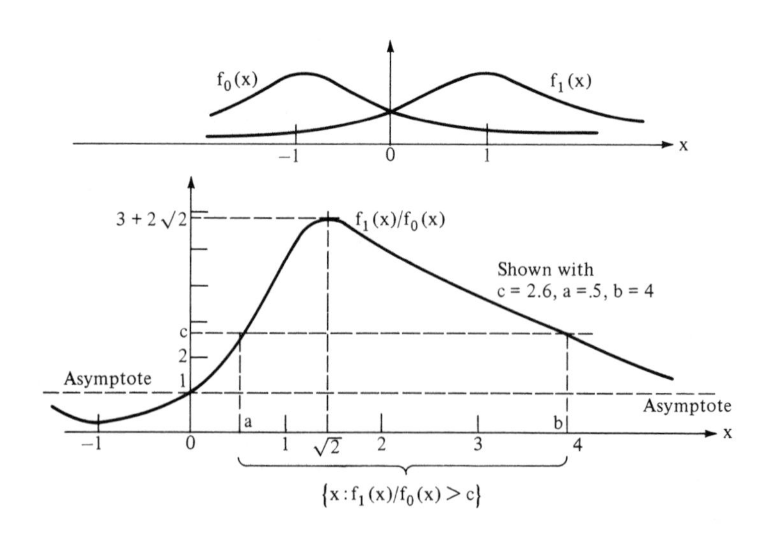

For n observations in the Cauchy case, the explicit form of the critical region is hard to draw. This is also connected with the complexity of the minimal sufficient statistic in the Cauchy case. In an example like 8.1(a) where the minimal sufficient statistic $Y = \sum_1^n X_i$ is simple, *we could also have proceeded by using* (Chapter 6) *our knowledge that we can always do as well with some test based on* Y *as with any given test*. Under H_0 (respectively, H_1), Y has density $f_0^*(y) = \frac{1}{(n-1)!} y^{n-1} e^{-y}$ $\left(\text{respectively, } f_1^*(y) = \frac{2^n}{(n-1)!} y^{n-1} e^{-2y}\right)$ for $y > 0$. The N-P lemma *for a test based only on* Y gives the following calculation of a UMP critical region:

$$f_1^*(y)/f_0^*(y) > c \Leftrightarrow y < c' \quad \text{(related to } c\text{)},$$

by a calculation like (8.9), so we obtain again the result of Example 8.1(a).

EXAMPLE 8.3. If $S = \{x : 3/2 > x > 0\}$ and $f_i(x)$ is uniform from $i/2$ to $1 + i/2$, we obtain the following picture: If $0 < c \neq 1$, the test is again essentially unique if we take $B = \{x : f_1(x)/f_0(x) \geq c\}$.

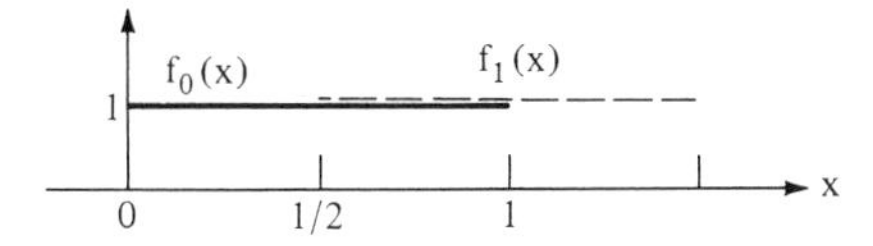

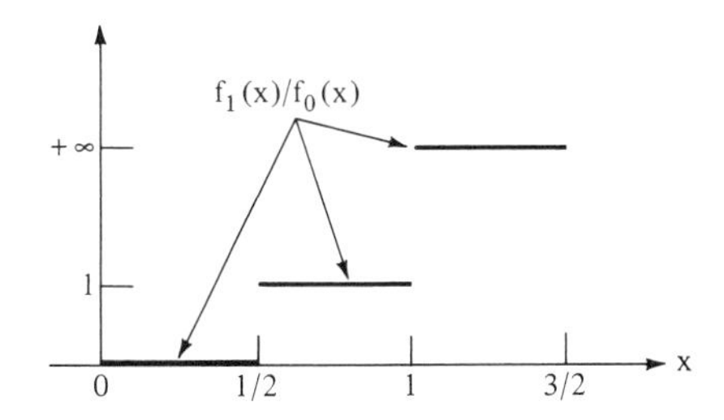

If $c = 0$, corresponding to a test of level 1/2 or of level $\alpha > 1/2$, we get for the latter the inadmissible tests described earlier, with risk points $(\alpha, 0)$ with $\alpha > q = 1/2$; the N-P test of size 1/2, $B = \{x : 1/2 < x < 3/2\}$, is better than these. The test of level $\alpha > 1/2$ is not unique; for example, a nonrandomized test of level 3/4 corresponds to any acceptance region formed by taking a subset of $\{x : 0 < x < 1/2\}$ of total length 1/4.

If $c = 1$, $f_1(x)/f_0(x) = c$ on the set $1/2 < x < 1$, of positive probability under both f_i. For a reason similar to that which yielded lack of uniqueness when $c = 0$ previously, we again have many possibilities satisfying (8.8). For example, when $\alpha = 1/8$, the two nonrandomized tests

$$\delta_2(x) = \begin{cases} 1 & \text{if} \quad x > 7/8, \\ 0 & \text{otherwise,} \end{cases}$$

$$\delta_3(x) = \begin{cases} 1 & \text{if} \quad 1/2 < x < 5/8 \quad \text{or} \quad x > 1, \\ 0 & \text{otherwise,} \end{cases}$$

are both MP, and so are the randomized tests

$$\delta_4(x) = \begin{cases} 1 & \text{if} \quad x > 1, \\ 1/4 & \text{if} \quad 1/2 \le x \le 1, \\ 0 & \text{if} \quad x < 1/2, \end{cases}$$

$$\delta_5(x) = \begin{cases} 1 & \text{if} \quad x > 1, \\ x - 1/2 & \text{if} \quad 1/2 \le x \le 1, \\ 0 & \text{if} \quad x < 1/2, \end{cases}$$

the last since $\int_{1/2}^{1} f_0(x) \cdot (x - 1/2)\, dx = 1/8$. All these tests have the same risk point $(1/8, 3/8)$ and agree (as they must by (8.8)) when $f_1(x)/f_0(x) \neq 1$.

EXAMPLE 8.4. In the preceding continuous case examples, randomization was not necessary (although it was possible in Example 8.3). Now suppose $S = \{(0,0), (0,1), (1,0), (1,1)\}$ and that $X = (X_1, X_2)$ where the X_i are iid Bernoulli random variables with parameter 1/3 under H_0 and 2/3 under H_1. Then $p_1(x_1, x_2)/p_0(x_1, x_2) = (\frac{2}{3})^{x_1+x_2}(\frac{1}{3})^{2-x_1-x_2}/(\frac{1}{3})^{x_1+x_2}(\frac{2}{3})^{2-x_1-x_2} = \frac{1}{4} \cdot 2^{x_1+x_2}$. For certain values of α there is a nonrandomized N-P test (e.g., if $\alpha = 5/9$, $B = \{(x_1, x_2) : x_1 + x_2 = 1 \text{ or } 2\}$), but for others there is not. For example, if $\alpha = 1/2$ this is the case; two possible N-P tests are

$$\delta_6(x_1, x_2) = \begin{cases} 1 & \text{if} \quad x_1 + x_2 = 2, \\ 0 & \text{if} \quad x_1 + x_2 = 0, \\ 1/8 & \text{if} \quad x_1 + x_2 = 1, \end{cases}$$

$$\delta_7(x_1, x_2) = \begin{cases} 1 & \text{if} \quad x_1 + x_2 = 2, \\ 0 & \text{if} \quad x_1 + x_2 = 0 \quad \text{or} \quad (x_1, x_2) = (0, 1), \\ 1/4 & \text{if} \quad (x_1, x_2) = (1, 0). \end{cases}$$

[Note that δ_7 does *not* depend only on the minimal sufficient statistic $X_1 + X_2$. (Similarly, in Example 8.3 where a minimal sufficient statistic T takes on three values describing which of the intervals $[0, 1/2)$, $[1/2, 1]$, or $(1, 3/2]$ X falls in, δ_5 does not depend on X simply through T.) But there *is* a level α MP test δ_0 (as there was δ_2, δ_3, or δ_4 in Example 8.3) which *does* depend only on T, as guaranteed by our discussion of sufficiency in Chapter 6. For $\alpha = 1/3$, the last lines of δ_6 and δ_7 are replaced by 1/2 and 1, so in that case δ_7 has the advantage of being *nonrandomized* although it does not depend only on $X_1 + X_2$.]

For common values of α and a reasonably large number of flips of the coin instead of two, one usually does not bother in practice to find a randomized test of level α, but instead finds a nonrandomized test of level close to α. Often the criterion of finding a (possibly randomized) N-P test of level α is modified by requesting "a *nonrandomized* test of as large a level $\leq \alpha$ as possible." Comfort is then taken in the fact that the probabilities of type I error are less than or equal to the specified α—the test is called *conservative*—which is misleading, since the other half of the story is that the power is also less than what it would have been for a nonrandomized N-P test of size slightly *larger* than α.

8.3. Composite Hypotheses: UMP Tests; Unbiased Tests

For composite hypotheses, the results are not always so nice as for simple hypotheses. This is because, if Ω_1 contains more than one element F_1, we cannot simply compare *numbers* $\beta_\delta(F_1)$ and select the δ in $\mathscr{C}'_\alpha$ which gives the largest such number, but must compare the *functions* $\beta_\delta(F)$, $F \in \Omega_i$, and they may cross and not yield a uniformly best one.

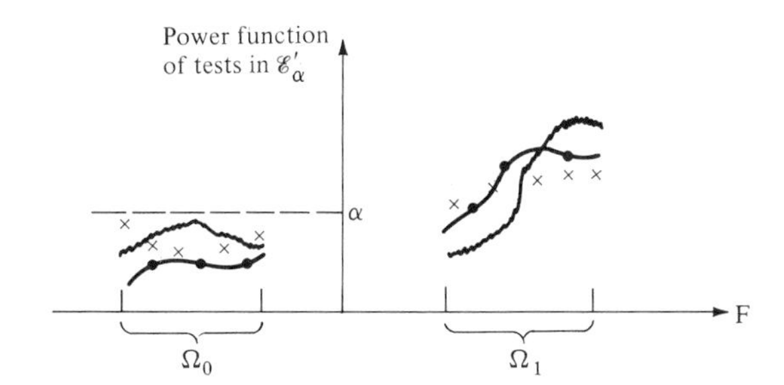

UMP Tests

There are a few cases in which UMP tests do exist, the most important of which we now discuss. *Throughout the remainder of Section* 8.3, Ω *will be a subset of* $\Omega^* = \{\theta : a < \theta < b\}$ and S is a subset of the reals, and we assume that the density or probability function $f_\theta(x)$ is such that

$$\text{for each pair of values } \theta_1 \text{ and } \theta_2 \text{ in } \Omega^*, \text{ if} \\ \theta_1 < \theta_2 \quad \text{then} \quad \frac{f_{\theta_2}(x)}{f_{\theta_1}(x)} \quad \text{is a nondecreasing function of } x \text{ on } S. \tag{8.11}$$

(We shall vary the subset Ω of Ω^* in the course of our discussion, thus

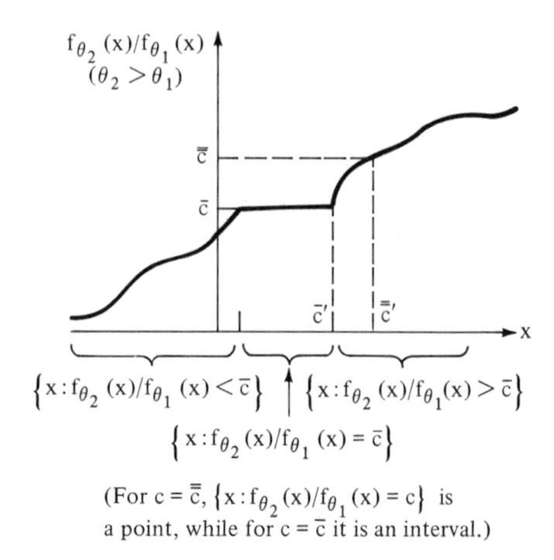

(For $c = \bar{\bar{c}}$, $\{x : f_{\theta_2}(x)/f_{\theta_1}(x) = c\}$ is a point, while for $c = \bar{c}$ it is an interval.)

considering various testing problems.) Such an Ω^* is called a *monotone likelihood ratio* (MLR) family (as it is also if "nondecreasing" is replaced by "nonincreasing" in (8.11)). The set $\{x : f_{\theta_2}(x)/f_{\theta_1}(x) > c\}$ for fixed θ_1 and θ_2 will be an interval $\{x : x > c'\}$, as shown. If Ω consists of two specified values θ_1 and θ_2 in Ω^*, with $\theta_1 < \theta_2$, and if we were to test

$$H_0 : \theta = \theta_1 \text{ against } H_1 : \theta = \theta_2,$$

it follows that, supposing for simplicity that randomization is unnecessary, the N-P test of level α is given by the critical region $B = \{x : x > c'\}$ where c' is chosen so that

$$\int_{c'}^{\infty} f_{\theta_1}(x)\, dx = \alpha.$$

The important point is that the structure of this test ($x > c'$), and in fact its exact description (the constant c'), *does not depend on θ_2 but only on θ_1, so long as $\theta_2 > \theta_1$*. Consequently, if $\theta_1 = 3$, *whether $\theta_2 = 5$ or $\theta_2 = 6$, we would get the same test* for a given α. Call this test δ^*.

Now consider $\Omega = \{\theta : \theta_1 \le \theta < b\}$ where θ_1 is a specified value in Ω^*, and suppose it is desired to test

$$H_0 : \theta = \theta_1 \quad \text{against } H_1' : \theta > \theta_1,$$

so that Ω_1 is composite. If any test δ' of level $\le \alpha$ has power greater than that of δ^* at some value $\theta_2' > \theta_1$, δ' would be better than δ^* for testing between the simple hypotheses

$$H_0 : \theta = \theta_1, \qquad H_1'' : \theta = \theta_2'$$

(when Ω consists of the two values θ_1 and θ_2'), and this is impossible since δ^* is N-P for H_0 versus H_1''. Since no such θ_2' and δ' exist, we conclude that δ^* *is UMP of* H_0 *versus* H_1', *among tests of level* α.

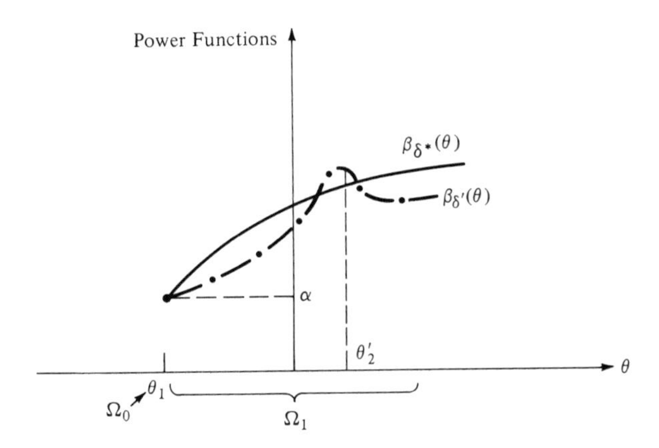

It is not hard to show that K-D densities are MLR, as are certain others (illustrated in Problems 8.3(a), 8.4(a), 8.5–8.7).

It is also easy to show that for an MLR family satisfying (8.11), $\beta_{\delta^*}(\theta)$ is nondecreasing. Hence δ^* also has size α when $\Omega = \Omega^*$, θ_1 is specified, and we are testing

$$H_0' : \theta \leq \theta_1 \quad \text{versus} \quad H_1' : \theta > \theta_1,$$

and δ^* is a UMP level α test of this composite hypothesis versus composite alternative.

This δ^* is often called a "*UMP one-sided test.*"

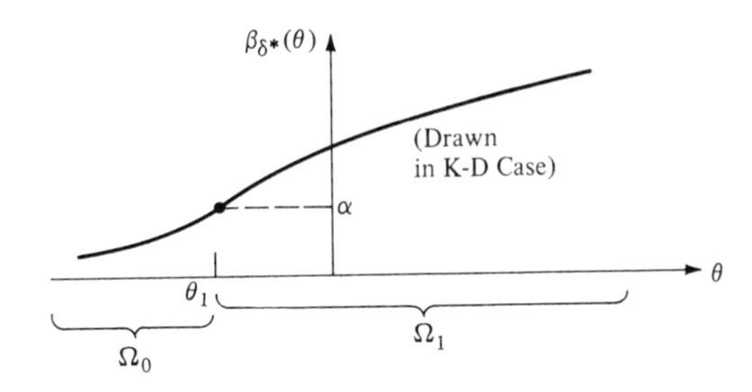

On the other hand, the figure shows that for testing

$$H_0 : \theta = \theta_1 \quad \text{versus} \quad H_1''' : \theta < \theta_1$$

the test δ^* would be terrible (in fact it is the uniformly *least* powerful test of level α), and a similar analysis shows that $B^{\#} = \{x : x < c''\}$ is UMP. (Details are to be carried out in Problem 8.5(b).)

In non-MLR examples, a UMP one-sided test will not generally exist. [This is illustrated in the Cauchy case of Problems 8.1 and 8.2.]

Unbiasedness

For testing, with $\Omega = \Omega^*$ and θ_1 specified,

$$H_0 : \theta = \theta_1 \quad \text{versus} \quad \tilde{H}_1 : \theta \neq \theta_1, \quad a < \theta < b,$$

there is no UMP test of level α in most MLR (in *all* K-D) cases, since δ^* and $B^{\#}$ will each have highest power on one side of θ_1 and will cross. [An exceptional example where a *UMP two-sided test exists* is given in Problem 8.6.]

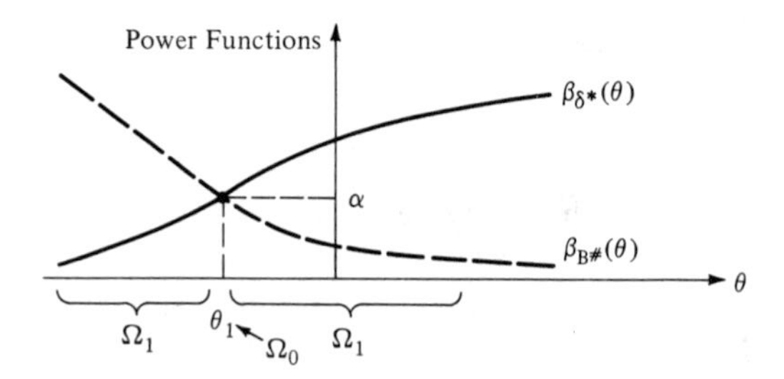

In other two-sided cases (H_0 versus $\tilde{H}_1$) people classically reduced the class $\mathscr{C}_\alpha$ still further to $\mathscr{D}_\alpha = \{$*unbiased* tests of level $\alpha\}$. A test δ is said to be *unbiased of level* α if it is of level α and if $\beta_\delta(F) \geq \alpha$ for $F \in \Omega_1$. The rationale for this restriction is that, if $\beta_\delta(F_1) < \alpha$ for some $F_1 \in \Omega_1$, it would be more probable to vote for H_0 when F_1 is true than when some F's in Ω_0 (those yielding the level α) are true, which seems unreasonable to many people. [Long after the origin of this definition of unbiased tests by Neyman and Pearson, having no clear relationship with the "unbiasedness" of estimation theory, a general concept including both of these was discovered by E. L. Lehmann. We need not pursue this here.]

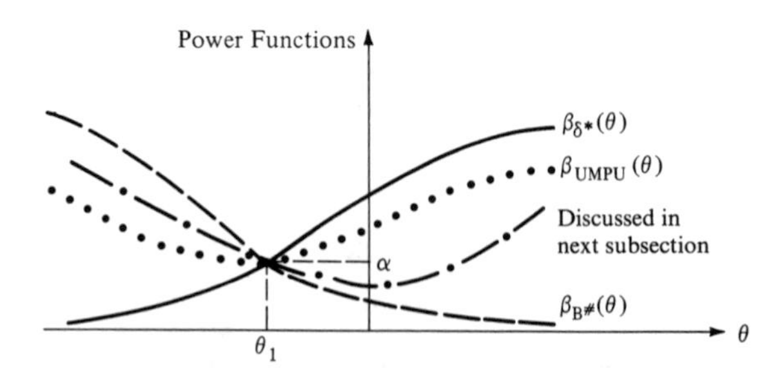

It turns out that in continuous (so as not to worry about randomization) K-D cases there is a unique UMP test among tests in $\mathscr{D}_\alpha$, called the uniformly most powerful unbiased (UMPU) test of level α. Its critical region is of the form $\{x : x < c_1 \text{ or } x > c_2\}$ where c_1 and c_2 are determined so that

$$\int_{c_1}^{c_2} f_{\theta_1}(x)\,dx = 1 - \alpha,$$
$$\int_{c_1}^{c_2} \frac{\partial}{\partial\theta} f_\theta(x)\bigg|_{\theta_1} dx = 0, \tag{8.12}$$

the last being equivalent to $\frac{\partial}{\partial\theta}\beta_\delta(\theta)\Big|_{\theta_1} = 0$. (Of course, neither $B^{\#}$ nor δ^* is unbiased for H_0 versus $\tilde{H}_1$.) Often (8.12) is approximated in practice by choosing c_1 and c_2 so as to satisfy the simpler conditions

$$\int_{-\infty}^{c_1} f_{\theta_1}(x)\,dx = \int_{c_2}^{+\infty} f_{\theta_1}(x)\,dx = \alpha/2. \tag{8.13}$$

Although the first condition of (8.12) is then satisfied, the second is not, except in special cases (e.g., f_θ normal with mean θ and variance 1). Nevertheless, the test of (8.13) is often enough easier to find than that of (8.12), and departs so little from being unbiased, that it is satisfactory in applications.

It is not hard to show that, in the cases discussed earlier, in which a UMP test δ^* exists, δ^* is also unbiased. Thus, "UMPU" is a less good (more easily satisfied) property than "UMP."

In more complex (non-MLR) cases UMP unbiased tests will also not exist.

Relationship to the Criteria of Chapter 4

In Section 8.2 and in the discussion in Chapter 4 of the geometry of risk points, we studied, for testing between two simple hypotheses, the relationship among N-P tests and (i) Bayes, (ii) admissible, and (iii) minimax tests.

We now turn to the relationship, in *composite* hypothesis testing, between these earlier notions and that of a level α test δ^* which is UMP among all tests of level α (or among all tests of level $\leq \alpha$). For simplicity and definiteness we consider the continuous case K-D setting of Section 8.3 for testing $H_0' : \theta \leq \theta_1$ (or, similarly, $H_0 : \theta = \theta_1$) against $H_1' : \theta > \theta_1$. [It will help the reader to compare the statements of (i), (ii), and (iii), which follow, with the almost identical conclusions we already know from Section 8.2 to hold for testing between the simple hypotheses $H_0 : \theta = \theta_1$ and $H_1 : \theta = \theta_2$, with θ_1 and θ_2 specified and $\theta_1 < \theta_2$, as described in the subsection on UMP tests.]

(i) First, considering the *risk* functions (8.2) of all tests, over all of Ω^* and without any restriction on "level," it can be shown that the class of *all UMP tests of all levels* coincides with the admissible tests.

(ii) Also, for any prior distribution ξ on Ω^*, the unique Bayes procedure is one of these tests (with its level α depending on ξ).

(iii) Finally, the minimax problem for the W and r of (8.1) and (8.2) is somewhat degenerate (unlike the problem for simple hypotheses), since $\lim_{\theta\downarrow\theta_1} r_\delta(\theta_1) = \lim_{\theta\downarrow\theta_1} P_\theta\{\delta \text{ makes decision } d_0\} = P_{\theta_1}\{\delta$ makes decision

$d_0\} = 1 - r_\delta(\theta_1)$ (the second equality by the continuity in θ of $f_\theta(x)$ in the K-D case). This means that $r_\delta(\theta_1+) + r_\delta(\theta_1) = 1$ (note the diagram below (8.5) with $b = \theta_1$), so that every procedure δ has maximum (more precisely, supremum) risk $\geq \max(r_\delta(\theta_1), r_\delta(\theta_1+)) \geq 1/2$. Thus, the UMP test of level 1/2 is seen to be minimax. (But there are many inadmissible minimax tests, e.g., $\delta(x) \equiv 1/2$, whereas for $\theta = \theta_1$ against $\theta = \theta_2$ there is only one minimax test.) To obtain a more realistic result, we depart from the W of (8.1), and suppose the nonnegative loss from making a wrong decision is an arbitrary $\begin{Bmatrix} W(\theta, d_1) & \text{if} & \theta \leq \theta_1 \\ W(\theta, d_0) & \text{if} & \theta > \theta_1 \end{Bmatrix}$. To avoid trivialities, we suppose neither $W(\theta, d_1)$ nor $W(\theta, d_2)$ is identically zero. (For example, $W(\theta$, wrong $d) = |\theta - \theta_1|$ may reflect the right quality.) The loss from making a correct decision is 0. Then the risk function of δ is

$$r_\delta(\theta) = \begin{cases} W(\theta, d_1)\beta_\delta(\theta) & \text{for} \quad \theta \leq \theta_1, \\ W(\theta, d_0)[1 - \beta_\delta(\theta)] & \text{for} \quad \theta > \theta_1. \end{cases}$$

It turns out that the conclusions of (i) and (ii) are still valid (modified only in that there are additional, inadmissible Bayes procedures, if ξ assigns all probability to a set for which $W(\theta, d_1) = W(\theta, d_2) = 0$). In particular, for most reasonable W's (such as the $W(\theta$, wrong $d) = |\theta - \theta_1|$ mentioned previously), there will now be a *unique* minimax procedure, no longer so easy to find as in the case of the W of (8.1) described earlier in this paragraph, but to be found as one of the UMP tests for some level α (which depends on W).

We turn next to corresponding questions for the *two-sided* continuous case K-D setting of the subsection on unbiasedness, for testing $H_0 : \theta = \theta_1$ against $\tilde{H}_1 : \theta \neq \theta_1$. Corresponding to (i) and (ii), the class of Bayes tests corresponding to all prior distributions, which again coincides with the admissible tests, consists of all critical regions of the form $\{x < c_1 \quad \text{or} \quad x > c_2\}$ as described just before (8.12), but now with c_1 and c_2 arbitrary, that is, *without either condition of* (8.12) *being applied*. In particular, if $c_1 = -\infty$ or $c_2 = +\infty$ we get the UMP *one-sided* tests discussed earlier (each of which performs very well for half of $\tilde{H}_1$ and very poorly for the other half). If we consider for a fixed value α only those tests which have level α, then by varying c_1 and c_2 between $-\infty$ and $+\infty$ while keeping the first relation of (8.12) satisfied, we obtain, as shown in the diagram illustrating (8.12), other tests such as the one whose power function is there labeled "discussed in next subsection." Any two tests of this infinite family (all with $\beta_\delta(\theta_1) = \alpha$) have power functions which cross only at θ_1. One of them is the UMPU test which satisfies the second condition of (8.12), too.

We can also vary W in the manner described in (iii) for the one-sided setting, obtaining analogous results. In particular, there will be a c_1 and c_2, which depend on W and which yield a minimax test. Except in certain examples in which f_θ possesses appropriate symmetry, this minimax test need satisfy neither (8.13) ("equal tails") nor the second condition of (8.12) ("unbiasedness").

Such a symmetric example will be discussed in Section 8.7 in the subsection on problems and tests.

8.4. Likelihood Ratio (LR) Tests

The likelihood ratio (LR) test of $H_0 : F \in \Omega_0$ versus $H_1 : F \in \Omega_1$ rejects H_0 if

$$\frac{\max_{F \in \Omega_0} p_F(x)}{\max_{F \in (\Omega_0 \cup \Omega_1)} p_F(x)} < k \tag{8.14}$$

where k is adjusted to give the right significance level. This is analogous to the ML method of estimation; in fact, if $\hat{t}$ is the ML estimator of F *assuming we know only that* $F \in \Omega$, and $\hat{t}_0$ is the ML estimator *assuming we know that* $F \in \Omega_0$, then (8.14) is $p_{\hat{t}_0}(x)/p_{\hat{t}}(x) < k$.

If Ω_0 and Ω_1 are simple, this yields essentially the N-P tests. Otherwise this intuitive approach to test construction can yield very bad procedures in some examples, just as the ML construction of estimators can.

The classical *normal theory* tests to be studied later are derived in many books by this LR approach. It is more satisfactory to obtain them as UMP or UMPU tests (see preceding discussion) or as UMP invariant tests (discussed later) whenever they have such properties, since the LR method in general is no more justifiable than is the ML method.

For large sample sizes certain good properties hold for LR tests, just as for ML estimators.

8.5. Problems Where n Is To Be Found

Many testing problems are phrased (as other statistical problems can also be phrased) in terms of finding the minimum sample size which achieves a specified upper bound on expected loss due to making an incorrect decision, rather than in terms of a given S (and, thus, a given sample size). We now illustrate this.

EXAMPLE 8.5. Suppose n is to be chosen so that we observe $X_1, \ldots, X_n$, iid with distribution $\mathcal{N}(0, 1)$ under H_0 and $\mathcal{N}(.2, 1)$ under H_1. How large need n be to obtain a test of level .05 and power $\geq .9$? (This "choice of n" can be thought of as a problem of experimental design—see Chapter 3.) We know the N-P test is of the form "reject H_0 if $\sum_1^n X_i > c$." Since $\sum_1^n X_i$ is $\mathcal{N}(0, n)$, it is easily computed that $c = 1.65\sqrt{n}$, since $\Phi(1.65) = .95$. The power is $P_1\left\{\sum_1^n X_i > 1.65\sqrt{n}\right\} = P_1\left\{\frac{\sum_1^n X_i - .2n}{\sqrt{n}} > 1.65 - .2\sqrt{n}\right\} = 1 - \Phi(1.65 - .2\sqrt{n})$, since $\left(\sum_1^n X_i - .2n\right)\Big/\sqrt{n}$ is $\mathcal{N}(0, 1)$ under H_1. Since

$\Phi(-1.28) = .1$ we must set $1.65 - .2\sqrt{n} \leq -1.28$, so $n = 215$ is the smallest sample size which works.

An alternative formulation to that of Example 8.5 treats the sum $E_F W(F, t(X_1, \ldots, X_n)) + nC$ as the risk function, where C is the cost per observation; risk functions are then compared as we have discussed earlier. This is usually less attractive to practitioners than the formulation of Example 8.5, in large part because of the common difficulty of specifying units of W relative to C.

8.6. Invariance

There are many problems which can be treated by the invariance approach. The main change from invariance as discussed in estimation (Section 7.3) is that the transformations $\{g_k, h_k, j_k\}$ must be such that $j_k(d_i) = d_i$ and such that if $\theta \in \Omega_i$ then $h_k(\theta) \in \Omega_i$. As before, if X has a df with parameter value θ, then $g_k(X)$ has a df with parameter value $h_k(\theta)$. The distributions in each Ω_i remain in the same Ω_i when transformed by any of the h_k.

Without going into more detail, let us treat an example. Suppose X_1, X_2 are independent normal random variables with variance 1 and means θ_1, θ_2 where $\Omega_0 = \{(\theta_1, \theta_2): -\infty < \theta_1 = \theta_2 < +\infty\}$ and $\Omega_1 = \{(\theta_1, \theta_2): -\infty < \theta_1 < \theta_2 < \infty\}$. Both Ω_i are composite. The transformation $(X_1, X_2) \to (X_1', X_2') \stackrel{\text{def}}{=} (X_1 + c, X_2 + c)$ "leaves the problem invariant" since (X_1', X_2') has a law in Ω_0 if (X_1, X_2) did and has a law in Ω_1 if (X_1, X_2) did. An invariant *test* should come to the same conclusion about whether or not the random variables X_i have equal means, regardless of shift in origin; an *invariant test* (which in the general setup of the previous paragraph is defined to satisfy $\delta(g_k x) = \delta(x)$ for all g_k and x) here is defined to satisfy $\delta(x_1 + c, x_2 + c) = \delta(x_1, x_2)$ for all x_1, x_2, c. Putting $c = -x_2$, we obtain $\delta(x_1, x_2) = \delta(x_1 - x_2, 0)$: *an invariant test depends only on* $X_2 - X_1 = Y$ (*say*). The df of Y is $\mathcal{N}(0, 2)$ under H_0 and $\mathcal{N}(\phi, 2)$ with $\phi = \theta_2 - \theta_1 > 0$ under H_1. We know (from the discussion of UMP tests) a UMP test of this one-sided hypothesis, based on Y, is "reject H_0 if $Y > c$," where c is chosen to give level α. This test is hence termed a "*UMP level α invariant test*," meaning "a UMP test among all level α invariant tests." We will treat more such examples in Section 8.7.

8.7. Summary of Common "Normal Theory" Tests

This section is intended to relate the earlier results of Chapter 8 to procedures encountered most frequently in the literature.

Distributions

The following summarizes the most important distribution theory for common normal theory tests. Details can be found in Wilks, Scheffé, Graybill-

Mood, etc. The lack of details in this section conforms with our intention to deemphasize the computation of probabilities when such computations are not elementary and add nothing to the understanding of the principles of statistical inference being discussed. In fact, the derivation of these distributions is carried out in detail in many probability texts without any indication of why they are of interest, or with a routine application of the distributions to the construction of tests which are in no way justified as having any good properties. *All of these distributions are of interest only because the level and power of certain tests, described and justified in the next subsection, or the risk of certain procedures of Chapter 9, can be expressed in terms of these distributions.*

Let $Y_1, Y_2, \ldots$ be iid $\mathcal{N}(0, 1)$ (normally distributed with mean 0 and variance 1).

(i) Then $\sum_1^n Y_i^2$ is said to have a *χ^2-distribution (chi-squared) with* n *degrees of freedom*, abbreviated χ_n^2. If U has a χ_n^2 distribution and b is constant, we will say that bU has a *$b\chi_n^2$-distribution*. The distribution of $\sum_1^n (Y_i + c_i)^2$ turns out to depend on the c_i's only through $\lambda = \sum_1^n c_i^2$ and is called the *noncentral χ^2-distribution with* n *degrees of freedom and noncentrality parameter λ*, sometimes abbreviated $\chi_n^2(\lambda)$; the case $\lambda = 0$ described in the previous sentence is sometimes referred to as the *central χ_n^2-distribution*. The sum of two independent random variables U_i, if U_i is $\chi_{m_i}^2$-distributed $(i = 1, 2)$, is $\chi_{m_1+m_2}^2$.

(ii) The distribution of $(Y_{n+1} + \gamma)\Big/\left[\sum_1^n Y_i^2/n\right]^{1/2}$ (the ratio of a $\mathcal{N}(\gamma, 1)$ variable to the square root of n^{-1} times an independent χ_n^2-variable is called the *noncentral t-(or "Student's") distribution with* n *degrees of freedom and noncentrality parameter* γ, abbreviated $t_n(\gamma)$; the case $\gamma = 0$ is called *central* (or simply the *t_n-distribution* if no ambiguity can arise).

(iii) The distribution of $\left[\sum_1^n (Y_i + c_i)^2/n\right]\Big/\left[\sum_{n+1}^{n+m} Y_i^2/m\right]$ (ratio of two independent χ^2-variables, each divided by its degrees of freedom, the denominator central) is called the *F-distribution with* n *numerator and* m *denominator degrees of freedom and noncentrality parameter λ*, abbreviated $F_{n,m}(\lambda)$; here λ is as defined in (i); the case $\lambda = 0$ is again referred to as *central* (or simply $F_{n,m}$).

Warning: In some tables the "noncentrality" parameter is $\gamma/2$ in (ii) or $\lambda/2$ in (iii).

[There are also F-distributions with noncentrality in the denominator, which will not concern us.]

(iv) Note that the square of a $t_n(\gamma)$ variable has an $F_{1,n}(\gamma^2)$ distribution.

All of the preceding *central* densities have fairly simple expressions; the *noncentral* densities are usually expressed in power series.

Problems and Tests

We shall see that the preceding distributions arise in the following typical examples, in all cases the central distribution being used to compute the significance level; in each case the distributions used in computing the level or power are listed briefly, just after a short description of the problem. In Problems A and B we shall indicate the methods for justifying the tests; there are similar justifications in C, D, E, F, where we shall give fewer details.

Problem A: One- or Two-Sided Test about the Mean of a Normal Distribution with Known Variance—(Standard) Normal Law. Here $X_1, X_2, \ldots, X_r$ are iid $\mathcal{N}(\theta, 1)$ with either $\Omega = \{\theta : \theta \geq 0\}$ or $\{\theta : -\infty < \theta < +\infty\}$. In the former case we test $H_0 : \theta = 0$ against $H_1' : \theta > 0$; in the latter case we test $H_0 : \theta = 0$ against $\tilde{H}_1 : \theta \neq 0$.

Since the sufficient statistic $T = \sum_1^r X_i$ (for either Ω) has the K-D form $\mathcal{N}(r\theta, r)$ when X_1 is $\mathcal{N}(\theta, 1)$, and we can reduce consideration to tests depending only on T, the discussion of K-D families in the subsection on UMP tests, and arithmetic like that of Example 8.5 show that the one-sided critical region

$$\{r^{-1/2}T > c'\}, \quad \text{where} \quad 1 - \Phi(c') = \alpha, \tag{8.15}$$

where Φ is again the standard normal df, *is UMP of level* α for H_0 against H_1'. The power function of this test is

$$P_\theta\{r^{-1/2}T > c'\} = P_\theta\{r^{-1/2}(T - r\theta) > c' - r^{1/2}\theta\} = 1 - \Phi(c' - r^{1/2}\theta), \tag{8.16}$$

the last equality following from the fact that $r^{-1/2}(T - r\theta)$ is $\mathcal{N}(0, 1)$ when X_1 is $\mathcal{N}(\theta, 1)$.

For testing H_0 against the *two-sided* alternative $\tilde{H}_1$, Section 8.3 shows that there is no UMP level α test. The commonly used level α critical region is

$$\{r^{-1/2}|T| > c_2\} \quad \text{where} \quad 1 - \Phi(c_2) = \alpha/2. \tag{8.17}$$

We shall now indicate four ways in which this test can be justified:

(i) In (8.12) let f_θ be the density of $r^{-1/2}T$ when X_1 is $\mathcal{N}(\theta, 1)$. Our critical region is of the form discussed there with $c_1 = -c_2$. The second line of (8.12) is easily seen to be satisfied (with $\theta_1 = 0$), either by detailed computation or by the symmetry about 0 of the $\beta(\theta)$ for the region (8.17). Thus, (8.17) is UMPU of level α.

(ii) The distribution of $-T$ satisfies H_0 (respectively, $\tilde{H}_1$) if that of T does. A development analogous to that of the example of Section 8.6 thus leads to consideration of the transformation $(T, \theta) \to (-T, -\theta)$ (as described in Section 7.3) and an *invariant* procedure is one for which $\delta(T) = \delta(-T)$, that is, which depends on the value of T only through $|T|$. The density function of $|T| = U$ (say) is easily seen to depend on θ only through $|\theta|$. Thus, if we reduce consideration to invariant procedures, we must test

$H_0^*: |\theta| = 0$ against $\tilde{H}_1^*: |\theta| > 0$ based on U. The family of densities of U for $|\theta| \geq 0$ is a MLR family [proving this is not difficult]; Section 8.3 thus implies that the critical region $\{U > \bar{c}\}$ of level α is UMP of H_0^* against $\tilde{H}_1^*$, which means that it is *UMP invariant of level* α (UMP among invariant tests of level α) of the original H_0 against $\tilde{H}_1$. And $\{U > c\}$ is the test of (8.17) with $\bar{c} = r^{1/2}c_2$.

(iii) The test of (8.17) also has *minimax* properties. For example, if as in the subsection on the relation of criteria in Chapter 4, we depart from our assumption of (8.1) by letting $w(|\theta|) \geq 0$ denote the loss incurred by making the wrong decision when θ is the true parameter value, then it can be shown that, among all tests δ of level α, the test (8.17) minimizes the maximum of the risk function over Ω_1 (which maximum is $\max_{\theta \neq 0}[1 - \beta_\delta(\theta)]w(|\theta|)$). Alternatively, doing away with the restriction to tests of level α, one can consider the maximum of the risk function over *all* of Ω, namely, $\max\{w(0)\beta_\delta(\theta), \max_{\theta \neq 0}[1 - \beta_\delta(\theta)]w(|\theta|)\}$, and this is minimized for δ of the form (8.17) for some c_2 (related to w now, rather than to α).

(iv) For any prior distribution ξ on $\{-\infty < \theta < +\infty\}$ which is *symmetric about* 0, an appropriate choice of c_2 (depending on ξ) makes (8.17) a *Bayes test*.

The discussion of these four criteria (i), (ii), (iii), (iv) did not arise in connection with the UMP level α test (8.15) of H_0 against the *one-sided* alternative H_1' because a ump level α test δ^* is *automatically* (i) UMPU of level α; (ii) UMP invariant of level α under any possible collection of transformations under which δ^* is invariant and which leave the problem invariant (the only transformation leaving the problem invariant in the present case being the identity, which means that invariance, which imposed a restriction on δ in the case $\tilde{H}_1$, imposes none here in the case H_1'); (iii) minimax in either of the senses described under (iii) (the second, for an appropriate choice of α or c' depending on w); and (iv) Bayes relative to any prior distribution ξ (for an appropriate choice of c' depending on ξ). (See the subsection on the relation of Chapter 4 criteria regarding (iii) and (iv); (i) was mentioned at the end of the subsection on unbiasedness.)

The power function of the test of (8.17) is (computing as in (8.16))

$$\begin{aligned} &P_\theta\{r^{-1/2}T > c_2\} + P_\theta\{r^{-1/2}T < -c_2\} \\ &\quad = 1 - \Phi(c_2 - r^{1/2}\theta) - \Phi(-c_2 - r^{1/2}\theta). \end{aligned} \tag{8.18}$$

The relations (8.17) and (8.18) are simple enough. We shall nevertheless rewrite them, for future reference, so that we can see later that Problem D can be viewed essentially as a generalization of the present H_0 against $\tilde{H}_1$. The event $\{r^{-1/2}|T| > c_2\}$ of (8.17) is of course equivalent to

$$r^{-1}T^2 > c_2^2, \tag{8.19}$$

which can also be written $\{[r^{-1/2}(T - r\theta) + r^{1/2}\theta]^2 > c_2^2\}$; since $r^{-1/2}(T - r\theta)$

is $\mathcal{N}(0,1)$ when θ is the true parameter value, we conclude from (i) of the subsection on distributions that (8.19) is equal to the probability that a $\chi_1^2(r\theta^2)$ variable exceeds c_2^2; in particular, (8.17) is achieved by choosing c_2^2 so that the probability is α that a central χ_1^2 variable (namely, the left side of (8.19)) exceeds c_2^2.

We note that, if $X_1^\#, X_2^\#, \ldots, X_r^\#$ are iid $\mathcal{N}(\theta, \sigma^2)$ *where* σ^2 *is known*, then for testing $H_0^\# : \theta = L$ against $H_1^\# : \theta > L$ or $\tilde{H}_1^\# : \theta \neq L$ where L is specified, the transformation $X_i = (X_i^\# - L)/\sigma$ reduces the problem to that of testing H_0 against H_1' or $\tilde{H}_1$ as described previously.

Problem B: Same as Problem A but with Unknown Variance—t-Distribution. With $r \geq 2$, $X_1, X_2, \ldots, X_r$ are again iid $\mathcal{N}(\theta, \sigma^2)$, but now $\Omega = \{(\theta, \sigma^2) : \theta \geq 0, \sigma > 0\}$ or $\Omega = \{(\theta, \sigma^2) = -\infty < \theta < +\infty, \sigma > 0\}$. That is, σ *is also unknown.* The descriptions of H_0, H_1', and $\tilde{H}_1$ are written as in Problem A, but the difference in meaning should be understood; for example, H_0 was simple in Problem A, but is now composite since "$\theta = 0$" means "$\theta = 0, 0 < \sigma < +\infty$."

In this problem (for either Ω) a minimal sufficient statistic is $(T_1, T_2) \stackrel{\text{def}}{=} \left(\bar{X}_r, \sum_1^r (X_i - \bar{X}_r)^2\right)$. When X_1 is $\mathcal{N}(\theta, \sigma^2)$, the joint density of (T_1, T_2) can be stated as

$$\begin{aligned} &\text{(i) } T_1 \text{ is } \mathcal{N}(\theta, \sigma^2/r), \\ &\text{(ii) } T_2/\sigma^2 \text{ is } \chi_{r-1}^2(0), \\ &\text{(iii) } T_1 \text{ and } T_2 \text{ are independent.} \end{aligned} \tag{8.20}$$

Of these, (i) is the standard result on the distribution of the sample mean from a normal distribution. As for (ii), we first note for comparison, upon putting $Y_i = (X_i - \theta)/\sigma$ and using the subsection on distributions, that $\sum_i (X_i - \theta)^2/\sigma^2$ is $\chi_r^2(0)$. But $T_2 = \sum_1^r (X_i - \bar{X}_r)^2/\sigma^2$. One can make a linear (actually, orthogonal) transformation $\mathscr{L}$ (say) from the iid $\mathcal{N}(0,1)$ random variables $\{Y_1, \ldots, Y_r\}$ to independent normal random variables $\{Y_1', Y_2', \ldots, Y_{r-1}', r^{1/2}\bar{Y}_r\}$, which also turn out to be iid $\mathcal{N}(0,1)$, and such that $T_2 = \sigma^2 \sum_1^{r-1} (Y_i')^2$. (See, e.g., Scheffé or Wilks.) This shows that T_2/σ^2 is $\chi_{r-1}^2(0)$. This relationship between $\sum_1^r (X_i - \theta)^2$ and $\sum_1^r (X_i - \bar{X}_r)^2$ should recall our discussion of $Q_{\min}$ in least squares (linear estimation) theory (Sec. 5.4); we had $E_{\theta,\sigma^2} \sum_1^r (X_i - \theta)^2 = r\sigma^2$ but $E_{\theta,\sigma^2} \sum_1^r (X_i - \bar{X}_r)^2 = (r-1)\sigma^2$, reflecting the "loss of one degree of freedom due to estimating θ by $\bar{X}_r$." This is consistent, in present terms, with the fact that the $\chi_r^2(0)$ and $\chi_{r-1}^2(0)$ distributions have means r and $r-1$, respectively. (Of course, $Q_{\min}/\sigma^2$ will not have a χ^2 distribution if the X_i's are not normal.)

As for (8.20) (iii), it is actually a property special to the normal df. It follows from the independence of $Y_1', \ldots, Y_{r-1}', \bar{Y}_r$ resulting from the transformation $\mathscr{L}$ described previously.

By sufficiency, we can restrict attention to tests based on (T_1, T_2). Thus, in the one-sided case we are reduced to testing

$$H_0 : T_1, T_2 \text{ independent}; \qquad T_1 : \mathscr{N}(0, \sigma^2/r);$$
$$T_2 : \sigma^2 \chi_{r-1}^2(0), \quad \text{for some unknown } \sigma > 0,$$

against

$$H_1' : T_1, T_2 \text{ independent}; \qquad T_1 : \mathscr{N}(\theta, \sigma^2/r);$$
$$T_2 : \sigma^2 \chi_{r-1}^2(0), \quad \text{for some (unknown) } \theta > 0 \quad \text{and} \quad \sigma > 0.$$

It is not very difficult to write down prior distributions which produce a variety of forms of Bayes procedures; there is no characterization so simple as that of Problem A. There are a number of properties in terms of which one can justify the "t-test" described in (8.28). Rather than to list them all (as in the two-sided case of Problem A), we shall describe one justification, largely because an analogous justification is the simplest with which to work in most cases of the remaining problems (D–F) described later.

That justification is in terms of *invariance* (Section 7.3). If c is any positive value, then the random variables $T_1' = cT_1$, $T_2' = c^2 T_2$ (which correspond to $(X_1', \ldots, X_r') = (cX_1, \ldots, cX_r)$, discussed further later in this section) satisfy H_0 for some $\sigma > 0$ (namely, c times that of T_1, T_2) if and only if (T_1, T_2) do. And they satisfy H_1 for some $\mu > 0$ and $\sigma > 0$ (both c times those of T_1, T_2) if and only if (T_1, T_2) do. That is, this is a *scale-parameter problem* similar to the scale-parameter estimation problems of Section 7.3. Two experimenters using the same data, recorded in different units, should both reach the same conclusion about accepting or rejecting H_0 if they use the same nonrandomized procedure δ. That is, such an *invariant procedure* δ must satisfy

$$\delta(\tau_1, \tau_2) = \delta(\tau_1', \tau_2') = \delta(c\tau_1, c^2\tau_2) \tag{8.21}$$

for all possible data values (τ_1, τ_2) of (T_1, T_2) and transformation coefficients $c > 0$ (and, consequently, for all data values (τ_1', τ_2') of (T_1', T_2')). In particular (employing the technique used in Section 7.3 to obtain the form of an invariant procedure), (8.21) must hold with $c = \tau_2^{-1/2}$; that is,

$$\delta(\tau_1, \tau_2) = \delta(\tau_1/\sqrt{\tau_2}, 1); \tag{8.22}$$

conversely, it is easy to show that any δ satisfying (8.22) also satisfies (8.21). We conclude, from (8.22), that *any scale-invariant procedure depends on* $(\mathrm{T}_1, \mathrm{T}_2)$ *only through* $\mathrm{T}_1/\sqrt{\mathrm{T}_2}$; or, equivalently (multiplying by a known constant), through

$$T_{r-1}^* \stackrel{\text{def}}{=} r^{1/2} T_1/\sqrt{T_2/(r-1)}. \tag{8.23}$$

We now follow a development like that of (ii) in the two-sided case of Problem A. When X_1 is $\mathscr{N}(\theta, \sigma^2)$ we can write (even though we don't know θ

or σ)

$$T^*_{r-1} = \frac{[r^{1/2}\sigma^{-1}(T_1 - \theta) + r^{1/2}\sigma^{-1}\theta]}{[T_2/\sigma^2(r-1)]^{1/2}}. \tag{8.24}$$

Since by (8.20) the numerator and denominator of (8.24) are independent, the numerator being $\mathcal{N}(r^{1/2}\sigma^{-1}\theta, 1)$ and the T_2/σ^2 of the denominator being $\chi^2_{r-1}(0)$, we see from (ii) of the subsection on distributions that

$$T^*_{r-1} \quad \text{has a} \quad t_{r-1}(r^{1/2}\sigma^{-1}\theta) \text{ distribution}. \tag{8.25}$$

Note that $r^{1/2}\sigma^{-1}\theta$ is 0 under H_0 and its range is the positive reals under H'_1. That means that, *if we restrict attention to scale-invariant tests*, we are faced with testing, on the basis of T^*_{r-1},

$$\begin{aligned} &H^*_0 : T^*_{r-1} \quad \text{has a } t_{r-1}(0) \text{ distribution against} \\ &H'^*_1 : T^*_{r-1} \quad \text{has a } t_{r-1}(\gamma) \text{ distribution for some unknown } \gamma > 0. \end{aligned} \tag{8.26}$$

It turns out that

$$\{\text{the } t_{r-1}(\gamma) \text{ densities}, \quad 0 \le \gamma < +\infty\} \text{ is an MLR family}. \tag{8.27}$$

(The family of (8.27) is not K-D, and (8.27) is not elementary.) Hence, from (8.26), (8.27), and the subsection on UMP tests, a UMP level α critical region of H^*_0 against H^*_1 (based on T^*_{r-1}) is

$$T^*_{r-1} > c_\alpha \quad \text{where} \quad P_{0,\sigma^2}\{T^*_{r-1} > c_\alpha\} = \alpha; \tag{8.28}$$

since the $t_{r-1}(0)$ distribution of T^*_{r-1} when $\theta = 0$ *does not depend on* σ^2, the value of c_α for which the probabilistic condition of (8.28) is satisfied can be determined from tables of the $t_{r-1}(0)$ df even though σ is unknown.

Hence (again in analogy with the development of (ii) in Problem A) the test (8.28) is *UMP among level* α *invariant tests of the original* H_0 *against* H'_1. As in the subsection on UMP tests, H_0 can be replaced by $H'_0 : \theta \le 0$ in this statement. And $H_0 : \theta = \theta_1$ (or $H'_0 : \theta \le \theta_1$) against $H'_1 : \theta \ge \theta_1$ with θ_1 specified can be transformed to the preceding form by replacing X_i everywhere by $X_i - \theta_1$. Moreover, for testing $H_0 : \theta = 0$ against $H'''_1 : \theta < 0$, the analogue of $B^\#$ of the subsection on UMP tests is easily seen to be $T^*_{r-1} < -c_\alpha$; the central t-density is symmetric.

T^*_{r-1} is called *Student's t-statistic with* r − 1 *degrees of freedom*, and (8.28) is *Student's one-sided t-test of level* α. "Student" was a pseudonym of the British statistician Gosset, who feared that his employer (the brewer Guinness) might disapprove of articles' appearing under his real name.

The preceding test can also be arrived at as a UMP invariant test from the original $X_1, \ldots, X_r$ without first reducing the problem by sufficiency to (T_1, T_2). But then additional transformations to $(X_1, \ldots, X_r) \to (cX_1, \ldots, cX_r)$ are required to reduce consideration to tests depending on (8.23).

The test (8.28) is more often derived as the *LR test* (Section 8.4), or is obtained by an intuitive development from the test of (8.15) for the random variables $X_i^{\#}$ of the last paragraph of Problem A by substituting for σ (known in Problem A but unknown here) the estimate $[T_2/(r-1)]^{1/2}$ of σ. But neither of these derivations justifies the test in terms of any precise property of goodness.

In the case of the two-sided alternative $\tilde{H}_1 : \theta \neq 0$ to $H_0 : \theta = 0$, the analogue of (8.17) is now

$$|T_{r-1}^*| > c_{\alpha/2} \tag{8.29}$$

where, consistent with the notation of (8.28), the probability that a $t_{r-1}(0)$ random variable exceeds $c_{\alpha/2}$ is $\alpha/2$. If one considers, as in (ii) of Problem A, the transformation $(X_1, \ldots, X_r) \to (-X_1, \ldots, -X_r)$ (or $(T_1, T_2) \to (-T_1, T_2)$) as well as the transformations of (8.21), one can show that (8.29) is *UMP invariant of level* α. Moreover, the ("for-future-reference") analogue here of (8.19) is to rewrite (8.29) as

$$(T_{r-1}^*)^2 > c_{\alpha/2}^2 \tag{8.30}$$

and to observe on squaring (8.24) and using (iv) of the subsection on distributions that $(T_{r-1}^*)^2$ has an $F_{1,r-1}(0)$ distribution under H_0; we shall refer to this under Problem E.

Finally, the power function of (8.28) or (8.29) can be obtained from the noncentral $t_{r-1}(\gamma)$ distribution of T_{r-1}^* described in (8.25) (or the latter, in the form (8.30), from the noncentral $F_{1,r-1}(\gamma^2)$ distribution). These are analytically less simple than (8.16) or (8.17), but there are limited tables of them. (See Scheffé's book for references.)

When r is large, $T_2/(r-1)$ will be close to the true σ^2 with high probability, and from (8.24) we see then that T_{r-1}^* is approximately normal with unit variance. Hence, when r is large the c_α of (8.28) is approximately the c' of (8.15), and the power function of the test (8.28) can be approximated by (8.16); similarly for (8.29) and (8.17)–(8.18).

Problem C: One- or Two-Sided Test about the Variance of a Normal Distribution with Mean (i) *Known or* (ii) *Unknown—χ^2 Distribution* (*Central Only, Even for Power*).

(i) *Mean known*: Suppose $X_1, \ldots, X_r$ are iid with *known mean* θ_1 and unknown variance σ^2. With σ_1^2 a specified value, we want to test $H_0 : \sigma^2 = \sigma_1^2$ against either $H_1' : \sigma^2 > \sigma_1^2$ or $\tilde{H}_1 : \sigma^2 \neq \sigma_1^2$. Thus, Ω is either $\{\sigma^2 : \sigma^2 \geq \sigma_1^2\}$ or else $\{\sigma^2 : \sigma^2 > 0\}$. In either case a minimal sufficient statistic is

$$T' = \sum_1^r (X_i - \theta_1)^2; \tag{8.31}$$

and, if σ^2 is the true unknown parameter value, T'/σ^2 has the $\chi_r^2(0)$ distribution by (i) of the subsection on distributions, where we then defined the $\sigma^2\chi_r^2(0)$ distribution to be the distribution of T'. In terms of (8.31), the one-sided hypothesis H_0 against H_1' can be rewritten

$$H_0 : T'/\sigma_1^2 \quad \text{has the } \chi_r^2(0) \text{ distribution}$$

$$\text{against} \tag{8.32}$$

$$H_1' : T'/\sigma_1^2 \quad \text{has the } k\chi_r^2(0) \text{ distribution for some unknown } k > 1.$$

It is now much simpler than in the case (8.27) to show that

$$\{\text{the } k\chi_r^2(0) \text{ distributions}, \quad 0 < k < \infty\} \quad \text{is an MLR family}, \tag{8.33}$$

since in this case the MLR family is even K-D, namely, a family of. Γ-densities differing only in scale parameter. A consequence of (8.33) and the subsection on UMP tests (with $\theta = \theta_1$ there becoming $k = 1$ here) is that the test

$$T'/\sigma_1^2 > c' \quad \text{where} \quad P\{(\chi_r^2(0) \text{ random variable}) > c'\} = \alpha \tag{8.34}$$

is *UMP of level* α. The power function of this test is

$$\begin{aligned} P_{\sigma^2}\{T'/\sigma_1^2 > c'\} &= P_{\sigma^2}\{T'/\sigma^2 > c'\sigma_1^2/\sigma^2\} \\ &= P\{(\chi_r^2(0) \text{ random variable}) > c'\sigma_1^2/\sigma^2\}, \end{aligned} \tag{8.35}$$

since T'/σ^2 has the $\chi_r^2(0)$ distribution when σ^2 is the true variance.

The analogue of $B^{\#}$ of the subsection on UMP tests, now for testing $H_0 : \sigma^2 = \sigma_1^2$ against $\sigma^2 < \sigma_1^2$, is $T'/\sigma_1^2 < c''$, where c'' is obtained from the lower tail of the $\chi_r^2(0)$ distribution; c'' cannot be obtained from the c' of (8.34) in the manner of Problems A and B, since of course the χ^2 densities are not symmetric.

For testing H_0 against $\tilde{H}_1$, there is no transformation (analogous to $X_i \to -X_i$ in Problem A or B) with which to reduce the problem further from that of consideration of T'. Often one uses (8.33) and the development of the subsection on unbiasedness to obtain the *UMPU level* α *critical region* $\{T'/\sigma_1^2 < c_1 \quad \text{or} \quad > c_2\}$, by satisfying (8.12) with the $k\chi_r^2(0)$ density for f_θ there and $\theta_1 = 1$ in (8.12). More often, this test is approximated by using (8.13).

(ii) *Mean unknown*: With $X_1, \ldots, X_r$ iid $\mathcal{N}(\theta, \sigma^2)$, with either $\Omega = \{-\infty < \theta < +\infty, \quad \sigma^2 \geq \sigma_1^2\}$ or $\Omega = \{-\infty < \theta < +\infty, \quad \sigma^2 > 0\}$, the statistic (T_1, T_2) of Problem B is again minimal sufficient. Both the problem of testing $H_0 : \theta = \theta_1$ against $H_1' : \theta > \theta_1$ and that of testing $H_0 : \theta = \theta_1$ against $\tilde{H}_1 : \theta \neq \theta_1$ are invariant under $(X_1, \ldots, X_r) \to (X_1 + b, \ldots, X_r + b)$, where b is an arbitrary real number, since the $X_i + b$ are i.i.d. normal with unknown mean and with the same variance as the X_i. Since T_2 is unchanged if X_i is replaced by $X_i + b$ (and $\bar{X}_r$, by $\bar{X}_r + b$), the transformation in terms of the sufficient statistic is $(T_1, T_2) \to (T_1 + b, T_2)$. Hence, an invariant δ now satisfies (in place of (8.21))

$$\delta(\tau_1, \tau_2) = \delta(\tau_1 + b, \tau_2) \tag{8.36}$$

for all τ_1, τ_2, and b. Putting $b = -\tau_1$, we have $\delta(\tau_1, \tau_2) = \delta(0, \tau_2)$; that is, *an invariant procedure* (among procedures depending on (T_1, T_2)) *depends only on* T_2. The remaining development is much simpler than that of (8.23)–(8.28) of Problem B, where we encountered a new family of densities (8.25) compared with what we had in Problem A, when we went from testing about θ with σ known to testing with σ unknown. For, now, when we go from case (i) of Problem C in testing about σ with θ known to the present case (ii) in testing about σ with θ unknown, we simply go from consideration of the T' of (8.31) to consideration of T_2, *and, as stated in* (8.20) (ii), T_2 *has a* $\sigma^2\chi^2_{r-1}(0)$ *distribution*, in place of the $\sigma^2\chi^2_r(0)$ distribution of T'. Thus, if we restrict consideration to the "location-invariant" procedures we have described by (8.36), *all considerations of case* (i) *of Problem C* may be applied by simply replacing T' and r there by T_2 and $r - 1$ here.

Problem D: Model I of ANOVA with Variance Known—Normal or χ^2 *Distribution.* Suppose X_{ij}, $1 \le j \le r_i$, $1 \le i \le k$, are independent normal random variables, X_{ij} distributed $\mathcal{N}(\phi_i, \sigma^2)$. *This was called the one-way ANOVA in Chapter* 5 (without assuming normality in the linear estimation considered there). We assume σ^2 known. (This will be changed in Problem E.) We shall use the notation $X_{i\cdot} = r_i^{-1}\sum_j X_{ij}$ and $X_{\cdot\cdot} = \left(\sum_i r_i\right)^{-1} \sum_{i,j} X_{ij}$, just as in Chapter 5.

First suppose $k = 2$. [When $r_1 = r_2 = 1$, the example of this paragraph is that of Section 8.6.] For testing $H_0 : \phi_1 = \phi_2$ (which is composite since the common value of the ϕ_i's can be any real value) against $H_1' : \phi_1 - \phi_2 > 0$, a minimal sufficient statistic is the pair $(X_{1\cdot}, X_{2\cdot})$. The problem can be seen to be invariant under transformations $(X_{1\cdot}, X_{2\cdot}) \to (X_{1\cdot} + c, X_{2\cdot} + c)$, and an invariant procedure is then seen to depend only on $X_{1\cdot} - X_{2\cdot}$; or, equivalently, on

$$(X_{1\cdot} - X_{2\cdot})/(r_1^{-1} + r_2^{-1})^{1/2}\sigma. \tag{8.37}$$

The random variable of (8.37) is normal with unit variance and with mean 0 under H_0 and >0 under H_1'. *This reduced problem in terms of* (8.37) *is that of Problem A (one-sided) with* $r^{1/2}T$ *there replaced by* (8.37). Hence, the UMP test among invariant tests of level α is that of (8.15), with $r^{1/2}T$ replaced by (8.37).

Similarly, still with $k = 2$, for testing $H_0 : \phi_1 = \phi_2$ against $\tilde{H}_1 : \phi_1 - \phi_2 \ne 0$, the preceding reduction by invariance to tests based on (8.37) leaves us with the second (two-sided) case of Problem A, and we derive the test (8.17), again with $r^{1/2}T$ replaced by (8.37); for example, the additional transformation of (ii) in Problem A yields this test as UMP among level α tests invariant under both this transformation and those used in obtaining (8.37). We now recall the development of (8.19). Since $X_{\cdot\cdot} = (r_1X_{1\cdot} + r_2X_{2\cdot})/(r_1 + r_2)$, a simple computation shows that $X_{1\cdot} - X_{\cdot\cdot} = r_1^{-1}(X_{1\cdot} - X_{2\cdot})/(r_1^{-1} + r_2^{-1})$, with an analogous expression for $X_{2\cdot} - X_{\cdot\cdot}$; hence, the square of (8.37) can be written as

$$\sum_{i=1}^{2} r_i(X_{i\cdot} - X_{\cdot\cdot})^2/\sigma^2, \tag{8.38}$$

and thus this exceeds c_2^2 (as the form (8.19) demands) if

$$\sum_{i=1}^{k} r_i(X_{i\cdot} - X_{\cdot\cdot})^2/\sigma^2 > \bar{c}, \tag{8.39}$$

where $k = 2$ and $\bar{c} = c_2^2$. We recall from the discussion of (8.19) that $\bar{c}$ can be obtained from the $\chi_1^2(0)$ distribution, and the power is obtained from the noncentral χ_1^2 distribution.

We have developed this unnecessarily complex form (8.39) of the test of (8.17) (with (8.37) for $r^{1/2}T$) when $k = 2$ in order to motivate the corresponding test for general k. Specifically, for testing $H_0 : \phi_1 = \phi_2 = \cdots = \phi_k$ against $\tilde{H}_1$: at least two ϕ_i's are unequal, *the critical region* (8.39) *is generally used.* The test can be justified by an invariance argument (the transformations being more complicated than they were for $k = 2$). The left side of (8.39) has a $\chi_{k-1}^2(0)$ distribution under H_0, from which $\bar{c}$ can be determined to yield a test of specified level α.

If it is desired to test only equality of a subset of the ϕ_i's, say $\phi_1 = \phi_2 = \cdots = \phi_s$, the left side of (8.39) is replaced by $\sum_{i=1}^{s} r_i\left(X_{i\cdot} - \sum_{j=1}^{s} r_j X_{j\cdot}\Big/\sum_1^s r_j\right)^2\Big/\sigma^2$, and the χ^2 statistic has $s - 1$ degrees of freedom. If $s = 2$, we can use the 1- or 2-tailed normal test, instead.

[The fact that there are $k - 1$ degrees of freedom instead of k is analogous to the presence of $r - 1$ in place of r in (8.20) (ii). If, instead, the null hypothesis were H_0^L : all ϕ_i equal a *specified* value L, against the alternative $\tilde{H}_1^L$: at least one ϕ_i differs from L, the usual test would be (8.39) with $X_{\cdot\cdot}$ replaced by L, and the $\chi_k^2(0)$ distribution would be used to determine $\bar{c}$. An alternative intuitive motivation for the development of the actual (8.39) is to begin by noting that this modification,

$$\sum_{1}^{k} r_i(X_{i\cdot} - L)^2/\sigma^2 > \bar{c}, \tag{8.40}$$

when $k = 1$, is *exactly* the test of (8.17) for the setting described in the last paragraph of Problem A. Next, one proceeds by suggesting the same form (8.40) for general k, remarking that this test can actually be justified by an invariance argument, or that it seems intuitively reasonable because each summand $r_i(X_{i\cdot} - L)^2$ has expectation σ^2 under H_0^L and a greater value under $\tilde{H}_1^L$, so that under the latter the power can be hoped to be larger than the level. Finally, if L is unknown (so that H_0^L becomes H_0), one argues intuitively that L in (8.40) should be estimated by $X_{\cdot\cdot}$, yielding (8.39). This intuitive argument may help the reader in obtaining a feeling that (8.39) is reasonable. It seems less satisfactory than the actual development of (8.39), since it does not even give a genuine justification in terms of an optimum property when $k = 2$, and it omits the one-sided case when $k = 2$.]

The power of the test (8.39) can be obtained upon deriving the fact that the left side of (8.39) has the noncentral $\chi_{k-1}^2(\lambda)$ distribution with $\lambda = \sum_1^k r_i(\phi_i - \phi_\cdot)^2/\sigma^2$ where $\phi_\cdot = (\sum r_i)^{-1} \sum r_i \phi_i$. This should be thought of as an extension of the result for $k = 2$; the power function when $k = 2$ is obtained from our reduction below (8.37) to the test of (8.19) (with (8.37) for $r^{1/2}T$), and from the discussion of power just below (8.19).

Note that although Problems C and D both use the central χ^2 distribution to compute levels, the power function computation uses only the central distribution in case C, but the noncentral in case D. Remark (b) of (ii) of Problem F will discuss this further.

When $k > 2$, there is no simple analogue of the one-sided test described just after (8.37). For example, "reasonable" tests of H_0 against $H_1' : \phi_1 > \phi_2 > \cdots > \phi_k$ can be given, but none with a simple natural optimum property has been obtained.

Other more complex ANOVA problems, such as the two-way ANOVA, have a similar treatment to that of the present Problem D. Typically, one is interested in testing that all "row effects" are equal, or that all "column effects" are equal, or both of these facts.

In the applied literature, the term *ANOVA problem* or *ANOVA test* usually refers *not* to a problem of the present case D, but rather to one of case E, where (more realistically) σ^2 is unknown.

Problem E: Model I ANOVA (Usual) with Variance Unknown—t or F Distribution. We now have the various problems of Problem D, *but with σ^2 unknown.*

When $k = 2$, a minimal sufficient statistic now consists of $(X_{1\cdot}, X_{2\cdot}, Q_{\min})$, where $Q_{\min}$ is the residual sum of squares of Chapter 5 (with expectation $(r_1 + r_2 - 2)\sigma^2$) when the present setting is considered instead as one for linear estimation of the ϕ_i's; that is,

$$Q_{\min} = \sum_{i=1}^{k} \sum_{j=1}^{r_i} (X_{ij} - X_{i\cdot})^2, \tag{8.41}$$

and $Q_{\min}$ has the $\sigma^2 \chi^2_{n-k}(0)$ distribution with $n = \sum_1^k r_i$, analogous to (8.20) (ii). For testing $H_0 : \phi_1 = \phi_2$ against $H_1' : \phi_1 > \phi_2$, the location-invariance argument leading to (8.37) in Problem D, together with the scale-invariance argument of (8.21)–(8.22), reduces the problem to that of (8.26), with T^*_{r-1} there replaced by

$$(X_{1\cdot} - X_{2\cdot})/(r_1^{-1} + r_2^{-1})^{1/2} [Q_{\min}/(r_1 + r_2 - 2)]^{1/2}. \tag{8.42}$$

The statistic (8.42) has the $t_{r_1+r_2-2}(0)$ distribution under H_0 and a corresponding noncentral t distribution under H_1'. The test (8.28) with T^*_{r-1} replaced by (8.42) is *UMP invariant.*

Similarly, and still with $k = 2$, for testing H_0 against $\tilde{H}_1 : \phi_1 \neq \phi_2$, we obtain a UMP invariant test by replacing T^*_{r-1} by (8.42) in (8.29). Rewriting this as in (8.30), and using the arithmetic which yielded (8.38), we can rewrite this test as

$$(k-1)^{-1} \sum_{i=1}^{k} r_i (X_{i\cdot} - X_{\cdot\cdot})^2 \Bigg/ \left[Q_{\min} \Bigg/ \left(\sum_1^k r_i - k \right) \right] > \bar{c}. \tag{8.43}$$

When $k = 2$ we have already seen below (8.30) that the left side of (8.43) has an $F_{1,\sum r_i - k}$ distribution, central under H_0, from which the choice of $\bar{c}$ and the power function can be obtained.

For $k > 2$, paralleling the development of (8.39) in Problem D for general k, we now use (8.43) for testing $H_0 : \phi_1 = \phi_2 = \cdots = \phi_k$ against $\tilde{H}_1$: at least two ϕ_i's are unequal. (The inclusion of the factor $(k-1)^{-1}(\sum r_i - k)$ is not really essential; it is included here because the appropriate tables consider the left side of (8.43) with this factor present.) The left side of (8.43) now has the $F_{k-1,\sum r_i - k}(\lambda)$ distribution (see (iii) of the subsection on distributions), central under H_0 and with $\lambda = \sum_1^k r_i(\phi_i - \phi.)^2/\sigma^2$ under $\tilde{H}_1$. *This is the usual "ANOVA" test one finds in the literature.* It is UMP among tests with an appropriate invariance property. An alternative, but *intuitive*, derivation of this test, is to replace σ^2 (which is now unknown) on the left side of (8.39) by its unbiased estimator $Q_{\min}/(\sum r_i - k)$; the result, multiplied by the constant $(k-1)^{-1}$, is the left side of (8.43). Or, as one finds the motivation described in many applied books, the numerator and denominator of (8.43) each have expectation σ^2 under H_0, but the numerator has a larger expectation under $\tilde{H}_1$, so that it is reasonable to hope that the critical region of (8.43) will be good for distinguishing between H_1 and H_0.

Just as in Problem D, if we only want to test that a subset of the population means are equal, say $\phi_1 = \phi_2 = \cdots = \phi_s$ where $s < k$, we replace the numerator of (8.43) by $(s-1)^{-1} \sum_{i=1}^{s} r_i \left(X_{i\cdot} - \left(\sum_1^s r_j \right)^{-1} \sum_{j=1}^{s} r_j X_{j\cdot} \right)^2$, and the numerator degrees of freedom number is $s - 1$. Note that the denominator is unchanged. If $s = 2$, we can use a 1- or 2-tailed t-test, instead.

In more complex ANOVA problems (e.g., 2-way ANOVA), the assumptions usually made again make the denominator $Q_{\min}/(n-h)$ of usual tests (in the notation of Chapter 5) have a $[\sigma^2/(n-h)]\chi^2(0)$ distribution. The numerator of the F-test for testing whether or not p specified linearly independent estimable parametric functions (see Chapter 5) are all 0, is $p^{-1}\hat{t}'R^{-1}\hat{t}$ where $\hat{t}$ is the p-vector of B.L.U.E.'s of the p parametric functions and $\sigma^2 R$ is their covariance matrix. (We saw in Chapter 5 that R is known, even though σ^2 is unknown.) In the one-way ANOVA, $p = k - 1$ and $\theta_1 - \theta_k$, $\theta_2 - \theta_k$, $\ldots$, $\theta_{k-1} - \theta_k$ are 0 under $\tilde{H}_0$.

Problem F: (i) *One- or Two-Sided Test about Equality of the Variances of Two Normal Distributions, with Means Known or Unknown*; *also*, (ii) *Model II of ANOVA—F-Distribution (Central, Only, Even for Power)*.

(i) We suppose $X_{11}, \ldots, X_{1r_1}, X_{21}, \ldots, X_{2r_2}$ are independent, with X_{ij} distributed $\mathcal{N}(\phi_i, \sigma_i^2)$. For brevity we treat only the case where the ϕ_i's are *unknown*; it should be obvious from Problem C how to replace $X_{i\cdot}$ below by ϕ_i with the gain of a degree of freedom for each of $\bar{T}_1$ and $\bar{T}_2$ below, if the ϕ_i are known.

For testing $H_0 : \sigma_1^2 = \sigma_2^2$ against $H_1' : \sigma_1^2 > \sigma_2^2$, a sufficient statistic is

$(X_{1}., X_{2}., \bar{T}_1, \bar{T}_2)$, where $\bar{T}_i = \sum_j (X_{ij} - X_{i}.)^2$ has a $\sigma_i^2 \chi^2_{r_i - 1}(0)$ distribution. An invariance argument reduces consideration to tests based on $\bar{T}_1/\bar{T}_2$, and (iii) of the subsection on distributions and an MLR argument like that running from (8.32) to (8.35) yields

$$
\begin{aligned}
&(r_1 - 1)^{-1}\bar{T}_1/(r_2 - 1)^{-1}\bar{T}_2 > c_{r_1 - 1, r_2 - 1}(\alpha), \quad \text{where} \\
&P\{(F_{m,n}(0) \text{ random variable}) > c_{m,n}(\alpha)\} = \alpha,
\end{aligned}
\tag{8.44}
$$

as a UMP invariant test of level α, with power function

$$
\begin{aligned}
&P_{\sigma_1^2, \sigma_2^2}\{(r_1 - 1)^{-1}\bar{T}_1/(r_2 - 1)^{-1}\bar{T}_2 > c_{r_1 - 1, r_2 - 1}(\alpha)\} \\
&\quad = P\{(F_{r_1 - 1, r_2 - 1}(0) \text{ random variable}) > \sigma_2^2\sigma_1^{-2}c_{r_1 - 1, r_2 - 1}(\alpha)\}.
\end{aligned}
\tag{8.45}
$$

The usual two-tailed critical region for testing $H_0 : \sigma_1^2 = \sigma_2^2$ against $\tilde{H}_1 : \sigma_1^2 \neq \sigma_2^2$ is the approximately UMPU level α test based on $\bar{T}_1$ and $\bar{T}_2$, obtained by using (8.13) with f_{θ_1} in (8.13) replaced by the $F_{r_1 - 1, r_2 - 1}(0)$ distribution:

$$
(r_1 - 1)^{-1}\bar{T}_1/(r_2 - 1)^{-1}\bar{T}_2 \begin{cases} > c_{r_1 - 1, r_2 - 1}(\alpha/2) \\ \text{or} \\ < c_{r_1 - 1, r_2 - 1}\left(1 - \dfrac{\alpha}{2}\right). \end{cases}
\tag{8.46}
$$

Tables often give only "upper tail" values $c_{m,n}(\alpha)$ of (8.44) or $c_{m,n}(\alpha/2)$ of (8.46), for commonly employed small values of α or $\alpha/2$. To compute $c_{r_1 - 1, r_2 - 1}\left(1 - \frac{\alpha}{2}\right)$ of (8.46), one then uses the fact (from (iii) of the subsection on distributions) that, if U is $F_{r_1 - 1, r_2 - 1}(0)$, then U^{-1} is $F_{r_2 - 1, r_1 - 1}(0)$; this implies that the $c_{r_1 - 1, r_2 - 1}\left(1 - \frac{\alpha}{2}\right)$ of (8.46) can be computed from the fact that $c_{r_1 - 1, r_2 - 1}\left(1 - \frac{\alpha}{2}\right) = 1/c_{r_2 - 1, r_1 - 1}(\alpha/2)$.

Some erroneous applications have been caused by this listing of only upper tail probability critical values $c_{m,n}(\alpha)$, which values are sometimes tabled by listing m as "greater mean square degrees of freedom" and n as "lesser mean square degrees of freedom." The "mean squares" are the statistics $(r_i - 1)^{-1}\bar{T}_i$, and the recipe of the previous sentence implies using their ratio *with the larger mean square in the numerator*. Thus, for testing whether or not $\sigma_1^2 = \sigma_2^2$, users of these tables use the F-statistic $(r_1 - 1)^{-1}\bar{T}_1/(r_2 - 1)\bar{T}_2$ if this is ≥ 1, and use $(r_2 - 1)^{-1}\bar{T}_2/(r_1 - 1)^{-1}\bar{T}_1$ if that is > 1; this results in the test of (8.46) if, in the former case, one rejects H_0 if the F-statistic exceeds $c_{r_1 - 1, r_2 - 1}(\alpha/2)$, and if, in the latter, it exceeds $c_{r_2 - 1, r_1 - 1}(\alpha/2)$. Phrased in this way, one is using only an "upper tail" *in the tables*, and consequently it is easy to make the mistake of thinking of this as a one-tailed (8.44) test of H_0 against H_1'; the error is thus sometimes made of trying to obtain a level α test by rejecting H_0 when the ratio of greater to lesser mean square exceeds an α-critical point ($c_{r_1-1,r_2-1}(\alpha)$ or $c_{r_2-1,r_1-1}(\alpha)$ in the respective previous cases). *But in reality this yields the test of* (8.46) *with the desired level α there replaced by 2α instead*, as the discussion of this paragraph shows.

[The discussion of the previous paragraph does not arise with the ANOVA tests of (8.43) and (8.47), which are genuinely one-tailed.]

(ii) *Model II of ANOVA*: The simplest illustration of this model is that of Problem E *modified by regarding the ϕ_i's as random variables* in the following way: $\phi_1, \phi_2, \ldots, \phi_k$ are iid $\mathcal{N}(\mu, \bar{\sigma}^2)$ where μ and $\bar{\sigma}$ are unknown and $-\infty < \mu < +\infty$ and $0 \leq \bar{\sigma}^2 < +\infty$; then, *given* the values of the ϕ_i's, the conditional joint distribution of the X_{ij}'s is that of case E: the X_{ij}'s are conditionally independent and X_{ij} is $\mathcal{N}(\phi_i, \sigma^2)$. *We suppose all* r_i*'s are equal.* (Otherwise, the development is less simple than that of (8.47), later.) Note that, if this conditional density of the X_{ij}'s is multiplied by the density of the ϕ_i's and the latter are integrated out to give the joint *unconditional* density of the X_{ij}'s alone, the latter are seen not to be independent. Note also that $\bar{\sigma}^2$ is allowed to be 0, in which case the ϕ_i's are all equal (and equal to μ), with probability one.

The hypothesis to be tested here is $H_0 : \bar{\sigma}^2 = 0$ against $\tilde{H}_1 : \bar{\sigma}^2 > 0$. Viewing this in terms of the ϕ_i's, these parameters are equal under H_0 and unequal under $\tilde{H}_1$ (with probability one). What, then, is the difference between the meaning here and that in Problem E, where in model I of ANOVA H_0 and $\tilde{H}_1$ seemed to have the same meaning for the ϕ_i's? The answer is that the ϕ_i's are no longer thought of in Model II merely as unknown constants as in Model I, but as random variables with a specific form of distribution. (Such a change in the distribution of the X_{ij}'s, as described for our model in the previous paragraph, would generally result in a new test. See, however, Remark (a) later, regarding the special problem treated here.) From a practical point of view, inference in Model I was concerned with the specific ϕ_i's at hand; in Model II it is concerned with the $\mathcal{N}(\mu, \bar{\sigma}^2)$ population that produced those ϕ_i's rather than the particular ϕ_i's themselves.

For example, suppose ϕ_i is the true value of an index of intelligence of the ith individual chosen at random from a very large population—so large compared with k that we can regard the individuals as being chosen independently of each other. There is a normal distribution of intelligences within the population (an assumption made all too often by education experts), namely, the $\mathcal{N}(\mu, \bar{\sigma}^2)$ distribution, whose parameters are unknown. We would like to know whether all individuals in the population are equally intelligent, which would be the case when $\tilde{H}_0 : \bar{\sigma}^2 = 0$ is true. But we cannot observe directly the ϕ_i's of the k chosen individuals. Rather, we can give them intelligence tests, and, *given* the ith individual chosen (and hence his or her unknown ϕ_i), his or her performance score X_{ij} on the jth test he or she takes is normally distributed about his or her actual intelligence ϕ_i, having the $\mathcal{N}(\phi_i, \sigma^2)$ distribution. Note that $\bar{\sigma}^2$ is the variance *among individuals*, and σ^2 is the variance *among performances of any given individual*. Since $\text{var}(X_{ij}) = \sigma^2 + \bar{\sigma}^2$ has these two "components," we sometimes call this Model II the *components of variance model*. (In much of the literature, "Model II" or "components of variance" is used when all means of interest are treated in the preceding fashion. If, in the 2-way analysis of variance of Section 5.4 the β_j were treated as unknown constants, and the α_i were regarded to be $\mathcal{N}(0, \bar{\sigma}^2)$, this would be called a *mixed model*.)

Suppose instead that we had k fertilizers to be compared in their efficacy for growing a particular variety of corn, where ϕ_i is the expectation of the weight X_{ij} of the jth plant using fertilizer i. This (with the usual normality assumption) is an example of the Model I of ANOVA of Problem E. It would make little sense to think of the k fertilizers as being chosen at random from a normal population of fertilizers—the technological development of fertilizers is simply not of that nature. So Model II would be inappropriate. We can only make an inference about the possible equality of *these* k *fertilizers*, not about the nature of some population of fertilizers from which these k are a sample. In contrast, regarding the possibility of treating a Model II problem as though it were a Model I problem, we see in the example of the previous paragraph that it is perfectly legitimate to ignore the random sampling which produced the k individuals, to regard the people (and hence their ϕ_i's) as being given, and to seek inference only about *these* k *individuals* (exactly as we did about k fertilizers), accepting or rejecting the hypothesis that these k individuals have equal intelligence without making any statement about a population from which they came; that is, we could use Model I as in Problem E.

These examples illustrate the important fact that one can always apply Model I in place of Model II, but not conversely. Model II requires the additional presence of a normal population of ϕ_i's. But Model II also yields inference about a larger population and not just about the k individuals at hand.

By sufficiency and appropriate invariance considerations applied to the joint density of the X_{ij}'s, one obtains (recalling that we assumed $r_i = r_1$ for all i)

$$(k-1)^{-1} r_1 \sum_{i=1}^{k} (X_{i\cdot} - X_{\cdot\cdot})^2 / [(r_1 - 1)k]^{-1} \sum_{i,j} (X_{ij} - X_{i\cdot})^2 > c_{k-1,(r_1-1)k}(\alpha) \tag{8.47}$$

as UMP level α invariant test, where $c_{m,n}(\alpha)$ is as defined in (8.44). In fact, the numerator of the left side of (8.47) can be shown to be distributed as $(k-1)^{-1}(\sigma^2 + r_1\bar{\sigma}^2)$ times a $\chi^2_{k-1}(0)$ random variable, and the denominator is independent of the numerator and is $[(r_1 - 1)k]^{-1}\sigma^2$ times a $\chi^2_{(r_1-1)k}$ random variable. (In this computation the ϕ_i's are integrated out.) Thus, (8.47) can be thought of, in the manner of (8.44), as testing $H_0 : \sigma^2 + r_1\bar{\sigma}^2 = \sigma^2$ against $H_1' : \sigma^2 + r_1\bar{\sigma}^2 > \sigma^2$ (this being a way of rewriting, in a manner that parallels the $H_0 : \sigma_1^2 = \sigma_2^2, H_1' : \sigma_1^2 > \sigma_2^2$ of (8.44), the $H_0 : \bar{\sigma} = 0, H_1' : \bar{\sigma} > 0$ of the present problem), based on the ratio of independent quadratic estimators analogous to the $(r_i - 1)\bar{T}_i$ of (8.44). The power function of the test (8.47) can be shown, in analogy with (8.45), to be

$$P\{(F_{k-1,(r_1-1)k}(0) \text{ random variable}) > [\sigma^2/(\sigma^2 + r_1\bar{\sigma}^2)]c_{k-1,(r_1-1)k}(\alpha)\}. \quad (8.48)$$

Some Further Remarks on This Test.

(a) The critical region (8.47) is identical with that of (8.43) used in Model I. This is due to the simplicity of the one-way model with all r_i's equal, and in more complex settings the two models will *not* generally yield the same test. [In fact, even in the one-way ANOVA setting with *unequal* r_i's, it is no longer so clear that (8.43) is the "obvious test" to use, and the left side of (8.43) is no longer even distributed as a multiple of an F random variable.] Some settings (e.g., 2-way ANOVA) in which the ϕ_i's in one set (e.g., rows) are considered as unknown constants as in Model I, and the ϕ_i's in another set (e.g., columns) are random variables as in Model II, are called *mixed models.*

(b) Note that although the levels of both (8.43) and (8.47) are computed by using a central F-distribution, *the power functions differ in being obtained from the noncentral F-distribution for* (8.43) *and from the central F-distribution for* (8.47). [Since by (a) the two critical regions are the same, this difference in the power function computations can be thought of as follows: if the power function of (8.43) is thought of as a *conditional* probability depending on the value of $\lambda = r_1 \sum (\phi_i - \phi.)^2/\sigma^2$, and if expectation is then taken with respect to the random variable λ in terms of the $\mathcal{N}(\mu, \bar{\sigma}^2)$ distribution of the ϕ_i's in Model II, then one obtains the power function for (8.47). Similarly, the noncentral χ^2 distribution appears in Problem D power computations, whereas only the central χ^2 distribution appears in Problem C; if Model I of ANOVA with σ^2 known (Problem D), and with all r_i's equal, were changed into Model II with σ^2 known by having the ϕ_i's normally distributed as in the present section, then the power function of (8.39) of Problem D would become one similar to (8.35) for the test (8.34) of Problem C.]

(c) Although the ϕ_i's are random variables, Model II is not merely a Bayesian approach to Model I, since $\bar{\sigma}^2$, μ, and σ^2 are unknown. [A Bayesian might also specify a prior distribution on $\mu, \sigma^2, \bar{\sigma}^2$, but this would in general not yield the test (8.47).]

(d) The hypotheses considered here can be altered, for any specified value $J > 0$, to $H_0 : \bar{\sigma}^2/\sigma^2 \leq J$ against $H_1 : \bar{\sigma}^2/\sigma^2 > J$, with analogous results. But there are probably few experimental situations in which an educator has a meaningful value of J with which he or she wants to compare $\bar{\sigma}^2/\sigma^2$.

PROBLEMS

Part I is concerned with Sections 8.1–8.6; part II is concerned with Section 8.7.

I. *Problems on Sections* 8.1–8.6. Suggested: At least one of 8.1–8.4 and one of 8.9–8.11.

8.1. Suppose X is a real random variable with Cauchy pdf

$$f_{\theta;X}(x) = 1/\pi[1 + (x - \theta)^2].$$

(a) In a problem like that of Example 8.2 of Section 8.2 it is desired to test the simple hypothesis $H_0: \theta = 0$ against the simple alternative $H_1: \theta = 3$. Draw a graph of g_3, where $g_\theta(x) \overset{\text{def}}{=} f_{\theta;X}(x)/f_{0;X}(x)$ and show that the N-P critical region $\{x: g_3(x) \geq 10\}$ is the finite interval $A^* = \{x: 3 \leq x \leq 11/3\}$. Find the level and power of this test. $\left[\textit{Hint: } \dfrac{d}{dx}\tan^{-1} x = 1/(1 + x^2).\right]$ This example illustrates the faultiness of the intuitive argument, "since f_3 is just f_0 shifted 3 units to the right, we should vote for f_3 when X is large; i.e., 'good' critical regions will be of the form $\{x: x \geq c\}$".

(b) Draw a graph also of $g_1(x) = f_{1;X}(x)/f_{0;X}(x)$. It is desired to find an N-P test of $H_0: \theta = 0$ against $\bar{H}_1: \theta = 1$, of the same level as the test of part (a). Without necessarily finding that test explicitly, show that it cannot be the same one as that given by A^* of (a), by showing that there is no c' such that $\{x: g_1(x) \geq c'\} = A^*$, so that A^* cannot be an N-P test of *any* level for testing $H_0: \theta = 0$ against $\bar{H}_1: \theta = 1$. Conclude that a "UMP one-sided test" of the level obtained in (a) does not exist here (as it does in some cases mentioned in Chapter 8) for testing $\theta = 0$ against $\theta > 0$.

8.2. Work Problem 8.1, but, in place of $\theta = 3$ under H_1, and $\{x: g_3(x) \geq 10\}$, use one of the following:

(a) $\theta = 2$ and $\{x: g_2(x) \geq 5\}$. Here $A^* = [2, 3]$.
(b) $\theta = 4$ and $\{x: g_4(x) \geq 5\}$. Here $A^* = [3, 7]$.
(c) $\theta = 1$ and $\{x: g_1(x) \geq 2\}$. Here $A^* = [1, 3]$. Use $\theta = 2$ and g_2 in (b).

8.3. [(a) and (b) are easy; (c) is only slightly harder.] Suppose $X = (X_1, \ldots, X_n)$ where the iid X_i's have absolutely continuous case pdf

$$f_\theta(x_1) = \begin{cases} \theta(x_1 + \theta)^{-2} & \text{if } x_1 > 0, \\ 0 & \text{if } x_1 \leq 0. \end{cases}$$

Here $\theta > 0$.

(a) For $n = 1$, show that an N-P critical region of $H_0: \theta = 2$ against the *simple alternative* $\bar{H}_1: \theta = \theta_1$, where θ_1 is a *specified positive value* < 2, is of the form $\{X_1 \leq c_\alpha\}$; state how c_α depends on α if the test is to be of level α. Since this test does not depend on θ_1, conclude that it is a UMP level α test of H_0 against the composite $H_1': 0 < \theta < 2$.

(b) Is the test of (a) a UMP level α test of H_0 against $\theta \neq 2$ (two-sided alternatives)?

(c) For $n = 2$, show that an N-P critical region of H_0 against $\bar{H}_1: \theta = 1$ is of the form $\dfrac{x_2 + x_1 + 3}{(x_1 + 1)(x_2 + 1)} \geq c'$, and that this region cannot coincide with the N-P critical region of H_0 against $\bar{\bar{H}}_1 : \theta = 1/2$ except in the trivial cases $\alpha = 0$ or 1. Hence, does a UMP test of H_0 against $H_1' : 0 < \theta < 2$ exist? [Moral: The conclusions of (a) and (c) show that an Ω which is "nice" for one n may not be nice for another.]

8.4. Work Problem 8.3 with the substitution of the density of Problem 6.3(vi) for f_θ, $-1 \le \theta \le 1$. In (a) and (b), use $H_0 : \theta = 0$ against $\bar{H}_1 : \theta = \theta_1$, θ_1 *specified* and <0. In (c), $\bar{H}_1 : \theta = -1$ yields N-P critical region $(1 - x_1)(1 - x_2) \ge c'$, to be compared with the critical region for $\bar{\bar{H}}_1 : \theta = -\frac{1}{2}$: does there exist a UMP test of H_0 against $H_1' : \theta < 0$?

8.5. (a) As an example of a monotone likelihood ratio (MLR) family which is not of K-D type, show that the rectangular distribution from $\theta - \frac{1}{2}$ to $\theta + \frac{1}{2}$ (for one observation), $-\infty < \theta < \infty$, is MLR. Find a UMP test of level $\frac{1}{4}$ for testing $\theta = 0$ against $\theta > 0$ in this case. [*Note*: For testing $\theta = 0$ against, e.g., $\theta = 1/2$, observe that there are many possible N-P critical regions of level 1/4 of the form $(-\infty, -.5) \cup A$, but essentially only one choice of A yields a test that is also N-P against every other $\theta > 0$. (The inclusion of $(-\infty, -.5)$, of probability 0 under each $\theta \ge 0$, does not matter in part (a) but is needed to give the largest possible N-P region of level 1/4; in part (b), it is obvious that no smaller region, obtained by excluding part of $(-\infty, -.5)$, could possibly be UMP for the two-sided hypothesis, since it would have smaller power than this largest region, for some $\theta < 0$.)]

(b) In the K-D case as well as in the example of (a), show that if $\Omega = \{\theta : a < \theta < b\}$ and $a < \theta_0 < b$, then a UMP test of $\theta = \theta_0$ against $\theta > \theta_0$ is not UMP for $\theta = \theta_0$ against $\theta = \theta_1$ (specified) $<\theta_0$. [*Hint*: A demonstration exactly like that of this chapter concerning UMP one-sided tests, and which it is unnecessary to repeat, shows the structure of an N-P test of the latter hypothesis.] Hence, show that in these cases there do not exist UMP "two-sided" tests of $\theta = \theta_0$ against $\theta \ne \theta_0$. [The next problem, 8.6, gives an example in which such tests do exist.]

8.6. As we have seen, there rarely exist UMP "two-sided" tests. Here is an example in which such a test exists: Let $X_1, X_2, \ldots, X_n$ be iid with common. pdf

$$f_\theta(x) = \begin{cases} e^{-(x-\theta)} & \text{if } x \ge \theta, \\ 0 & \text{otherwise.} \end{cases}$$

Here $\Omega = \{\theta : -\infty < \theta < \infty\}$. It is desired to test the hypothesis that $\theta = \theta_0$ (specified) against the alternative $\theta \ne \theta_0$.

(a) Show that the critical region

$$B = \{(x_1, \ldots, x_n) : \min_{1 \le i \le n} x_i \begin{cases} < & \theta_0 \\ & \text{or} \\ > & \theta_0 + n^{-1} \log \frac{1}{\alpha} \end{cases}$$

is a UMP test of size α. [*Hint*: Show that this region (i) is of size α, (ii) is MP of $\theta = \theta_0$ against every $\theta = \theta_1$ (specified) $>\theta_0$, (iii) is MP of $\theta = \theta_0$ against $\theta = \theta_1'$ (specified) $<\theta_0$.] [Note also that this is not the only UMP critical region of $\theta = \theta_0$ against $\theta < \theta_0$, since any set of the form $\{\min x_i < \theta_0\} \cup A$, where $P_{\theta_0}\{A\} = \alpha$, satisfies (iii). However, it can be shown that the test B considered in the problem is the *only* test of this latter form which also satisfies (ii) for *every* $\theta_1 > \theta_0$.]

(b) Letting $Y_i = e^{-x_i}$ and $\varphi = e^{-\theta}$ in (a), conclude the existence of a UMP two-sided test for testing $\varphi = \phi_0$ against $\phi \neq \phi_0$ when the observations are independent and identically distributed, uniformly from 0 to ϕ, with $\Omega = \{\phi : \phi > 0\}$.

(c) If the transformation $Y_i = e^{-x_i^r}$, $\varphi = e^{-\theta r}$ is used instead of that of (b), what conclusion is reached? Here $r > 0$ is specified.

8.7. Refer to the remarks on invariance in Section 8.6. Suppose X and Y are independent exponentially distributed random variables with means θ_1 and θ_2, respectively. Here $\Omega = \{(\theta_1, \theta_2) : 0 < \theta_1 \leq \theta_2 < \infty\}$. It is desired to test $H_0 : \theta_2 = \theta_1$ against $H_1 : \theta_2 > \theta_1$ (thus, both hypotheses are composite).

(a) Show that for every $c > 0$, the problem is invariant under the transformations

$$g_c(x, y) = (cx, cy)$$

$$h_c(\theta_1, \theta_2) = (c\theta_1, c\theta_2)$$

$$j_c(d_i) = d_i \quad \text{for} \quad i = 1, 2.$$

(b) Show that any invariant test depends only on $Y/X = U$ (say).

(c) Writing $\theta_2/\theta_1 = \phi$, find the density function of U and show that it depends only on ϕ. Show that the hypothesis can be written as $H_0 : \phi = 1$ against $H_1 : \varphi > 1$.

(d) Show that the densities $f_{U;\phi}$ of (c) are an MLR family and thereby find, among all tests based on U, a UMP test of size α of H_0 against H_1. Conclude that this is thus a UMP invariant test (that is, UMP among invariant tests) for the original problem based on X and Y.

8.8. Suppose that $p_0(x) = 1/3$ for $x = 0, 1, 2$; $p_1(1) = 1/3$, $p_1(2) = 2/3$; $p_2(0) = 1/3$, $p_2(2) = 2/3$. Show that, for testing X has law p_0 against $H_1 : X$ has law p_1 or p_2, there is a UMP test of level 1/3, but not of level 2/3. [Randomization doesn't help.]

8.9. [Important type of problem!] The random variables X_i are iid Poisson random variables with $E_\theta X_i = \theta$. It is desired to test $H_0 : \theta = 1$ against $H_1 : \theta = .5$ with a test of level approximately .01 and power approximately .98 or more. Use the Central Limit Theorem and an approach like that of Section 8.5 to determine, approximately, the smallest sample size n for which a test based on $X_1, X_2, \ldots, X_n$ will achieve these goals. [A difference of this setup from that of Example 8.5 in Section 8.5 is that $\sum_1^n X_i$ no longer has the same variance under H_0 and H_1.]

8.10. Work Problem 8.9 with the Poisson law replaced by the exponential law with mean θ, $f_{\theta;X_1}(x_1) = \theta^{-1}e^{-x_1/\theta}$ for $x_1 > 0$.

8.11. Much of Section 7.6 has an analogous development for hypothesis testing. In the present problem we illustrate the use of simple and highly efficient (although not asymptotically 100 percent efficient) sequences, in terms of alternatives close to H_0. The pdf f_θ is that of Problem 8.3, in which it was seen that tests with exact optimum properties are hard to write down. (a) For testing $H_0 : \theta = 2$ against

$H_1' : 0 < \theta < 2$, use the asymptotic law (Section 7.6) of $Z_{1/2,n}$ to find constants c_n such that the critical region $\{Z_{1/2,n} \leq c_n\}$ has level $\to \alpha$ as $n \to \infty$. (b) For each fixed $\delta > 0$, find the limit $(n \to \infty)$ of the power of the preceding test at the alternative $\theta = 2 - \delta n^{-1/2}$. [*Answer*: Of the form $\Phi(A + B\delta)$, where A and B are constants.] (c) Find the smallest n for which the preceding test, of level approximately .02, has power .9 at $\theta = 1.97$. [*Note*: It can be shown that this test has asymptotic efficiency .75, the same value as for the corresponding estimator in Problem 7.6.7 (d) with $h = 1$. This means that the N-P test of $\theta = 2$ against $\theta = 1.97$ would require about 3/4 as many observations to achieve the same level and power. More efficient (but more complex) tests can be found, paralleling the results for estimation in Section 7.6.]

II. *Problems on Section* 8.7: Suggested: One of 8.12–8.14, perhaps one or more of 8.16–8.20.

Note: In some problems the same data are used, for brevity, in several different tests. In practice one should not do this blindly, since the tests are not independent; and when the outcome of one test motivates the performing of another, the level of the combined tests is not very simple to compute. Unfortunately, it is not uncommon to see an incorrect assessment of the meaning of such a collection of tests.

8.12. The model and data are those of Problem 5.15(d), with the following additional assumption: *The* Y_{ij} *are normally distributed.* (Without this assumption, we could still find B.L.U.E.'s, but the present forms of parametric tests depend on the normality assumption.) This should help you, both in connecting the present developments with the earlier ones and also in requiring less computation in the examples since you have already done much of the computing.

(i) [*Not hard.*] Find the usual level .05 critical region for testing each of the following (use t-, F-, χ^2 tables in the back of a statistics text); in each part, state which case of the subsection on problems and tests is the model, which critical region of that section you are using, and which decision you would make for the data of Problem 5.15(d). Also, in "practical terms" regarding nutrients, what question is being asked in each case?

(a) $$H_0 : \theta_1 = \theta_2 \quad \text{against} \quad H_1 : \theta_1 > \theta_2.$$

(b) $$H_0 : \theta_1 = \theta_2 \quad \text{against} \quad H_1 : \theta_1 \neq \theta_2.$$

(c) $$H_0 : \theta_1 = \theta_2 = \theta_3 \quad \text{against} \quad H_1 : \text{not } H_0.$$

(d) $$H_0 : \sigma_F^2 = 2{,}000 \quad \text{against} \quad H_1 : \sigma_F^2 > 2{,}000.$$

(e) How would each of the tests (a), (b), (c) be altered if you *knew* $\sigma_F^2 = 2{,}000$?

(ii) "*Model II*" *of ANOVA*. Suppose the specification of part (i) is viewed, instead, as describing only the *conditional distribution of the* Y_{ij} *given the values of the* θ_i, and that we now complete this specification by considering θ_0 to be an unknown constant and the θ_i themselves ($i = 1, 2, 3$) as *independent normal random variables* with means 0 and common unknown variance σ_{1F}^2. This is an example of "Model II."

(a) Show that the "usual" (Section 8.7) level .05 critical region for testing $H_0 : \sigma_{1F}^2 = 0$ against $H_1 : \sigma_{1F}^2 > 0$ is the same as that of part (i) (c).

(b) What is the difference in practical interpretation of the tests of (i) (c) and (ii) (a) in terms of fertilizers? When would it be reasonable to use the (ii) (a) interpretation, and when would it be unreasonable?

(c) What is the difference between the distribution of the F-statistic in (i) (c) and (ii) (a), and consequently on what parameters do the power functions depend in the two cases?

8.13. Suppose X_{ij} ($i = 1, 2, 3; j = 1, 2$) are independent and (close enough to) normal, with common unknown variance σ^2 and expectations

$$EX_{ij} = \theta_0 + \theta_i. \tag{8.49}$$

Here X_{ij} is the number of seconds laboratory rat i takes to get through maze j. The values taken on by X_{ij} are

j \ i	1	2	3
1	33	32	42
2	42	37	46

(i) Find the usual level .05 critical region for testing each of the following (use tables in any book); in each part, state which case of the subsection on problems and tests is the model, which critical region you are using, and which decision you would make for the data given above.

(a) $H_0: \theta_1 = \theta_2$ against $H_1: \theta_1 > \theta_2$.

(b) $H_0: \theta_1 = \theta_2$ against $H_1: \theta_1 \neq \theta_2$.

(c) $H_0: \theta_1 = \theta_2 = \theta_3$ against H_1: not H_0.

(d) $H_0: \sigma^2 = 4$ against $H_1: \sigma^2 < 4$.

(e) How would each of (a), (b), (c) be altered if σ^2 were known to equal 5?

(ii) *"Model II" of ANOVA*. Suppose we instead view the specification up to and including (8.49) as giving the *conditional* distribution of the X_{ij} given the θ_i's and consider θ_0 to be the unknown constant and θ_i ($i = 1, 2, 3$) to be independent and normal with means 0 and common variance σ_1^2 (an example of Model II).

(a) Show that the "usual" (Section 8.7) level .05 critical region for testing $H_0: \sigma_1^2 = 0$ against $H_1: \sigma_1^2 > 0$ is the same as that of problem (i)(c).

(b) What is the difference of interpretation of the tests of (i)(c) and (ii)(a) in terms of rats?

(c) On what parameters do the power functions of (i)(c) and (ii)(a) depend? How, and in terms of what distributions, are these two power functions computed?

8.14. This illustrates the care needed for "large numbers of observations" to yield small errors. In more complex examples this phenomenon can be less transparent.

Let $X_1, X_2, \ldots, X_{21}$ (21 of them) and $Y_1, Y_2, \ldots, Y_n$ be independent normal random variables. The X_i's all have unknown mean μ_1 and unknown variance σ_1^2, and the Y_i's have unknown mean μ_2 and unknown variance σ_2^2.

(a) How large should n be in order that the standard level .05 test of $H_0: \sigma_1^2 = \sigma_2^2$ against $H_1: \sigma_1^2 = 4\sigma_2^2$ have power .95? (Work by trial and error from F-tables.) [*Hint*: Use the fact, noted in the discussion of Problem F in Section 8.7, that $P\{(F_{p,q}$ random variable) $< c\} = P\{(F_{q,p}$ random variable) $> 1/c\}$, where an "$F_{m,n}$ random variable" or "$b\chi_m^2$ random variable" denotes a random variable with the corresponding law, defined in the subsection on distributions.]

(b) If the desired power had been specified to be .9999, show that no n would have sufficed. [*Hint*: Note that, as $q \to \infty$, the $F_{p,q}$-distribution approaches the $p^{-1}\chi_p^2$-distribution, and that no value of n can be preferable to "knowing σ_2^2" (i.e., to "$n \to +\infty$"). Now consult χ_{20}^2-tables.]

8.15. With the model and data of Problem 5.20, assuming also that the Y_{rs} are normal:
(a) Test at the .05 level that music has no effect on weight, against its effect being favorable.
(b) Test that neither music nor amount of feed (in this range) has an effect, against all alternative possibilities (not necessarily a reasonable choice of alternative hypothesis, but often used for simplicity of using the F-test), at the .01 level.
(c) Test $\sigma_F^2 = 10$ against the alternative $\sigma_F^2 < 10$, at the .05 level.
(d) If it is *known* that $\sigma_F^2 = 9$, how are the tests of (a) and (b) altered?

8.16. With the model and data of Problem 5.23 and the assumption of normality of $\{Y_{rs}\}$, test at the .05 level the hypothesis that $\phi_1 = \phi_2$ against the alternative $\phi_1 > \phi_2$ (i.e., that the effect of a month of sun exceeds that of a decimeter of water, over this range of those variables).

8.17. In Problem 5.25, assuming normality, test at the .05 level H_0 : response law is of first degree against H_1 : not H_0. How would the test change if one *knew* the response law were convex in the range studied?

8.18. In Problem 5.27, assuming normality, test at the .02 level whether or not the response law is a straight line. How would the test be changed if one *knew* the response law had nonpositive second derivative in the range studied?

8.19. In Problem 5.28, assuming normality, test whether or not there *is* seasonal variation, at the .01 level.

CHAPTER 9

Confidence Intervals

We discussed interval estimation briefly in Chapters 3 and 4. We shall recall briefly how interval estimation differs from the considerations of Chapters 5 and 7, in which D consisted of the possible values of some function φ of the unknown parameter value θ, the decision d_0 meaning "my guess is that the true value of $\varphi(\theta)$ is d_0." Thus d_0 was typically a *single point* of an interval D (i.e., for simplicity we think of the set of decisions as a real number interval), and these problems were termed ones of *point estimation.* If the decision is to be an interval of real numbers rather than a single point, the problem is called one of *interval estimation.* In a problem of this kind, a decision may be "my guess is that the table is between 4 and 4.3 feet long" rather than "my guess is that the table is 4.1 feet long." Thus, if we are discussing confidence intervals, D consists of a collection of intervals, and we may denote by $[a, b]$ that decision which consists of the set of real numbers between a and b inclusive; the decision in the example of the table would be $[4.0, 4.3]$.

A nonrandomized statistical procedure t can now be written in terms of a pair of real-valued functions $L(x)$ and $U(x)$ with $L(x) \leq U(x)$ for all x. If the two functions L, U are used and the outcome of the experiment is $X = x'$, then the decision $[L(x'), U(x')]$ is made. We write $t = [L, U]$ for this procedure, thinking of t as a function whose values are intervals. (Equivalently, t is often thought of as the pair of functions L, U.)

The classical development of the subject of interval estimation made no use of loss functions but proceeded in the following way: Suppose the procedure $t = [L, U]$ is used to estimate $\varphi(\theta)$ and that the true (unknown) value of θ is θ_0. It might, by chance, turn out that the observed value x' of X is such that $L(x') \leq \varphi(\theta_0) \leq U(x')$ (in which case the guess $[L(x'), U(x')]$ is "correct"), or that it is such that the interval $[L(x'), U(x')]$ does not contain $\varphi(\theta_0)$ (so that the guess represented by the decision is "incorrect"). If $L(x') \leq \varphi(\theta_0) \leq U(x')$, we say that the interval $[L(x'), U(x')]$ *covers* the true value $\varphi(\theta_0)$, and the quantity

$$\gamma_t(\theta_0) = P_{\theta_0}\{L(X) \leq \varphi(\theta_0) \leq U(X)\} \tag{9.1}$$

is called the *probability of coverage* when the true value of θ is θ_0 and the procedure $t = [L, U]$ is used. The number

$$\bar{\gamma}_t = \min_{\theta \in \Omega} \gamma_t(\theta)$$

(with "infimum" replacing "minimum" if the latter is not attained) is called the *confidence coefficient* of the procedure t. The decision $t(x') = [L(x'), U(x')]$ is called a *confidence interval*, of which $L(x')$ and $U(x')$ are the *lower and upper confidence limits*. It is desirable to choose the procedure t so that $\bar{\gamma}_t$, and hence $\gamma_t(\theta)$ for each θ, is close to one, since $\gamma_t(\theta_0)$ is the probability of making a correct decision when $\theta = \theta_0$. However, this is not the only consideration; for, we can make $\bar{\gamma}_t = 1$ by letting the decision $[L(x'), U(x')]$ always consist of the set of all possible values of $\varphi(\theta)$, regardless of x', and this would be useless for practical problems. (In flipping a coin with unknown probability θ of coming up heads, this would mean ignoring the data and always making the "correct" decision "my guess is that $0 \leq \theta \leq 1$".)

What is really wanted is a compromise between the aim of having the interval include the true value of $\varphi(\theta)$ and having the interval short in length (so as to be of practical value). This could be achieved in the decision-theoretic approach to statistics by a suitable choice of the loss function. For example, the loss function

$$W(\theta, d) = k \cdot (\text{length of } d) + \begin{cases} 0 & \text{if} \quad \varphi(\theta) \quad \text{is in } d, \\ 1 & \text{if} \quad \varphi(\theta) \quad \text{is not in } d, \end{cases} \tag{9.2}$$

where k is a positive constant, weighs contributions both from making an incorrect decision and also from choosing d to be a long interval.

In classical textbooks, the subject is usually treated instead by first specifying a number like .95 and then choosing a t such that $\bar{\gamma}_t \geq .95$. This is analogous to specifying the level of a test in hypothesis testing. To stop here and not consider whether or not t tends to yield short intervals is analogous to choosing a test so as to have small level without ever considering power. Just as the better classical descriptions will indicate that a certain test, subject to a certain restriction on the level, has some optimum property in terms of power under the alternative hypothesis, so they will also indicate that a recommended confidence interval procedure will, subject to a certain restriction on the confidence coefficient, have some good property in terms of the length of the confidence interval or some other property. (General methods for constructing confidence intervals will be discussed later.)

An alternative to (i) considering length, subject to a restriction on coverage (as in the previous paragraph), or to (ii) combining penalties for noncoverage and for length (as in the paragraph before last), is (iii) to restrict consideration to procedures $t = [L, U]$ for which the *length* $U(x) - L(x)$ never exceeds some specified constant $2c$ (say), and then to use $1 - \gamma_t(\theta) = P_\theta\{\varphi(\theta) \notin [L(X), U(X)]\}$ as the risk function. We then may as well assume that $U(x) - L(x) = 2c$ *for all* x, since it cannot increase the risk if we broaden the interval whenever its length is $< 2c$. Now note that, if $t^*(x)$ is the midpoint of the interval $[L(x), U(x)]$ of length $2c$, then that interval contains the value $\varphi(\theta)$ if and only if $|t^*(x) - \varphi(\theta)| \leq c$. Hence, $1 - \gamma_{[L,U]}(\theta) = r_{t^*}(\theta)$, where the latter is the risk function of t^* for the problem of *point estimation* of $\varphi(\theta)$ with loss function

$$W(\theta, d) = \begin{cases} 0 & \text{if} \quad |\varphi(\theta) - d| \leq c, \\ 1 & \text{otherwise,} \end{cases}$$

examples of which were studied in Chapter 4, Sections 4.1–4.4.

Of these three approaches, (i) is most used by many practitioners, to whom a high probability of coverage is the principal goal. Both approaches (i) and (iii) can in some examples result in procedures t for which $\bar{\gamma}_t = 0$; this will be illustrated in Example 9.2.

The interpretation of the statement that "$L(x') \leq \varphi(\theta) \leq U(x')$" must be properly understood. *After* the experiment is performed and it has been observed that $X = x'$, the quantities $L(x')$ and $U(x')$ are ordinary real numbers, and the preceding statement that the true value of $\varphi(\theta)$ lies between these two numbers is either true or false. There is (if the approach is non-Bayesian) no longer any probability of .95 (for example) that it is true rather than false. This is similar to the situation in testing hypotheses: *after* the decision has been reached there, we no longer use the size or power to tell us the probability of this particular decision's being right, since it has already been made and is either right or wrong. Rather, size and power, as well as confidence coefficients, tell us probabilities of reaching certain decisions from the point of view we have in considering the problem *prior* to the observation of the chance variable X. Alternatively, one can think of the meaning of the confidence coefficient in terms of the law of large numbers: suppose, for simplicity, that $\gamma_t(\theta)$ is the same constant $\bar{\gamma}_t$ for all θ. (This often is the case in important practical examples.) Then, if the experimenter uses this same procedure t repeatedly in many independent experiments where Ω is the same, he has the same probability $\bar{\gamma}_t$ of a "success" on each experiment (i.e., of the confidence interval covering the true value of $\varphi(\theta)$ in each experiment, even though the true value of θ may vary from experiment to experiment). Hence, by the law of large numbers for Bernoulli trials, the probability is close to one that the *actual* proportion of correct statements he or she makes as a result of using the procedure t for many experiments, is close to $\bar{\gamma}_t$. (Later in this chapter we shall mention the "fiducial" and Bayesian interpretations of interval estimation.)

The computation of $\gamma_t(\theta)$ in (9.1) is usually carried out by rewriting the event $L(X) \leq \varphi(\theta) \leq U(X)$ so that the random variable X appears in the middle of the string of inequalities. For example, if X is a real random variable and if $\varphi(\theta) = \theta$, $L(X) = X - 5$, and $U(X) = X + 3$, we can rewrite

$$x - 5 \leq \theta \leq x + 3 \tag{9.3}$$

as

$$\theta - 3 \le x \le \theta + 5, \tag{9.4}$$

the first inequality of (9.3) being the second of (9.4) and vice versa. The probability of X taking on a value in the interval (9.4) can then be computed as

$$\int_{\theta-3}^{\theta+5} f_\theta(x)\,dx,$$

where $f_\theta(x)$ is the density of X when θ is the true parameter value. (People are sometimes confused by the appearance of θ in the middle term of the inequality of (9.1) or (9.3), into thinking of θ as a random variable instead of an unknown constant. It is no more random in this non-Bayesian development than is the number 4 in the inequality $Z \le 4 \le Z + 1$.)

EXAMPLE 9.1 (*Location Parameter Problem*). Suppose X_1 has density function $f_\theta(x) = \bar{g}(x - \theta)$, $-\infty \le x < +\infty$, where $\Omega = \{\theta : -\infty < \theta < +\infty\}$ and the function $\bar{g}$ (equal to f_0) is known. The problem is one of finding an interval estimator of θ. Suppose, on the basis of invariance considerations, we use an interval estimator of length $2c$ of the form

$$[L(X_1), U(X_1)] = [X_1 + b - c, X_1 + b + c] = t_{c,b}(X_1) \text{ (say).}$$

The probability of coverage can be computed in terms of the distribution function G corresponding to $\bar{g}$, using calculations like those of (9.3)–(9.4), as follows:

$$\begin{aligned} \gamma_{t_{c,b}}(\theta) &= P_\theta\{X_1 + b - c \le \theta \le X_1 + b + c\} \\ &= P_\theta\{-c - b \le X_1 - \theta \le c - b\} \\ &= G(c - b) - G(-c - b), \end{aligned} \tag{9.5}$$

the last since $X_1 - \theta$ has density f_0 when X_1 has density f_θ (which fact we have used previously, e.g., in Chapter 8).

Thus, $\gamma_{t_{c,b}}(\theta)$ does not depend on θ (reflecting the "invariance" of $t_{c,b}$), and thus (9.5) gives $\bar{\gamma}_{t_{c,b}}$.

If we fix c, then the arithmetic of maximizing $\bar{\gamma}_{t_{c,b}}$ with respect to b is like that of the corresponding Chapter 4 Bayes estimation problem treated in (4.1)–(4.4), if g is unimodal. In particular, if $\bar{g}$ is symmetric about 0 and unimodal ($g(x)$ nonincreasing for $x \ge 0$), then the choice $b = 0$ maximizes (9.5) and hence a best invariant procedure $t_{c,0}$ has confidence coefficient

$$\bar{\gamma}_{t_{c,0}} = G(c) - G(-c) = 2G(c) - 1,$$

the last since $\bar{g}$ is symmetric. [If $\bar{g}(x)$ is *strictly* decreasing for $x \ge 0$, then $b = 0$ is the unique maximizing value.]

Still assuming $\bar{g}$ symmetric and unimodal, suppose we now consider the approach of specifying the confidence coefficient γ and minimizing the length $2c$ among invariant procedures with at least this great a confidence coefficient. It is not hard to see that, if c_γ is the value such that $2G(c_\gamma) - 1 = \gamma$, then $t_{c_\gamma,0}$ is a best invariant procedure.

Finally, if we use the loss function (9.2) for symmetric unimodal $\bar{g}$, we have

$$k(2c) + 1 - \gamma_{c,0} = 2[kc + 1 - G(c)] \tag{9.6}$$

for the constant risk of $t_{c,0}$ (and at least this much for $t_{c,b}$ when $b \neq 0$). The derivative of (9.6) with respect to c is $2[k - \bar{g}(c)]$. This last is nondecreasing in c for $c \geq 0$, and hence (9.6) is minimized by

$$\begin{cases} c = 0 \quad \text{if} \quad \bar{g}(0) < k; \\ c = \text{any value } \bar{c}_k \text{ for which } \bar{g}(\bar{c}_k) = k, \quad \text{if there is such a } \bar{c}_k; \\ \text{otherwise, } c = \text{the value } \bar{\bar{c}}_k \text{ such that } \bar{g}(\bar{\bar{c}}_k -) > k > \bar{g}(\bar{\bar{c}}_k +). \end{cases} \tag{9.7}$$

The $t_{c,0}$ with this choice of c is optimum among invariant procedures. For example, if f_θ is the $\mathcal{N}(\theta, 1)$ density, then for $k \geq (2\pi)^{-1/2}$ it is best to take $t(X_1) = [X_1, X_1]$, a degenerate interval (point) of length 0 and coverage 0 (this case being of little practical interest); whereas if $k < (2\pi)^{-1/2}$, then $t(X_1) = [X_1 - [-\log(2\pi k^2)]^{1/2}, X_1 + [-\log(2\pi k^2)]^{1/2}]$ is best invariant. On the other hand, if f_θ is uniform from $\theta - \frac{1}{2}$ to $\theta + \frac{1}{2}$, then in the trivial case $k > 1$ we can again use the degenerate procedure $[X_1, X_1]$; if $k = 1$, each procedure $[X_1 - c, X_1 + c]$ for $0 \leq c \leq 1/2$ has the same risk (9.6), although length and coverage vary with c; if $0 < k < 1$, one must take $c = \bar{\bar{c}}_k = 1/2$ (length $= 1$, $\bar{\gamma}_t = 1$). In this uniform example $\bar{g}$ is not *strictly* decreasing for $x \geq 0$, and hence $t_{c,0}$ is not uniquely best; the computation of (9.5) shows that, for $0 \leq c < 1/2$, $t_{c,b}$ has the same length and coverage as $t_{c,0}$, provided only that $c + |b| \leq 1/2$.

We conclude Example 9.1 by remarking that the close relationship of the solutions to the three approaches of the preceding three paragraphs depends strongly on the restriction to invariant procedures in this location-invariant problem (into which the W of (9.2) fits). The length and coverage for each $t_{c,0}$ were constants, and that is why the family of procedures $\{[X_1 - c, X_1 + c], 0 \leq c < \infty\}$ yielded the optimum procedures $t_{c,0}$, $t_{c_\gamma,0}$, $t_{\bar{c}_k,0}$ for the three approaches.

EXAMPLE 9.2. Suppose X_1 is a real random variable with density

$$f_\theta(x) = \begin{cases} \theta^{-1} e^{-x/\theta} & \text{if} \quad x > 0, \\ 0 & \text{otherwise,} \end{cases}$$

where $\Omega = \{\theta : \theta > 0\}$. Let $\varphi(\theta) = \theta$. Suppose (again from invariance considerations, but now for a scale-parameter problem) $L(x) = c_1 x$ and $U(x) = c_2 x$, where c_1 and c_2 are positive constants with $c_1 < c_2$. Then for $t = [L, U]$ we have (putting $y = x/\theta$)

$$\begin{aligned}
\gamma_t(\theta) &= P_\theta\{c_1X \le \theta \le c_2X\} \\
&= P_\theta\{\theta/c_2 \le X \le \theta/c_1\} \\
&= \int_{\theta/c_2}^{\theta/c_1} \theta^{-1}e^{-x/\theta}\,dx \\
&= \int_{1/c_2}^{1/c_1} e^{-y}\,dy = e^{-(1/c_2)} - e^{-(1/c_1)}.
\end{aligned}$$

Thus, $\gamma_t(\theta)$ is the same for all θ, and $\bar{\gamma}_t = e^{-(1/c_2)} - e^{-(1/c_1)}$. Notice that, in this case of unknown *scale* parameter, the length $(c_2 - c_1)X$ of the confidence interval is not a constant (as it is in (9.3), or in the *location*-parameter problem of Example 9.1). This is necessary for any procedure t if $\bar{\gamma}_t$ is to be positive, since as $\theta \to \infty$ the density of X becomes very much spread out and the probability of an inequality like (9.3) being satisfied goes to zero. The *expected length* of the interval is

$$E_\theta[U(X) - L(X)] = E_\theta[(c_2 - c_1)X] = (c_2 - c_1)\theta.$$

Hence, if the confidence coefficient is specified in advance, for example as .80, then, as in the next paragraph, we may choose c_1 and c_2 so as to minimize $(c_2 - c_1)$ subject to the restriction $e^{-(1/c_2)} - e^{-(1/c_1)} = .80$. [Although we could deduce the linear form c_1x and c_2x of L and U from the invariance principle, the loss function (9.2) does not fit into the present scale-parameter example as it did into the location-parameter setup of Example 9.1; this is reflected in the nonconstancy of the risk function $E_\theta[U(X_2) - L(X_1)] + 1 - \gamma_t(\theta)$, in contrast with the constancy of the risk function of the location-invariant procedures of Example 9.1. However, if k is replaced by $k\theta^{-1}$ in (9.2), the loss function is "invariant" for the present example, and the risk function of $[L, U]$ is the constant $k(c_2 - c_1) + 1 - \bar{\gamma}_t$.]

We now solve the problem of minimizing $(c_2 - c_1)$ subject to $\bar{\gamma}_t = .8$. This is a trial-and-error solution, since by solving for c_2 in terms of c_1 in the equation $.8 = e^{-(1/c_2)} - e^{-(1/c_1)}$, substituting into $c_2 - c_1$ to yield $c_2 - c_1 = \psi(c_1)$ (say), and setting $\psi'(c_1) = 0$, we obtain a transcendental equation. (The first three trials began with column 1; thereafter, using better tables, with column 3.)

c_1	$1/c_1$	$e^{-(1/c_1)}$	$e^{-(1/c_2)} = .8 + e^{-(1/c_1)}$	$-1/c_2$	c_2	$\psi(c_1) = c_2 - c_1$
0+	$+\infty$	0	.8	.223	4.48	4.48
.25	4	.0183	.8183	.200	5.00	4.75
.10	10	.00005	.80005	.223	4.48	4.38
.1609	6.21461	.002	.80200	.22065	4.532	4.371
.14476	6.90776	.001	.80100	.22189	4.50674	4.3620
.13156	7.60091	.0005	.80050	.22252	4.49398	4.3624
.13898	7.19544	.00075	.80075	.22220	4.5004	4.3614

So $c_1 = .14$, $c_2 = 4.50$ is the solution to two decimal places. Although this result required fairly good tables, using five significant figures in the calculations to obtain two significant figures in c_1, note that little *penalty*—half of 1 percent in the value of $(c_2 - c_1)$—would have been incurred if we had stopped with the third trial. This is, of course, due to the vanishing of ψ' at the minimizing value c_1^* of c_1, which makes $\psi(c_1^* + \varepsilon) \approx \varepsilon^2\psi''(c_1^*)/2$ for small ε; that is, a moderate error ε in the minimizing value c_1 produces a much smaller error, of order ε^2, in $c_2 - c_1$. From a practical point of view, $c_1 = .10$, $c_2 = 4.48$ would have been good enough, although this c_1 is over 25 percent off from c_1^*.

Sometimes one is interested only in a lower (or upper) bound on $\varphi(\theta)$. For example, the confidence interval $c_1 x \leq \theta < +\infty$ in the preceding example is called a *one-sided (lower) confidence interval*, $L(x) = c_1 x$ being a "lower confidence limit or bound" with $\bar{\gamma}_t = 1 - e^{-(1/c_1)}$; hence, in Example 9.2, with $\bar{\gamma}_t$ specified as .8, we would have

$$c_1 = -1/\log(1 - .8) = .63.$$

This was a simpler choice of c_1 than that in the case of the two-sided bound of Example 9.2, in which it was easy to find *some* procedure $[c_1 X_1, c_2 X_1]$ with $\bar{\gamma}_t = .8$, but in which the existence of infinitely many such intervals made the choice of an *optimum* one require additional calculations. In effect, in the one-sided case we have taken the two-sided problem and set $c_2 = +\infty$, making the choice of c_1 unique.

If we had adopted the approach of adding the two types of loss in this one-sided case, the loss function (9.2) would no longer be appropriate in examples like this for which the "length" is $+\infty$. Among more suitable loss functions, one might use, when the estimating interval is $[a, +\infty]$, the loss

$$W(\theta, [a, +\infty)) = k|a - \theta| + \begin{Bmatrix} 1 & \text{if} & a > \theta \\ 0 & \text{if} & a \leq \theta \end{Bmatrix}.$$

This reflects the notion that, if a is taken too small in order to make $1 - \bar{\gamma}_t$ close to 0, the resulting interval is rather useless since the lower confidence bound is too far from the value θ being estimated; thus, in Example 9.2, the limiting case $a = 0$ yields the useless confidence statement "my guess is that $0 \leq \theta < +\infty$."

A common method of constructing confidence intervals, given in many books, is this: Suppose for simplicity that $\Omega = \{\theta : a \leq \theta \leq b\}$ and that we wish to estimate θ. Suppose we can find real increasing continuous functions g and h on Ω and a real statistic T such that for each fixed θ the numbers $g(\theta)$ and $h(\theta)$ satisfy

$$P_\theta\{g(\theta) \leq T(X) \leq h(\theta)\} \geq \lambda, \tag{9.8}$$

where λ is a positive constant. Let $\bar{L}$ and $\bar{U}$ be the inverse functions of h and g, respectively, on the sets where these are defined. Extend the definition of $\bar{L}$ and $\bar{U}$ by setting $\bar{L} = a$ for those T where $\bar{U}$ but not $\bar{L}$ was originally defined,

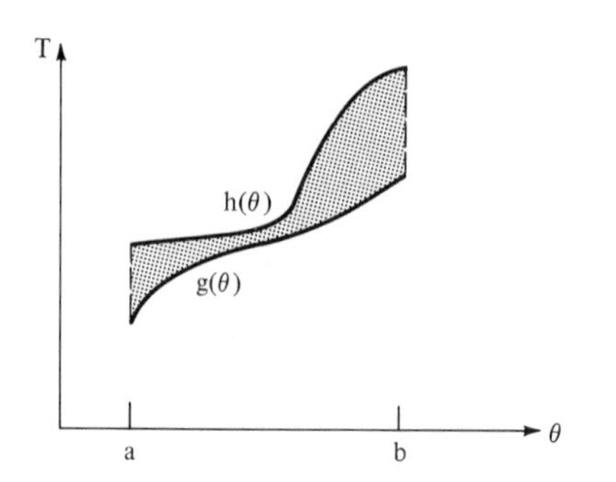

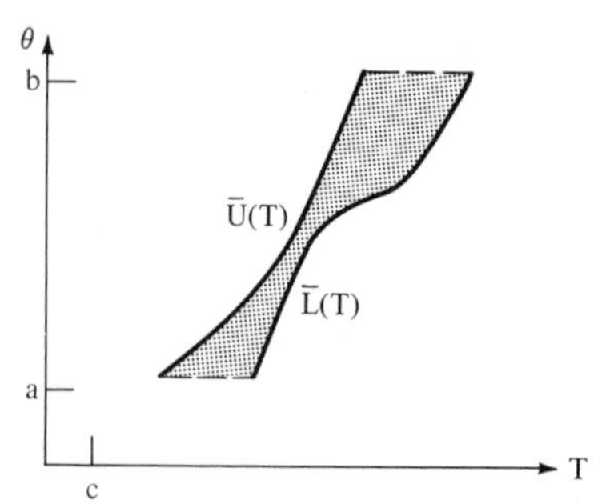

and $\bar{U} = b$ for those T where $\bar{L}$ but not $\bar{U}$ was originally defined (dashed lines in diagram). For $a \leq \theta \leq b$, it is clear that the point (θ, T) lies in the (shaded) region between (or on) the graphs of g and h in the first diagram if and only if (T, θ) lies between (or on) the graphs of the extended $\bar{L}$ and $\bar{U}$ in the second diagram. That is, $g(\theta) \leq T \leq h(\theta)$ if and only if $L(T) \leq \theta \leq U(T)$. (In the earlier discussion, this is the correspondence between (9.4) and (9.3).) Thus, from (9.8), the pair (L, U) defined by $L(x) = \bar{L}(T(x))$ and $U(x) = \bar{U}(T(x))$ give a confidence interval for θ of confidence coefficient at least λ, except for one minor complication: If T does not lie between $g(a)$ and $h(b)$ (e.g., $T = c$ in the second diagram above), the procedure takes the confidence interval to be the empty set. This lack of coverage in such cases is taken into account in the statement that the probability of covering the true θ is $\geq \lambda$, but it is disturbing to many practitioners to be told that the actual confidence set for the observed sample value turned out to be empty (so that θ can't possibly lie in it). For this reason the estimating covering set is usually restricted to be nonempty. Thus, for this usual definition of confidence intervals one makes sure that the interval from $g(a)$ to $h(b)$ includes all possible values of T. (It will also usually be that, if the confidence set is sometimes empty, as in the diagram, then a new $[\bar{L}, \bar{U}]$ can be found which reduces $\max_\theta E_\theta[\bar{U} - \bar{L}]$.)

If one of h (and, hence, $\bar{L}$) or g (and, hence, $\bar{U}$) is omitted in the preceding development, we obtain one-sided confidence intervals.

The point of the preceding method, illustrated in Examples 9.3 and 9.4 later, is that it is often easier to guess g and h satisfying (9.8) than to guess L and U starting from scratch. The method will not automatically lead to good procedures. (See the paragraph after next.)

In the general theory of confidence sets (where Ω is no longer necessarily 1-dimensional) the analogous nomenclature and technique are the following: a *confidence set procedure* is a function t from S into the subsets of Ω (nonempty if we adhere to the convention noted previously), used by "guessing that the true $\theta \in t(x)$" if $X = x$. The confidence coefficient is again $\bar{\gamma}_t = \min_{\theta \in \Omega} P_\theta\{\theta \in t(X)\}$. An often-used method of constructing such procedures is to find a function B from Ω into the collection of subsets of S such that $P_\theta\{X \in B(\theta)\} \geq \lambda$. (In the special case of the previous paragraph $B(\theta) = \{x : g(\theta) \leq t(x) \leq h(\theta)\}$.) For each x let $t(x) = \{\theta : x \in B(\theta)\}$. Since $x \in B(\theta)$ if and only if $\theta \in t(x)$, we conclude as before that the t obtained in this way is a confidence set procedure with $\bar{\gamma}_t \geq \lambda$.

In Neyman's development of the subject, the set $B(\theta_0)$ was often chosen, for each θ_0, to be the *acceptance region* of a test of level $\leq 1 - \lambda$ of $H_0 : \theta = \theta_0$. If that test has some good property (e.g., UMP or UMPU for each θ_0), the confidence set inherits a corresponding property in terms of making $P_{\theta_0}\{\theta_1 \in t(X)\}$ small for $\theta_1 \neq \theta_0$. This notion is described by Neyman in the word *shortest* (replacing *most powerful* for tests), but you should understand that Neyman's usage of this term does not refer to making the expected length or volume of the set $t(X)$ small in the manner described in the example treated earlier. In some simple cases, such as confidence intervals for the mean of a normal distribution, the two notions of "shortness" both apply to the same standard procedure.

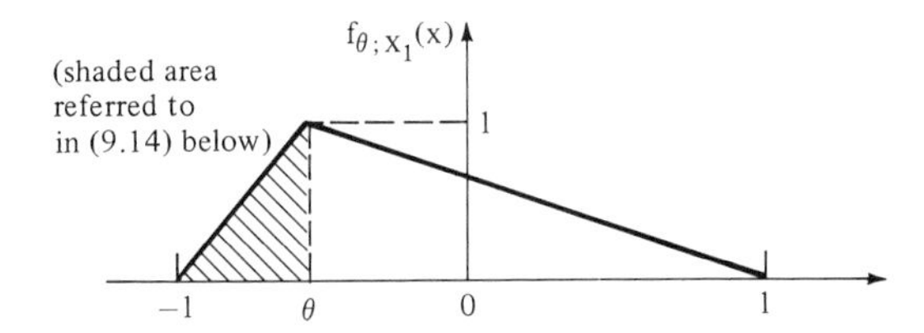

Example 9.3. Suppose X_1 has triangular density on $S = \{x : -1 \leq x \leq 1\}$ with mode $x = \theta$; that is,

$$f_{\theta;X_1}(x) = \begin{cases} (1 + x)/(1 + \theta) & \text{if} \quad -1 \leq x \leq \theta, \\ (1 - x)/(1 - \theta) & \text{if} \quad \theta \leq x \leq 1. \end{cases} \tag{9.9}$$

Here $\Omega = \{\theta : -1 \leq \theta \leq 1\}$ and the top (respectively, bottom) line of (9.9) is omitted if $\theta = -1$ (respectively, $\theta = 1$).

One can try to use the W of (9.2) and find a minimax procedure, or minimize $\max_\theta E_\theta[U(X_1) - L(X_1)]$ subject to $\bar{\gamma}_{[L,U]} \geq .9$, etc. The calculation of such a procedure is not easy, since there is no useful invariance consideration which will drastically reduce (as it did in Examples 9.1 and 9.2) the class of procedures to be considered; the present problem is invariant under the identity and the transformation $X_1 \to -X_1$, but this only reduces consideration to procedures for which $L(x) = -U(-x)$ (that is, $t(x) = -t(-x)$ in this *estimation* problem, by analogy with the $t(x) = t(-x)$ for the invariant procedures in the *hypothesis testing* Problem (A) (ii) of the subsection on problems and tests of Chapter 8).

Thus, we are led to try the method just described, using functions g and h, and we now construct three possible procedures, all with $T(x_1) = x_1$, to illustrate the method, with confidence coefficient .6. We shall see that a seemingly simple choice of g and h, in (i) in the following list, results in very difficult computations.

(i) It seems reasonable to try g and h, for simplicity, as the linear functions $A\theta \pm B$ where A and B are positive constants. Since $|\theta|$ cannot be >1, this really means using $g_1(\theta) = \max(A\theta - B, -1)$ and $h_1(\theta) = \min(A\theta + B, +1)$. These are increasing but not *strictly* increasing, so the diagram after (9.8) is modified to:

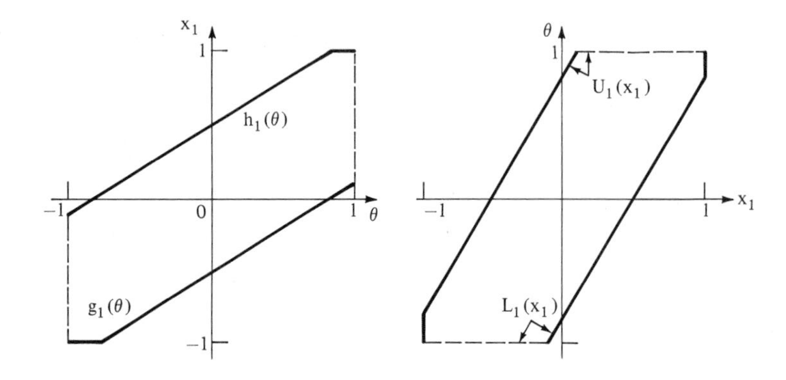

Corresponding to the horizontal portion of the graph of g_1, we get a vertical line segment at $X_1 = -1$, and $U_1(-1)$ must be taken to be the highest ordinate value. Thus,

$$
\begin{aligned}
L_1(x_1) &= \max(A^{-1}(x_1 - B), -1), \\
U_1(x_1) &= \min(A^{-1}(x_1 + B), +1).
\end{aligned}
\tag{9.10}
$$

To simplify the computation of the probabilistic expression of (9.8), let us suppose that $A - B \leq 1 \leq A + B$. This means $g_1(1) \leq 1$ and $g_1(-1) = -1$ and hence, from the form of g_1, we have $g_1(\theta) \leq \theta$ for $-1 \leq \theta \leq 1$. Consequently the first line of (9.9) yields

$$
\begin{aligned}
P_\theta\{X_1 \leq g_1(\theta)\} &= \int_{-1}^{g_1(\theta)} f_{\theta;X_1}(x)\,dx \\
&= \int_{-1}^{\max(A\theta - B, -1)} (1+\theta)^{-1}(1+x)\,dx \\
&= [\max(A\theta - B + 1, 0)]^2/2(1+\theta),
\end{aligned}
\tag{9.11}
$$

and similarly $P_\theta\{h_1(\theta) \leq X_1\} = [\max(1 - A\theta - B, 0)]^2/2(1-\theta)$, so that

$$
P_\theta\{g_1(\theta) \leq X_1 \leq h_1(\theta)\}
$$

$$
= \begin{cases}
1 - (1 - A\theta - B)^2/2(1-\theta) & \text{if} \quad \theta \leq (B-1)/A, \\
1 - (A\theta - B + 1)^2/2(1+\theta) - (1 - A\theta - B)^2/2(1-\theta) & \\
\quad \text{if} \quad B - 1 \leq A\theta \leq 1 - B, & \\
1 - (A\theta - B + 1)^2/2(1+\theta) & \text{if} \quad (1-B)/A \leq \theta.
\end{cases}
\tag{9.12}
$$

For fixed A and B one can compute the minimum over θ of each of the three lines of (9.12), and then $\bar{\gamma}_t$ is the minimum of these three minima. This is a slightly messy computation; moreover, one must then choose A and B to make $\bar{\gamma}_t \geq .6$; which of the possible pairs A, B one chooses could then be a matter of simplicity or (harder) of minimizing $\max_\theta E_\theta[U_1(X_1) - L_1(X_1)]$. This suggests trying another form of g and h, and we now do so.

(ii) Rather than trying to select "simple" g and h and then to compute (9.8) as in (i), we can try to select g and h to make that computation easy. One possibility is to choose g_2 and h_2 to satisfy

$$P_\theta\{X_1 \leq g_2(\theta)\} = .2 = P_\theta\{X_1 \geq h_2(\theta)\}. \tag{9.13}$$

For θ near -1, $g_2(\theta)$ will have to be $> \theta$ and the form (9.13) would require breaking $\int_{-1}^{g_2(\theta)} f_{\theta;X_1}(x)\,dx$ into two parts, so that it would be easier in such a case to compute the complementary probability. In fact, since this is so as long as $g_2(\theta) > \theta$, we first look for the value of θ such that $g_2(\theta) = \theta$, that is, such that

$$\begin{aligned} .2 &= P_\theta\{X_1 \leq \theta\} \\ &= (\theta + 1)/2 \quad \text{(area of shaded triangle in diagram of (9.9))}, \end{aligned} \tag{9.14}$$

or $\theta = -.6$. So we compute

$$\begin{cases} \text{if } \theta \geq -.6, \quad \text{then} \quad .2 = P_\theta\{X_1 \leq g_2(\theta)\} \\ \quad = \displaystyle\int_{-1}^{g_2(\theta)} (1+\theta)^{-1}(1+x)\,dx = [1 + g_2(\theta)]^2/2(1+\theta); \\ \text{if } \theta \leq -.6, \quad \text{then} \quad .8 = P_\theta\{X_1 \geq g_2(\theta)\} \\ \quad = \displaystyle\int_{g_2(\theta)}^{1} (1-\theta)^{-1}(1-x)\,dx = [1 - g_2(\theta)]^2/2(1-\theta); \end{cases} \tag{9.15}$$

and consequently

$$g_2(\theta) = \begin{cases} 1 - [1.6(1-\theta)]^{1/2} & \text{if} \quad -1 \leq \theta \leq -.6, \\ [.4(1+\theta)]^{1/2} - 1 & \text{if} \quad -.6 \leq \theta \leq 1. \end{cases} \tag{9.16}$$

This is an increasing function for $-1 \leq \theta \leq 1$, and its inverse, found by solving $x_1 = g_2(\theta)$ for θ (and recalling that $g_2(-.6) = -.6$), and by adjoining the "dashed line" portion described following (9.8), is

$$U_2(x_1) = \begin{cases} 1 - (1-x_1)^2/1.6 & \text{if} \quad 1 - \sqrt{3.2} \leq x_1 \leq -.6, \\ -1 + (1+x_1)^2/.4 & \text{if} \quad -.6 \leq x_1 \leq -1 + \sqrt{.8}, \\ 1 \quad \text{if} \quad -1 + \sqrt{.8} \leq x_1 \leq \sqrt{3.2} - 1. \end{cases} \tag{9.17}$$

By the symmetry of our construction, $L_2(x_1) = -U_2(-x_1)$. The resulting confidence interval procedure illustrates the phenomenon, mentioned following (9.8), of a sometimes-empty-interval: e.g., if $X_1 = -1/2$, the interval is $-1 \leq \theta \leq -13/32$ (since $U_2(-1/2) = -13/32$), but if $X_1 = -.8$

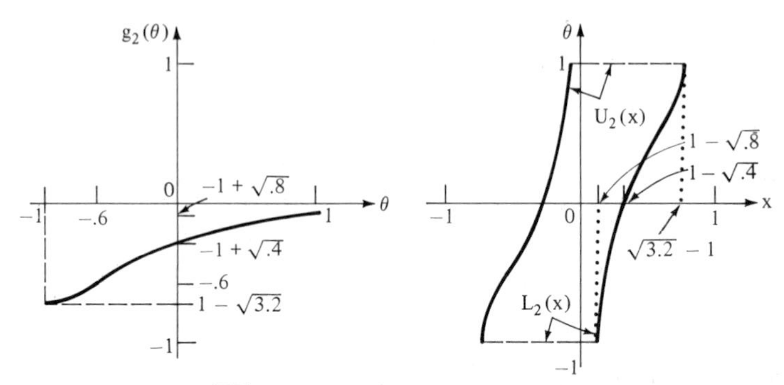

(which is $< 1 - \sqrt{3.2}$) the confidence set is empty. Nevertheless, despite the fact that "the true value of θ cannot be covered by the empty set" when $X_1 = -.8$, when we average probabilistically over all possible values of X_1 we still have $P_\theta\{L_2(X_1) \le \theta \le U_2(X_1)\} = .6$.

(iii) [More Complicated.] Again designing the form of g and h to make the computation of (9.8) simple, we now seek a choice which will make the confidence interval nonempty for all values of X_1. Examining g_2 and U_2 in the previous paragraph, we see that we must have $g(-1) = -1$ to avoid nonemptiness. We will also make a choice which illustrates the possibility of g and h having discontinuities (a common occurrence in cases in which X_1 is discrete, as we shall see in Example 9.4). We define $g_3(\theta)$ to satisfy

$$P_\theta\{X_1 \le g_3(\theta)\} = \begin{cases} 0 & \text{if} \quad -1 \le \theta < -.6, \\ .2 & \text{if} \quad -.6 \le \theta \le .6, \\ .4 & \text{if} \quad .6 < \theta \le 1, \end{cases} \tag{9.18}$$

and with $h_3(\theta) = -g_3(-\theta)$ the symmetry property $f_{\theta;X_1}(x) = f_{-\theta;X_1}(-x)$ implies that

$$P_\theta\{h_3(\theta) \le X_1\} = \begin{cases} .4 & \text{if} \quad -1 \le \theta < -.6, \\ .2 & \text{if} \quad -.6 \le \theta \le .6, \\ 0 & \text{if} \quad .6 < \theta \le 1, \end{cases} \tag{9.19}$$

so that (9.18) and (9.19) imply that the probabilistic expression of (9.8) equals .6. The computation of g_3 satisfying (9.18) is (integrating as in (9.15))

$$\begin{cases} \text{if } -1 \le \theta < -.6, \quad \text{then} \quad g_3(\theta) = -1; \\ \text{if } -.6 \le \theta \le .6, \quad \text{then} \quad g_3(\theta) = g_2(\theta) = [.4(1+\theta)]^{1/2} - 1; \\ \text{if } .6 < \theta \le 1, \quad \text{then} \quad .4 = P_\theta\{X_1 \le g_3(\theta)\} \\ \quad = [1 + g_3(\theta)]^2/2(1+\theta), \quad \text{or} \quad g_3(\theta) = [.8(1+\theta)]^{1/2} - 1. \end{cases} \tag{9.20}$$

(For the computation of the third line of (9.20), the analogue of the value $-.6$ obtained from (9.14), beyond which only one line of (9.9) need be used, is now $\theta = -.2$; since $\theta > -.2$ in the third line of (9.20), the first line of (9.9) is used for $f_{\theta;X_1}$ throughout the computation there.) We thus have

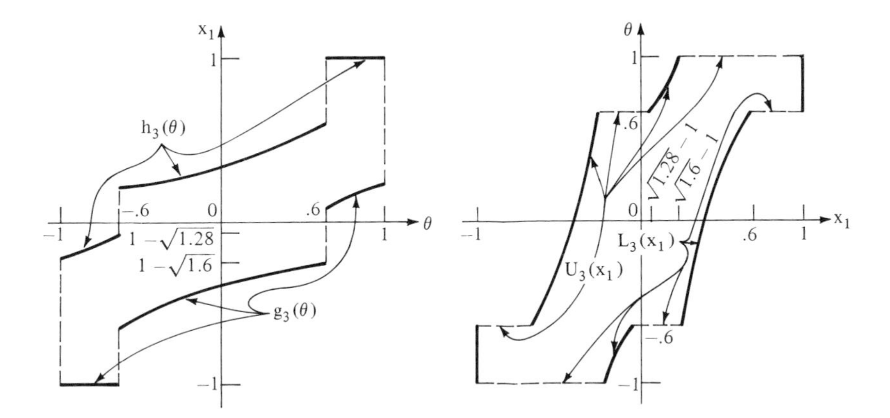

The appropriate definition of U_3 is obtained by adjoining to the graph of g_3 the vertical segments between its three components. Identifying the closed interval of ordinate values between the resulting values of $g_3(\theta)$ and $h_3(\theta)$, *including these vertical segments*, as the set $B(\theta)$ described in the third paragraph that follows (9.8), we obtain there $t(x) = [L_3(x), U_3(x)]$ with

$$-L_3(-x) = U_3(x)$$

$$= \begin{cases} -.6 \quad \text{if} \quad -1 \le x \le -.6, \\ -1 + (1+x)^2/.4 \quad \text{if} \quad -.6 \le x \le 1 - \sqrt{1.6}, \\ .6 \quad \text{if} \quad 1 - \sqrt{1.6} \le x \le \sqrt{1.28} - 1, \\ -1 + (1+x)^2/.8 \quad \text{if} \quad \sqrt{1.28} - 1 \le x \le \sqrt{1.6} - 1, \\ 1 \quad \text{if} \quad \sqrt{1.6} - 1 \le x \le 1. \end{cases} \tag{9.21}$$

This $[L_3(X_1), U_3(X_1)]$ gives a confidence interval with $\bar{\gamma}_{[L_3, U_3]} = .6$. For example, if $X_1 = .2$, we obtain the confidence set $-.6 \le \theta \le .8$.

EXAMPLE 9.4. [Also difficult.] In the discrete case one generally obtains g and h whose graphs contain horizontal line segments, and (on rotating the graph of g and h by 90°) this would seem to yield U and L with discontinuities. These discontinuities only appear because one looks at the domain of T as though it is a continuum; for the actual *discrete* X (or $T(X)$) there is simply an interval $[L(x), U(x)]$ of values of θ for each x in the *discrete* set of possible values of X. (For the continuous case of X, horizontal sections of g and h would produce actual discontinuities in L and U.) For example, suppose X_1, X_2, X_3, X_4 are iid Bernoulli random variables with $\Omega = \{\theta : 0 \le \theta \le 1\}$. Suppose it is desired to have $\bar{\gamma}_t \ge .8$ for estimating $\phi(\theta) = \theta$. We let T be the minimal sufficient statistic $\sum_1^4 X_i$.

We begin by finding, for each θ, what the *possible* sets $B(\theta) = \{\tau : g(\theta) \le \tau \le h(\theta)\}$ of values τ of T are, such that $P_\theta\{T \in B(\theta)\} \ge .8$. Note that T has binomial probability function with parameters 4 and θ.

First, the set $B(\theta) = \{0\}$ (that is, $g(\theta) = 0 = h(\theta)$) will suffice as long as $(1-\theta)^4 \geq .8$. Since $(1-\theta)^4$ is decreasing in θ and equals .8 when $\theta = .054$, we conclude that $B(\theta) = \{0\}$ suffices if $0 \leq \theta \leq .054$.

Next, since $(1-\theta)^4 + 4\theta(1-\theta)^3$ is also decreasing and equals .8 when $\theta = .212$, we can use $B(\theta) = \{0, 1\}$ (so that $g(\theta) = 0, h(\theta) = 1$) as long as $\theta \leq .212$. Since the use of the subset $\{0\}$ of $\{0, 1\}$, of the previous paragraph, when permissible, will yield a smaller t, we will use $\{0, 1\}$ only when $.054 < \theta \leq .212$.

Similarly, $(1-\theta)^4 + 4\theta(1-\theta)^3 + 6\theta^2(1-\theta) \geq .8$ as long as $0 \leq \theta \leq .419$, and we will only use the corresponding set $B(\theta) = \{0, 1, 2\}$ when $\theta > .212$, because of the availability of the subset $\{0, 1\}$ (previous paragraph) when $\theta \leq .212$. However, we may not decide to use $\{0, 1, 2\}$ for all θ satisfying $.212 < \theta \leq .419$, because of the next calculation.

That calculation shows not only that the set $\{0, 1, 2, 3\}$ can be used for θ near .419, but in fact that the smaller set $\{1, 2, 3\}$ can be used. In fact, we find that $P_\theta\{T \in \{1, 2, 3\}\} \geq .8$ for $.343 \leq \theta \leq .657$. Thus, we see that for $.343 \leq \theta \leq .419$ either $\{0, 1, 2\}$ or $\{1, 2, 3\}$ could be used as $B(\theta)$, and since neither is a subset of the other (and in fact they contain the same number of elements) there is no automatic preference between them. Thus, if we pick any constant ρ satisfying $.343 \leq \rho \leq .419$, we can use the set

$$B_\rho(\theta) = \begin{cases} \{0\} & \text{if } 0 \leq \theta \leq .054, \\ \{0, 1\} & \text{if } .054 < \theta \leq .212, \\ \{0, 1, 2\} & \text{if } .212 < \theta \leq \rho, \\ \{1, 2, 3\} & \text{if } \rho < \theta < 1 - \rho, \\ \{2, 3, 4\} & \text{if } 1 - \rho \leq \theta < .788, \\ \{3, 4\} & \text{if } .788 \leq \theta < .946, \\ \{4\} & \text{if } .946 \leq \theta \leq 1. \end{cases} \tag{9.22}$$

The last three and a half lines of (9.22), of course, follow from the first three and a half by the fact that the probability function of T under θ is the same as that of $4 - T$ under $1 - \theta$.

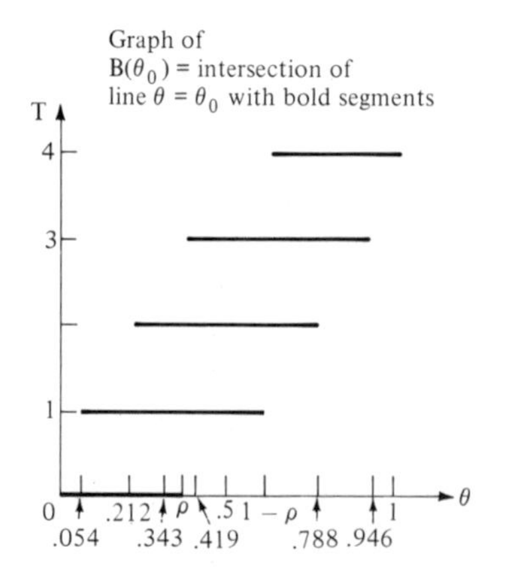

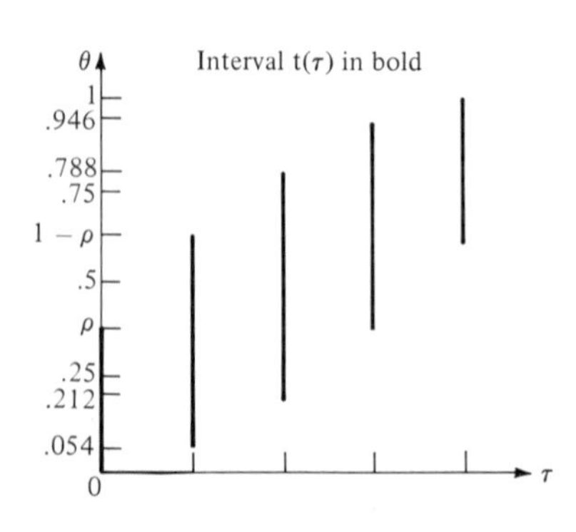

Thus, for example, if $T = 1$, the confidence statement is "$.054 < \theta \leq 1 - \rho$." The half-openness of this interval is a consequence of the form of the inequalities of (9.22), and it would be more usual to use the (larger) closed intervals in all cases:

$$t_\rho(\tau) = [L_\rho(\tau), U_\rho(\tau)] = \begin{cases} [0, \rho] & \text{if} \quad \tau = 0, \\ [.054, 1 - \rho] & \text{if} \quad \tau = 1, \\ [.212, .788] & \text{if} \quad \tau = 2, \\ [\rho, .946] & \text{if} \quad \tau = 3, \\ [1 - \rho, 1] & \text{if} \quad \tau = 4. \end{cases} \tag{9.23}$$

In constructing such a procedure, a development in a typical handbook would usually consider only one of the possible procedures t_ρ, corresponding to one of the limiting values $\rho = .343$ or $\rho = .419$. We have gone through the derivation of (9.23) to illustrate the multiplicity of satisfactory procedures in even such a simple discrete example. From a decision-theoretic point of view, ρ might be chosen to minimize $\max_\theta E_\theta[U_\rho - L_\rho]$, although the range of ρ is small enough here that such a computation is of doubtful practical value. Intuitively, one might argue that when $\theta = 1/2$ one tends to get the larger intervals $t_\rho(\tau)$ most often (binomial law with least probability that $T \in \{0, 4\}$), and that $T = 1$ or 3 is more likely than $T = 0$ or 4 in that case; hence, one would choose ρ to minimize $.946 - \rho$ (length when $T = 1$ or 3) rather than to minimize ρ (length when $T = 0$ or 4), namely, $\rho = .419$.

[For other values of $\bar{\gamma}_t$ and the sample size, a larger variety of possible confidence interval procedures may exist. For example, when $\bar{\gamma}_t$ is smaller, the set $\{1, 2\}$ may also sometimes be usable as $B(\theta)$. That cannot occur in the present case, in which the function $P_\theta\{T = 1 \text{ or } 2\} = 4\theta^3(1 - \theta) + 6\theta^2(1 - \theta)^2$ is easily computed to attain a maximum value of $.70 < .8$ (at $\theta = (3 - \sqrt{3})/2$).]

In problems in which there are n observations and there is no simple sufficient statistic (e.g., more than 1 observation in Example 9.1, with g the standard Cauchy density), we may find it convenient for large n to use the asymptotic theory of Section 7.6 to construct confidence intervals with approximately a specified coverage, again at the sacrifice of some efficiency (now measured in terms of length rather than variance). Thus, if $g(x) = 1/\pi(1 + x^2)$ in Example 9.1 with n iid observations, then (Example 7.16) $n^{1/2}(Z_{1/2,n} - \theta)$ is approximately $\mathcal{N}(0, \pi^2/4)$, where $Z_{1/2,n}$ is the sample median. Consequently, if $\Phi(c) = 1 - \alpha/2$ where Φ is the standard normal df, then $[Z_{1/2,n} - c\pi/2n^{1/2}, Z_{1/2,n} + c\pi/2n^{1/2}]$ is an interval with coverage probability approximately $1 - \alpha$. Even when there is a simple sufficient statistic, but we do not have a scale- or location-parameter problem, exact computations can be tedious, as we shall now discuss.

When Example 9.4 is modified by having a sample size n much greater than 4, a computation like that of Example 9.4 would be very lengthy. An approximate solution is given by using the Central Limit Theorem, in a manner slightly more complicated than the use of the normal approximation in the previous paragraph.

EXAMPLE 9.4(a). In the setup of Example 9.4, but with n iid Bernoulli random variables X_i and $T_n = \sum_1^n X_i$, we know that, for $0 < \theta < 1$, by the Central Limit Theorem, the random variable $[n\theta(1-\theta)]^{-1/2}(T_n - n\theta)$ is approximately $\mathcal{N}(0,1)$, and hence for large n

$$P_\theta\{-1.28 \le [n\theta(1-\theta)]^{-1/2}(T_n - n\theta) \le +1.28\} \approx .8. \tag{9.24}$$

The inequalities in braces in (9.24) can be rewritten $-1.28[n\theta(1-\theta)]^{1/2} + n\theta \le T_n \le 1.28[n\theta(1-\theta)]^{1/2} + n\theta$, but it is a waste of time to compute g and h in this way from (9.24), since what we really want is L and U, and we can obtain them by solving the inequalities of (9.24) directly for θ (consider either inequality, replace it by equality, square both sides, extract the appropriate quadratic root), giving

$$(L_n, U_n) = \frac{[2T_n + (1.28)^2] \pm 1.28[(1.28)^2 + 4n^{-1}T_n(n - T_n)]^{1/2}}{2[n + (1.28)^2]}, \tag{9.25}$$

where the + sign yields U_n and the $-$ sign yields L_n. When n is large $\bar{X}_n = n^{-1}T_n$ is close to θ with high probability; and hence, unless θ is close to 0 or 1, little error is likely to be produced by replacing the three expressions $(1.28)^2$ in (9.25) by zero, yielding

$$(L_n', U_n') \stackrel{\text{def}}{=} \bar{X}_n \pm 1.28[n^{-1}\bar{X}_n(1 - \bar{X}_n)]^{1/2}. \tag{9.26}$$

This could also have been obtained directly from (9.24) by replacing the unknown standard deviation of T_n there (namely, $[n\theta(1-\theta)]^{1/2}$) by its ML estimator $[n\bar{X}_n(1-\bar{X}_n)]^{1/2}$.

The adequacy of (9.26) requires larger n than that of (9.25). The closer the true value of θ is to 0 or 1, the larger n must be for the coverage probability of (9.25) or (9.26) to be close to .8. In fact, when $\theta = 0$ or 1, so that $\bar{X}_n = 0$ or 1 (respectively) with probability one, either (9.25) or (9.26) yields probability *one* of covering the true value of θ.

We next illustrate the fact that, even for 1-dimensional Ω and S, there may be an advantage of using a confidence *set* which is not an interval (but rather a union of intervals).

EXAMPLE 9.5. The setup is that of Example 9.1 except that, instead of assuming g unimodal, we consider the problem for which

$$g(x) = \begin{cases} .40 & \text{if} \quad 10 \le |x| \le 11, \\ .01 & \text{if} \quad 0 \le |x| < 10, \\ 0 & \text{otherwise}, \end{cases} \tag{9.27}$$

and $f_{\theta;X_1}(x) = g(x - \theta)$ as before. Suppose we want a coverage probability of .5. We begin by demonstrating that, even if we do restrict consideration to

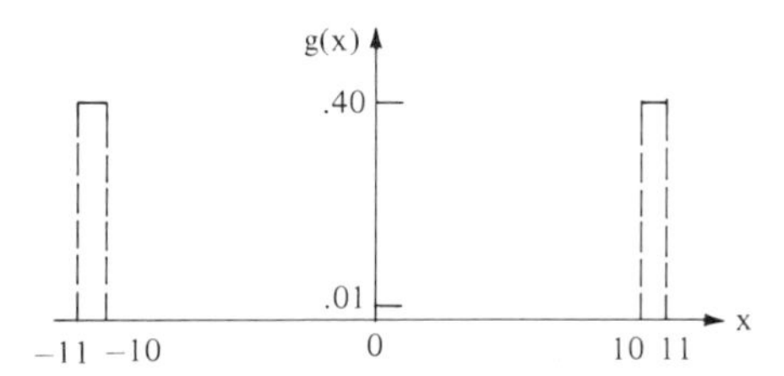

interval estimators rather than more general set estimators, the conclusions of Example 9.1 no longer hold for the present *nonunimodal* g. In fact, for $b = 0$ equation (9.5) yields $\gamma_{t_{c,0}} = .5$ when $c = 10.375$ (see the diagram accompanying (9.27), according to which $P_0\{X_1 > 10.375\} = .25$); equation (9.5) shows that the interval $t_{5.5,5.5}(X_1) = [X_1, X_1 + 11]$ also has coverage probability .5 and has length 11 compared with the length 20.75 of $[X_1 - 10.375, X_1 + 10.375]$. Thus, we see that the assertion two paragraphs after (9.5), that the choice $b = 0$ gives the smallest length $2c$, need not be true if g is not unimodal.

The procedures $t_{5.5,5.5}$ and $t_{5.5,-5.5}$ can be shown to be the invariant *interval* estimators of shortest length in this example. On the other hand, if we permit $t(X_1)$ to be a more general set (as in the description two paragraphs before (9.9)), then we can achieve a shorter total length. Indeed, if $t(X_1)$ is the union of two disjoint intervals $[X_1 + b_1 - c_1, X_1 + b_1 + c_1]$ and $[X_1 + b_2 - c_2, X_1 + b_2 + c_2]$, then a computation like that of (9.5) shows that the coverage probability is $\sum_{i=1}^{2} [G(c_i - b_i) - G(-c_i - b_i)]$. In the present example, if $c_1 = c_2 = .3125$ and $b_1 = -b_2$ is any value between 10.3125 and 10.6875 (so that $10 \leq b_1 - c_1 < b_1 + c_1 \leq 11$), then the diagram next to (9.5) shows that the coverage probability is .5. That is, for such a b_1,

$$P_\theta\{\theta \in [X_1 + b_1 - .3125, X_1 + b_1 + .3125]$$
$$\cup [X_1 - b_1 - .3125, X_1 - b_1 + .3125]\} = .5$$

for all real θ. This procedure, for any allowable b_1, yields a set whose total "length" is 1.25, much less than the length 11 of the best interval procedure of the previous paragraph. It can be shown that no other invariant set-valued procedure yields a shorter length than 1.25, and the multiplicity of allowable values of b_1 illustrates the lack of uniqueness of the best procedure. (In fact, there are procedures of total length 1.25, other than those described here, but the latter are the simplest.)

The practical result here is that we have achieved a shortening of length at the expense of allowing a more complicated form of decision (union of two intervals, in place of a single interval); whether or not the experimenter finds it appropriate to make such an allowance will depend on the circumstances of the application. Finally, when g is unimodal as in Example 9.1, it can be shown that one cannot improve on the length of $t_{c_\gamma,0}$ by using sets more general than intervals.

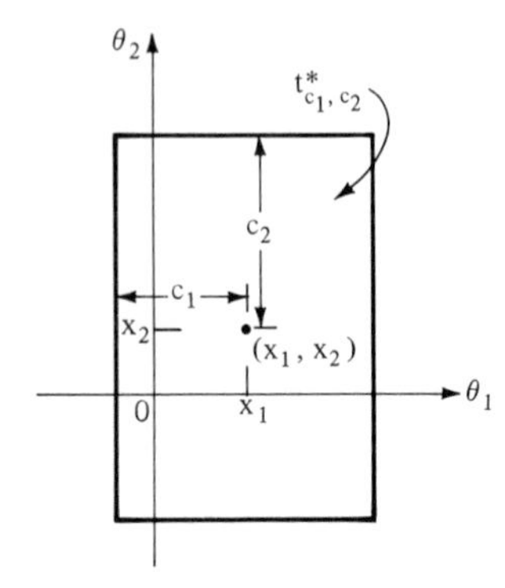

We now illustrate the construction of confidence sets when Ω is not 1-dimensional. The description two paragraphs before (9.9) still applies.

EXAMPLE 9.6. Suppose X_1 and X_2 are independent, with X_i having density $g_i(x_i - \theta_i)$ and $\Omega = \{(\theta_1, \theta_2): -\infty < \theta_1, \theta_2 < +\infty\}$. Here g_1 and g_2 are specified and each is assumed symmetric and unimodal as in Example 9.1. This is the 2-dimensional counterpart of Example 9.1 in the special case of independent components. It is clear from the calculation (9.5) and the independence of the X_i that the *rectangular* confidence set (with sides parallel to the coordinate axes)

$$\begin{aligned} &t^*_{c_1,c_2}(X_1, X_2) \\ &\quad = \{(\theta_1, \theta_2): X_1 - c_1 \leq \theta_1 \leq X_1 + c_1, X_2 - c_2 \leq \theta_2 \leq X_2 + c_2\} \end{aligned} \tag{9.28}$$

has coverage probability

$$\gamma_{t^*_{c_1,c_2}} = [2G_1(c_1) - 1][2G_2(c_2) - 1], \tag{9.29}$$

where G_i is the df corresponding to g_i. The *area* of this set is $4c_1c_2$, and if one were restricting consideration to such *rectangular* confidence sets one might choose c_1 and c_2, subject to $\gamma_{t^*_{c_1,c_2}}$ being the specified value, so as to minimize this area.

However, if we permit confidence sets which are not of the rectangular form (9.28), it may be possible to reduce the area further. This is illustrated in Problem 9.9(d), where the result to be proved is really that, if both g_i's are standard $\mathcal{N}(0, 1)$, then smaller area is achieved by a circular region $\{(\theta_1, \theta_2): (\theta_1 - X_1)^2 + (\theta_2 - X_2)^2 \leq c'\}$, of area $\pi c'$, than by a square region ((9.28) with $c_1 = c_2$, which is the best rectangular form in this case) with the same confidence coefficient. We now consider an arithmetically simpler illustration of a problem for which the best region of the form (9.28) can be improved upon.

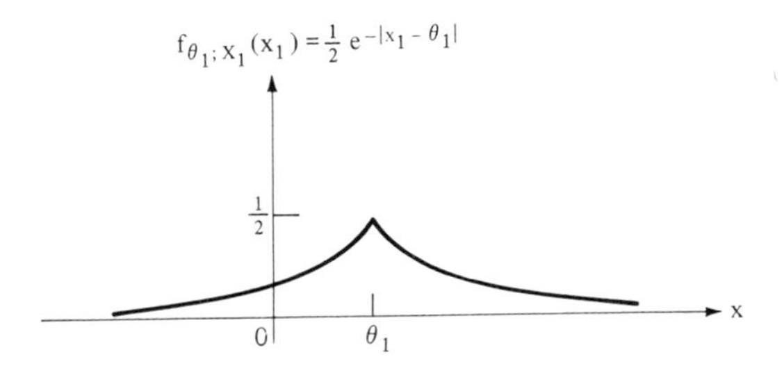

Suppose $g_1(x) = g_2(x) = \frac{1}{2}e^{-|x|}$ and we want coverage probability .8. It is not hard to show that the region of the form (9.28) with least area is the square with $c_1 = c_2 = -\log(1 - \sqrt{.8}) = 2.249$, with resulting area $4c_1^2 = 20.2$. On the other hand, consider the square region *with sides rotated* 45° *from the coordinate axes*,

$$t_{\bar{c}}(X_1, X_2) = \{(\theta_1, \theta_2) : |\theta_1 - X_1| + |\theta_2 - X_2| \le \bar{c}\}. \tag{9.30}$$

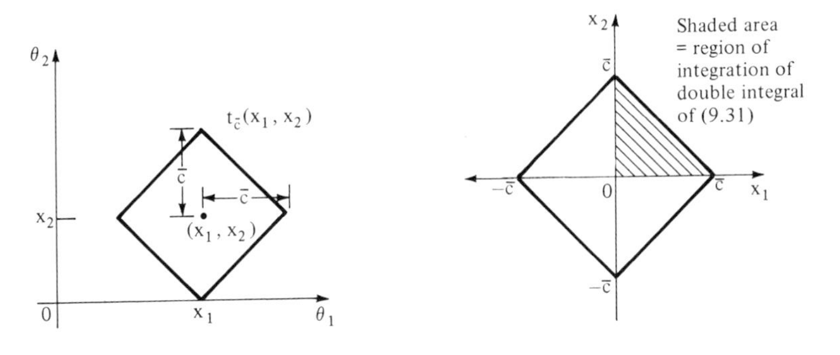

The probability of coverage is (by an argument like that of (9.5)) again constant and hence can be computed under $(\theta_1, \theta_2) = (0, 0)$. Using the symmetry of $\frac{1}{2}e^{-|x|}$ to break up the symmetric region of the first line of (9.31) into four sets of equal probability, we have

$$\begin{aligned}
\gamma_{t_{\bar{c}}} &= P_{0,0}\{|X_1| + |X_2| \le \bar{c}\} \\
&= 4P_{0,0}\{0 \le X_1; 0 \le X_2; X_1 + X_2 \le \bar{c}\} \\
&= 4\int_0^{\bar{c}}\int_0^{\bar{c}-x_2} \frac{1}{4} e^{-|x_1|-|x_2|}\,dx_1\,dx_2 \\
&= \int_0^{\bar{c}}\int_0^{\bar{c}-x_2} e^{-x_1-x_2}\,dx_1\,dx_2 = \int_0^{\bar{c}} e^{-x_2}(1 - e^{-\bar{c}+x_2})\,dx_2 \\
&= 1 - (\bar{c} + 1)e^{-\bar{c}}.
\end{aligned} \tag{9.31}$$

Setting this last expression equal to .8, we obtain (by trial and error) $\bar{c} = 2.994$ with corresponding area (for a square of side $\bar{c}\sqrt{2}$) $2\bar{c}^2 = 17.9$, an improvement over the area 20.2 of the best (square) region of the form (9.28), described previously.

[It can be shown that, for this g_1 and g_2, the region (9.30) actually yields smallest area among *all* invariant confidence set procedures with the same confidence coefficient, no matter what the shape.]

Bayes and Fiducial Intervals

The Bayesian interpretation of interval estimation fits into the discussion of Chapter 4. In fact, we considered there the point estimation problem with the loss function that is displayed two paragraphs after (9.2) (now setting $\phi(\theta) = \theta$ for simplicity) in (4.4) and computed the choice of d (that is, of the interval $[d - c, d + c]$) which gave a Bayes decision in an example involving the gamma density.

Often a Bayesian operates without such a notion of a particular loss function. Instead, given that $X = x$, he or she plots the posterior density of θ, say $g_\xi(\theta|x)$, for the given prior law ξ, and then chooses an estimating set $t(x)$ for which the posterior probability that $\theta \in t(x)$, namely, $\int_{t(x)} g_\xi(\theta|x)\,d\theta$, is some specified value. (Subject to this specification, t may be chosen to have smallest length.) This will often be intuitively satisfactory from a practical point of view; however, even in the setup of Example 9.1, the resulting $t(x)$ will not generally be Bayes relative to ξ for any simple W such as (9.2); this choice of $t(x)$ to give *constant posterior probability of coverage* is simply not the prescription obtained in Chapter 4 for a Bayes procedure relative to some W. Of course, the length of $t(x)$ will generally also be variable here, since the form of $g_\xi(\theta|x)$ changes with x.

Beginning instead with our development of confidence intervals, what is the interpretation of a confidence interval procedure like $t_{c_\gamma,0}$ of Example 9.1, from a Bayesian point of view? Since the probability $P_\theta\{\theta \in t_{c_\gamma,0}(X_1)\}$ is γ for each θ, it follows that, when θ has prior law ξ and $\gamma_t(\theta_0)$ is viewed as a *conditional* probability of coverage given that $\theta = \theta_0$, we have the *unconditional* probability of coverage under ξ also equal to γ. However, the Bayesian is often more interested in the posterior probability that $\theta \in t_{c_\gamma,0}(x_1)$, given that $X_1 = x_1$, and that probability will generally depend on x_1. Thus, by using $t_{c_\gamma,0}$ we do not satisfy the Bayesian described in the previous paragraph, who wanted an interval with specified posterior probability of coverage.

Nevertheless, the value γ of $t_{c_\gamma,0}$ does sometimes have approximately the right meaning for a Bayesian. Roughly speaking, if ξ is approximately uniform on a very large interval, then the probability is close to one that the value x_1 of X_1 turns out to be such that

$$\int_{t(x_1)} g_\xi(\theta|x_1)\,d\theta \approx \gamma.$$

R. A. Fisher discussed interval estimation in the framework of what he called "fiducial distributions." Roughly speaking, if $X = x$ Fisher would consider the likelihood function $\Lambda_x(\theta) = f_{\theta;X}(x)$ *as a function of* θ, with perhaps a slight modification to make $\int \Lambda_x(\theta)\,d\theta = 1$. Then, if $\int_{\tilde{L}(x)}^{\tilde{U}(x)} \Lambda_x(\theta)\,d\theta = \gamma$, Fisher would say that $[\tilde{L}(x), \tilde{U}(x)]$ was a *fiducial interval on* θ with *fiducial probability* γ (of coverage). In extensive discussions with Neyman and others, he treated $\Lambda_x(\theta)$ in calculations as if it were a density function of θ, but argued verbally that Λ_x was not to be thought of as an ordinary probability density. Few people claim thorough understanding of a logical development in this part of Fisher's work, and some doubt that he had one. In the location-parameter setup of Example 9.1 one obtains the same $[\tilde{L}, \tilde{U}]$ as the confidence interval $[L, U]$ whose coverage probability equals the fiducial probability. But there are more complicated examples in which Fisher's development leads to fiducial estimators which differ from Neyman's confidence intervals. If, in Example 9.1, there is a "diffuse prior law" of the type described in the previous paragraph, then fiducial probability and posterior probability are almost equal.

Problems

9.1. Suppose X is a real random variable with continuous case density

$$f_\theta(x) = \begin{cases} 2\theta^2/(x+\theta)^3, & 0 < x < \infty, \\ 0 & \text{otherwise.} \end{cases}$$

Here $\Omega = \{\theta : 0 < \theta < \infty\}$. [This family of densities was used in sufficiency and estimation examples in previous problems, e.g. 6.6(x) and 7.6.6 and 7.6.7.] It is desired to find a one-sided upper confidence limit on θ, with confidence coefficient equal to a specified value γ, $0 < \gamma < 1$.

(a) Following the development in Example 9.3 (the present problem being simpler arithmetically), for each possible value of θ find a value $g(\theta)$ such that $P_\theta\{X \geq g(\theta)\} = \gamma$. [*Answer*: $g(\theta) = (\gamma^{-1/2} - 1)\theta$.]

(b) Use the preceding to show that a confidence set with the desired properties is $t(x) = \{\theta : \theta \leq x/(\gamma^{-1/2} - 1)\}$. If $\gamma = .9$, what is the resulting confidence set if $X = 5$?

9.2. Suppose x has continuous case pdf

$$f_\theta(x) = \begin{cases} \dfrac{1+\theta x}{1+\frac{\theta}{2}}, & 0 \leq x \leq 1, \\ 0 & \text{otherwise.} \end{cases}$$

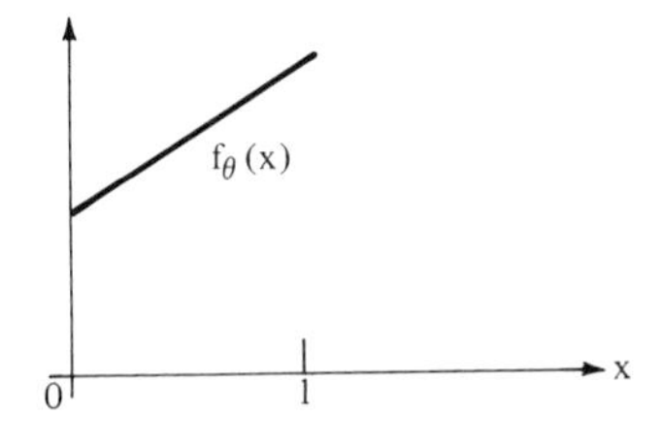

Here $\Omega = \{\theta : -1 \leq \theta < \infty\}$. It is desired to find a one-sided upper confidence limit on θ, with confidence coefficient equal to a specified value γ, $0 < \gamma < 1$.

(a) Following the development in Example 9.3 (the present problem being simpler arithmetically), for each possible value of θ find a value $g(\theta)$ such that $P_\theta\{X \geq g(\theta)\} = \gamma$. [*Answer*: $g(\theta) = \theta^{-1}[-1 + \sqrt{1 + (1 - \gamma)\theta(2 + \theta)}.]$ for $\theta \neq 0$, and $g(0) = 1 - \gamma$.]

(b) Use the preceding to show that a confidence set with the desired properties is

$$\begin{cases} \text{empty set of } \theta\text{'s} & \text{if } 0 \leq x < 1 - \sqrt{\gamma}; \\ \{\theta : -1 \leq \theta \leq 2[(1-\gamma) - x]/[x^2 - (1-\gamma)]\} & \text{if } 1 - \sqrt{\gamma} \\ & \quad \leq x < \sqrt{1-\gamma}; \\ \Omega & \text{if } x \geq \sqrt{1-\gamma}. \end{cases}$$

If $\gamma = .8$, what is the resulting confidence set if $X = .06$? If $X = .3$? If $X = .5$? This illustrates that the preceding method may lead to probabilistically valid procedures which can be unsatisfactory for certain values of X (the cases in which the confidence set is all of Ω or empty, for example).

9.3. The model is that of Problem 6.3(vi) with $n = 1$. It is desired to find a one-sided upper confidence limit on θ, with confidence coefficient 3/4. Following the development of Example 9.3 (harder than the present problem), for each possible value θ find $g(\theta)$ such that $P_\theta\{X \geq g(\theta)\} = 3/4$. [*Answer*: $g(\theta) = \theta^{-1}[-1 + (1 - \theta + \theta^2)^{1/2}]$, with $g(0) = -1/2$.] Use this to show that an upper confidence limit with the desired properties is

$$U(x) = \begin{cases} (1 + 2x)/(1 - x^2) & \text{if } 1 - \sqrt{3} \leq x \leq 0, \\ 1 & \text{if } 0 < x \leq 1, \end{cases}$$

with an empty confidence set if $x < 1 - \sqrt{3}$. [See final sentence of Problem 9.2.]

9.4. This illustrates the increased arithmetical work if X is discrete. Suppose $X = X_1$ has geometric *pdf*

$$P_\theta\{X_1 = x_1\} = (1 - \theta)\theta^{x_1} \quad \text{for} \quad x_1 = 0, 1, 2, \ldots, \tag{9.32}$$

where $\Omega = \{\theta : 0 \leq \theta < 1\}$ (with $P_0\{X_1 = 0\} = 1$). Find an upper confidence limit on θ of confidence coefficient .9 by finding a (discontinuous) g and (discontinuous) h as in the computations of Examples 9.3 and 9.4, as follows:

(a) find values $\theta_1, \theta_2, \ldots,$ such that

$$P_{\theta_i}\{X_1 \geq i\} = .9$$

$\left[\textit{Note}: \sum_{x=n}^{\infty} \theta^x = \theta^n/(1 - \theta).\right]$ [*Answer*: $\theta_i = .9^{1/i}$]

(b) Show that $P_\theta\{X_1 \geq i\}$ is increasing in θ for each fixed i.

(c) From (a) and (b) show that, if $\theta_0 = 0$ and

$$g(\theta) = i \quad \text{for} \quad \theta_i \leq \theta < \theta_{i+1}, \quad 0 \leq i < +\infty,$$

then

$$\min_{\theta \geq 0} P_\theta\{X_1 \geq g(\theta)\} = .9.$$

(d) From (c) and a diagram and discussion like that of the U of Example 9.3 (iii) and Example 9.4, or an equivalence like that of (9.3)–(9.4), show that if

$$U(x_1) = .9^{1/(x_1+1)} \quad \text{if} \quad x_1 = 0, 1, 2, \ldots,$$

then $\min_{\theta \geq 0} P_\theta\{\theta < U(X_1)\} = .9$.

(e) If $X_1, \ldots, X_n$ are iid with probability function (9.32) then the minimal sufficient statistic $T = \sum_1^n X_i$ has a negative binomial probability function. (Do not show this.) Even when n is small, the analogues of the θ_i must be obtained numerically and will not have a simple form as in (d); and when n is large the computation is very tedious. As an approximate solution for large n, use the Central Limit Theorem and the facts (which you need not prove) that $E_\theta X_1 = \theta/(1-\theta)$ and $\text{var}_\theta(X_1) = \theta/(1-\theta)^2$, to show that, when n is large, for $\theta > 0$

(i) $P_\theta\{T_n \geq g^*(\theta)\} \approx .9$, where $g^*(\theta) = [n\theta + 1.3\,(n\theta)^{1/2}]/(1-\theta)$;

(ii) the equation $g^*(\theta) = t$ has solution $\theta = U^*(t)$ where

$$U^*(t) = n^{-1}\left[\frac{-1.3 + [(1.3)^2 + 4t(1+n^{-1}t)]^{1/2}}{2(1+n^{-1}t)}\right]^2$$

[*Hint*: First write $\lambda = (n\theta)^{1/2}$ and substitute into $g^*(\theta) = t$, solving for λ];

(iii) g^* is increasing on $[0, 1)$, as is therefore U^* on $[0, \infty)$;

(iv) "$\theta \leq U^*(T_n)$" is an upper confidence bound statement with confidence coefficient approximately .9. [*Remark*: As $\theta \to 0$, the Central Limit Theorem requires increasingly large n for this to be a good approximation; when $\theta = 0$, since $U^*(0) = 0$, the coverage probability is 1.]

9.5. Suppose X is a real random variable with density function

$$f_\theta(x) = \begin{cases} e^{-(x-\theta)} & \text{if} \quad x \geq \theta, \\ 0 & \text{if} \quad x < \theta. \end{cases}$$

Here $\Omega = \{\theta: -\infty < \theta < \infty\}$. Consider the confidence interval given by $L(x) = x + a$, $U(x) = x + b$, where $a < b$ and a and b are constants.

(a) Show that $\gamma_t(\theta)$ is zero unless $a < 0$.

(b) If $a < 0$ and $b > 0$, show that $\gamma_t(\theta)$ has the same value as if a is kept the same and b is replaced by zero. Show that the second confidence interval is always shorter than the first. (Since the two procedures have the same confidence coefficient, this shows that we need never consider values $b > 0$.)

(c) If $a < b \leq 0$, show that $\gamma_t(\theta) = e^b - e^a$.

(d) For a specified value γ of $e^b - e^a$, show that the length of $b - a$ is minimized (over $a < b \leq 0$) by the choice $b = 0$, $a = \log(1-\gamma)$.

9.6. This is another illustration of a setting in which no invariant procedure is satisfactory. (See Sec. 7.3.) Suppose X is $\mathcal{N}(\mu, \sigma^2)$ where $\Omega = \{(\mu, \sigma^2): -\infty < \mu < \infty, \sigma^2 > 0\}$. There is only one observation and it is desired to estimate σ; thus, $D = \{d: d > 0\}$. The object is to put an upper confidence bound on σ, so that the decision d is considered to be this upper bound, and it is viewed as being "wrong" if it *underestimates* σ. Assuming W simple, this means $W((\theta, \sigma^2), d) = \begin{Bmatrix} 1 & \text{if} & d/\sigma < 1 \\ 0 & \text{if} & d/\sigma \geq 1 \end{Bmatrix}$.

(a) Show that this problem is invariant under the following "affine" transformation for each $b > 0$ and real c:

$$g_{(b,c)}(x) = bx + c,$$
$$h_{(b,c)}(\mu, \sigma^2) = (b\mu + c, b^2\sigma^2),$$
$$j_{(b,c)}(d) = bd.$$

Consequently, show that the procedure (upper confidence limit) t is invariant if and only if $t(bx + c) = bt(x)$ for all $b > 0$ and real c and x. Putting, in particular, $c = -bx$ (whatever b and x may be), show that the resulting equation implies, upon varying b, that $t(x) \equiv 0$. Since $0 \notin D$, this means *there is no invariant procedure.* Thus, the invariance approach leads nowhere.

(b) It may be felt that we should "be reasonable" and adjoin 0 to D, to permit the procedure $t^*(x) = 0$ of (a) to be used. Show that the procedure t^* is *uniformly worst.*

(c) Although no procedure can have risk identically 0 in this problem, show that the procedure $t_k(x) \overset{\text{def}}{=} kx$, for k sufficiently large, has $\bar{r}_{t_k}$ less than any preassigned $\varepsilon > 0$. [*Hint*: The crucial step is to show that $P_{\mu,1}\{|X| < k^{-1}\} \le P_{0,1}\{|X| < k^{-1}\}$ for $k > 0$ and all μ.]

[*Note*: The phenomenon of (a) and (b) cannot occur for sample size ≥ 2, in the present setting.]

9.7. Suppose $X = (X_1, \ldots, X_n)$ where the X_i are iid with common gamma density $\gamma(\theta, \alpha)$ of Appendix A. Here $\alpha > 0$ is known, and $\Omega = \{\theta : 0 < \theta < \infty\}$. Write $r = n\alpha$, and from Appendix A note what the density of the sufficient statistic $T_n = \sum_1^n X_i$ is. It is desired to estimate θ by an interval estimator whose upper and lower limits are in specified proportion $c > 1$, motivated by invariance considerations in this scale-parameter problem; these considerations then entail using an estimator of the form $[AT_n, cAT_n]$ for some choice of the constant A. Show that the choice of A that maximizes the probability that the interval will cover the true value of θ is $A = (c - 1)/rc \log c$. [The arithmetic is similar to that leading to (4.4), but the limits on the integral differ, as does the random variable (T_n, not θ).]

9.8. The spirit of this problem (highly efficient simple estimation for a complicated Ω) is that of 8.11, and the X_i are again iid with the density of Problem 7.6.7(d), $h = 1$, $f_\theta(x_1) = \theta(x_1 + \theta)^{-2}$ for $x_1 > 0$, with $\Omega = \{\theta : \theta > 0\}$. Since the method of confidence interval construction of Chapter 9 uses tests, and since efficient tests were seen in Problem 8.3 to be complicated to construct in this setting, for large n we base our method on $Z_{1/2,n}$, instead. Show (unless you already worked Problem 8.11) that $n^{1/2}(Z_{1/2,n} - \theta)$ is asymptotically $\mathcal{N}(0, 4\theta^2)$, and use this to give an approximate expression (in terms of Φ) for the confidence coefficient of the procedure $[(1 - cn^{-1/2})Z_{1/2,n}, (1 + cn^{-1/2})Z_{1/2,n}]$, where $c > 0$ is specified.

9.9. *Model I ANOVA confidence set on means.* Refer to Section 8.7 and the setup of Problem 8.12, which is assumed here. Write $\psi_1 = \theta_1 - \theta_2$ and $\psi_2 = 2\theta_3 - \theta_1 - \theta_2$. Parts (a)–(d) use "Model I" (i) of Problem 8.12.

(a) (*Easy*) For the data of Problem 5.15(d), find a 95 percent ($\gamma = .95$) confidence interval for ψ_1 in each of the two cases of Model I:

(i) σ_F^2 known to equal 2,000. [*Hint*: Use normal tables.]

(ii) σ_F^2 unknown. [*Hint*: Use "t"—("student"-) tables.]

(b) (*Slightly harder.*) Show that the B.L.U.E.'s t_1^* and t_2^* (say) of ψ_1 and ψ_2 are uncorrelated, and use this to conclude that

$$(t_1^* - \psi_1)^2/\mathrm{var}_F(t_1^*) + (t_2^* - \psi_2)^2/\mathrm{var}_F(t_2^*)$$

has a χ^2-distribution with two degrees of freedom. Use this result to find an *elliptical confidence region* on (ψ_1, ψ_2) for the data of Problem 5.15(d) in each of the two cases (i) and (ii) of (a).

(c) [*Optional.*] Show that, if (t_1, t_2, t_3) are any solution of the normal equations and $\bar{t}$ is their average, and $\bar{\theta}$ is the average of θ_1, θ_2, θ_3, then the confidence set of (b) (i) can be written in the more comprehensible symmetric form $\left\{\theta : \sum_1^3 (\theta_i - \bar{\theta} - t_i + \bar{t})^2 \le \text{constant}\right\}$.

(d) (*Easier than* (b)!) Show that, if U is a standard $\mathcal{N}(0, 1)$ random variable, $P\{|U| \le 2.237\} = \sqrt{.95}$. Use this to show that, in case (a) (i), the region

$$\{(\psi_1, \psi_2) : |\psi_1 - t_1^*| < 2.237\sqrt{\mathrm{var}(t_1^*)}, |\psi_2 - t_2^*| < 2.237\sqrt{\mathrm{var}(t_2^*)}\}$$

is a 95 percent rectangular confidence region. Compute its explicit form for the data of Problem 5.15(d). Compare the *area* of this region with that of b(i). [It can be shown that the region of (c) has smallest area among *rectangular* location-invariant regions, and that of (b) (i) has smallest area among *all* location-invariant regions.]

(e) "Model II" of the same setup. [In this part, the "$A \cdot \chi_n^2$-distribution" means the distribution of A times a variable with the χ_n^2 distribution, as in the subsection on distributions.] Refer to part (ii) of Problem 8.12, showing or recalling therefrom that the residual sum of squares $Q_{\min}$ still has the $(\sigma^2 \cdot \chi_{12}^2)$-distribution. [*Hint*: Its conditional distribution given $\theta_1, \theta_2, \theta_3$ is this for every value of $\theta_1, \theta_2, \theta_3$.] Also show that $\sum_i (X_{i\cdot} - X_{\cdot\cdot})^2$ has the $\left[\left(\frac{\sigma^2}{5} + \sigma_1^2\right) \cdot \chi_2^2\right]$-distribution. [*Hint*: Consider the distribution of the X_i's.] Find an upper confidence limit (one-sided confidence interval) on $\phi = \sigma_1^2/\sigma^2$ in cases (i) and (ii) of (a), using the same data.

APPENDIX A

Some Notation, Terminology, and Background Material

Abbreviations

iid = independent and identically distributed.

X iid$(n)L$ = "$X = (X_1, X_2, \ldots, X_n)$ where the X_j are iid, each with probability law L." [In usage, L will typically be replaced by a suitable specification such as "normal with mean 2 and variance 3," "Poisson with mean 4," etc. The word *law* is commonly used to mean the probability measure on the sample space. In the iid case this can conveniently be thought of as a specification of the pdf or df of X_j.]

df = (cumulative) distribution function.

pdf = probability density function.

The terminology pdf is used in both the *discrete case*, where the pdf p satisfies

$$P(A) = \sum_{x \in A} p(x)$$

and in the *absolutely continuous case*, where there is a pdf f such that

$$P(A) = \int_A f(x)\,dx.$$

In the following list of abbreviations of common probability laws, all of the pdf's can be found in most standard probability texts of the level of this book.

Discrete Laws, with Probability Function p

Abbreviation	Law
$\mathscr{B}(\theta)$	*Bernoulli* with parameter value $\theta = p(1) = 1 - p(0)$. Here $0 \leq \theta \leq 1$. This law has mean θ and variance $\theta(1-\theta)$.

Abbreviation	Law
$\mathscr{B}(n, \theta)$	*Binomial*, sum of n iid $\mathscr{B}(\theta)$ random variables: $$p(x) = \binom{n}{x}\theta^x(1-\theta)^{n-x} \quad \text{for} \quad x = 0, 1, \ldots, n.$$ This law has mean $n\theta$ and variance $n\theta(1-\theta)$.
$\mathscr{P}(\theta)$	Poisson with mean θ, where $\theta \geq 0$: $$p(x) = e^{-\theta}\theta^x/x! \quad \text{for} \quad x = 0, 1, \ldots$$ This law has mean θ and variance also θ.
$\mathscr{G}(\theta)$	Geometric with $p(0) = 1 - \theta$, where $0 \leq \theta < 1$: $p(x) = (1-\theta)\theta^x$ for $x = 0, 1, 2, \ldots$ This law has mean $\frac{\theta}{1-\theta}$. The sum of n iid $\mathscr{G}(\theta)$ random variables has a *negative binomial law*, referred to too rarely in the book to deserve a symbol.

Continuous Laws with Absolutely Continuous Case Density Function f

Abbreviation	Law
$\mathscr{N}(\mu, \sigma^2)$	Normal with mean μ (real) and variance $\sigma^2(>0)$: $$f(x) = (2\pi)^{-1/2}\sigma^{-1}e^{-(x-\mu)^2/2\sigma^2} = \sigma^{-1}\phi((x-\mu)/\sigma)$$ where ϕ is the "standard" $\mathscr{N}(0, 1)$ normal pdf. [The standard $\mathscr{N}(0, 1)$ df is often denoted by Φ]. If $\sigma^2 = 0$, the normal law is a degenerate discrete law assigning probability 1 to the value μ.
$\mathscr{R}(\theta_1, \theta_2)$	Rectangular, or uniform law, from θ_1 to θ_2: $$f(x) = \begin{cases} 1/(\theta_2 - \theta_1) & \text{if} \quad \theta_1 \leq x \leq \theta_2, \\ 0 & \text{otherwise.} \end{cases}$$ This law has mean $(\theta_1 + \theta_2)/2$ and variance $(\theta_2 - \theta_1)^2/12$.
$\gamma(\theta, \alpha)$	Gamma law with "scale parameter $\theta > 0$ and index parameter $\alpha > 0$": $$f(x) = \begin{cases} \dfrac{x^{\alpha-1}}{\Gamma(\alpha)\theta^\alpha}e^{-x/\theta} & \text{if} \quad x > 0, \\ 0 & \text{otherwise.} \end{cases}$$ Here $\Gamma(\alpha) = \int_0^\infty x^{\alpha-1}e^{-x}\,dx, \quad = (\alpha-1)! \quad$ if $\quad \alpha =$ positive integer. From the definition of Γ and the relation $\Gamma(\alpha) = (\alpha-1)\Gamma(\alpha-1)$ (proved by integrating by parts), one can prove that the $\gamma(\theta, \alpha)$ density has mean $\theta\Gamma(\alpha+1)/\Gamma(\alpha) = \theta\alpha$. For positive integral α, similarly, the rth moment is $\theta^r\alpha(\alpha+1)\,(\alpha+r-1)$. Consequently, the variance is $\theta^2\alpha$. Thus, $\gamma(\theta, 1)$ is the *exponential law* with mean θ. If X_i is $\gamma(\theta, \alpha_i)$, $i = 1, 2$, and the X_i are independent, $X_1 + X_2$ is $\gamma(\theta, \alpha_1 + \alpha_2)$. The law $\gamma(2, n/2)$ is often called the "chi-squared (χ^2) law with n degrees of freedom," discussed in Chapter 8. Occasionally the law obtained by translating the preceeding density occurs [density of $X + c$ where X has law $\gamma(\theta, \alpha)$].

Abbreviation	Law
$\beta(\alpha_1, \alpha_2)$	Beta law with parameters $\alpha_1, \alpha_2(>0)$: $$f(x) = \begin{cases} x^{\alpha_1 - 1}(1-x)^{\alpha_2 - 1}/B(\alpha_1, \alpha_2) & \text{for } 0 < x < 1, \\ 0 & \text{otherwise,} \end{cases}$$ where the "beta function" $B(\alpha_1, \alpha_2) = \int_0^1 x^{\alpha_1 - 1}(1-x)^{\alpha_2 - 1}\,dx = \Gamma(\alpha_1)\Gamma(\alpha_2)/\Gamma(\alpha_1 + \alpha_2)$. If X has this density, we see that $EX^r = B(\alpha_1 + r, \alpha_2)/B(\alpha_1, \alpha_2)$. In particular, the density has mean $\alpha_1/(\alpha_1 + \alpha_2)$ and variance $\alpha_1\alpha_2/(\alpha_1 + \alpha_2)^2(\alpha_1 + \alpha_2 + 1)$.

Linear Algebra

The reader of this book, especially of Chapter 5, is assumed to understand, both theoretically and computationally, the main ideas of finite dimensional (real) vector spaces, including dimension, matrix operations, rank, subspace spanned by a collection of vectors, projection, solution space of a set of linear equations, orthogonal decomposition of a space, length, inner product, and the Pythagorean principle.

We now summarize briefly, in matrix notation, some material used extensively in Chapter 5 on the notion of "covariance." In Section 5.4 there is a development in matrix notation that uses this material. The present development is background for that. We use a *prime* to denote *transpose*, and usually unprimed vectors are column vectors.

The expectation of a vector $X(N \times 1)$ or matrix $Z(r \times s)$ of random elements is defined as the corresponding vector or matrix of expectations. Thus, if Z has elements Z_{ij}, EZ is the matrix of elements EZ_{ij}.

For scalar random variables X and Y, we write $\operatorname{cov}(X, Y)$ for their covariance. For a vector $W(N \times 1)$, we write $\operatorname{Cov}(W)$ for its $N \times N$ *covariance matrix*, whose diagonal elements are the values $\operatorname{var}(W_i)$ and whose off-diagonal elements are the values $\operatorname{cov}(W_i, W_j)$.

Write $EW = \mu$ (both $N \times 1$). Since WW' is $N \times N$ with $(i,j)^{\text{th}}$ element $W_i W_j$, we obtain

$$\operatorname{Cov}(W) = E\{(W - \mu)(W - \mu)'\}.$$

For a random variable W_1, we know $\operatorname{var}(a_1 W_1 + b_1) = a_1^2 \operatorname{var}(W_1)$ if a_1 and b_1 are constants. Similarly,

$$\operatorname{cov}((a_1 W_1 + b_1), (a_2 W_2 + b_2)) = a_1 a_2 \operatorname{cov}(W_1, W_2).$$

If α and β (both $N \times 1$) are constant N-vectors, the scalar random variables $Z_1 = \alpha' W$ and $Z_2 = \beta' W$ have covariance

$$\begin{aligned}
\operatorname{cov}(Z_1, Z_2) &\overset{\text{def}}{=} E\{(\alpha' W - E\{\alpha' W\})(\beta' W - E\{\beta' W\})\} \\
&= E\{\alpha'(W - \mu)(W - \mu)'\beta\} \\
&\qquad [\text{since } \beta' W = W'\beta \quad \text{and} \quad \alpha'\mu = E\{\alpha' W\}, \text{etc.}] \qquad (*) \\
&= \alpha' E\{(W - \mu)(W - \mu)'\}\beta \quad [\text{since } E\{\alpha' T\} = \alpha' E\{T\}, \text{etc.}] \\
&= \alpha' \operatorname{Cov}(W)\beta.
\end{aligned}$$

Similarly, if the random $(M \times 1)$ vector $U = AW(M \times N$ times $N \times 1)$ one obtains

$$\underset{M \times M}{\operatorname{Cov}(U)} = \underset{(M \times N)}{A} \; \underset{(N \times N)}{\operatorname{Cov}(W)} \; \underset{(N \times M)}{A'} \;. \qquad (**)$$

In fact, if α and β are the ith and j^{th} rows of A, then $(*)$ is simply the $(i,j)^{\text{th}}$ element of the matrix equation $(**)$.

APPENDIX B

Conditional Probability and Expectation, Bayes Computations

Some of this material is used in the computation of Bayes procedures in Chapter 4.

(i) In some problems in Chapter 4 *no knowledge of conditional probability in the continuous case is required.* This includes Problem 4.5, which you can carry out with S and ξ discrete while still allowing D to be the interval containing all the values $a \leq d \leq b$ even though ξ assigns all probability to only a finite subset of values in $\Omega = \{\theta : a \leq \theta \leq b\}$. Also, no knowledge of conditional probability in the continuous case is required in the development preceding Example 4.1, which is valid also in the continuous case with no mention of conditional probability. For example, in the "k-urn" setting of Problem 4.4 with a general loss function, the expression $h_\xi(x, d) = \sum_F W(F, d)_{P_{F;X}}(X)P_\xi(F)$ in the development referred to when X has continuous case density function $f_{i;X}(x)$ under F_i and when ξ assigns all probability to $\{F_1, \ldots, F_k\}$, is simply $h_\xi(x, d_j) = \sum_i W(F_i, d_j)f_{i;X}(x)p_\xi(F_i)$; and the characterization of Bayes procedures merely compares these quantities (with simplification possible for the simple W as described in part of the problem). Also, the same expression is valid if Ω is represented as a Euclidean set on which there is an a priori *density* $f_\xi(\theta)$. For then $R_t(\xi) = \int r_t(\theta)f_\xi(\theta)\,d\theta$ and the derivation in Chapter 4 headed "Computation of Bayes Procedures" proceeds as before except with f_ξ for p_ξ, yielding $\int W(\theta, d)p_{\theta;X}(x)f_\xi(\theta)\,d\theta$ as the expression for $h_\xi(x, d)$. Here $p_{\theta;X}(x)$ could be the discrete case or continuous case pdf of X. Example 4.2 is also worked *without* using conditional probability. It is only in carrying the derivation preceding Example 4.2 over to the case for which F or X is not discrete that the notion of conditional probability in the continuous case arises.

(ii) Briefly, the situation is this: $P\{A|B\}$ can only be defined by $P\{A \cap B\}/P\{B\}$ when the denominator $P\{B\}$ is nonzero. If a real random variable X has density function $f_X(x)$, positive and continuous at $x = x_0$, the event $\{x_0 - \varepsilon < X < x_0 + \varepsilon\}$ has positive probability if $\varepsilon > 0$. So $P\{A|x_0 - \varepsilon < X < x_0 + \varepsilon\}$ can be defined as earlier. For "nice" events A this last conditional probability will approach a limit as $\varepsilon \to 0$, and we *define* this limit to be $P\{A|X = x_0\}$. (Since $P\{X = x_0\} = 0$ we could not have used the $P\{A \cap \{X = x_0\}\}/P\{X = x_0\}$ definition here.) This makes perfectly good intuitive sense, since "knowing" that an observation takes on the values 1.37528 isn't much different information from knowing it takes on a value between 1.37527 and 1.37529, in most applications.

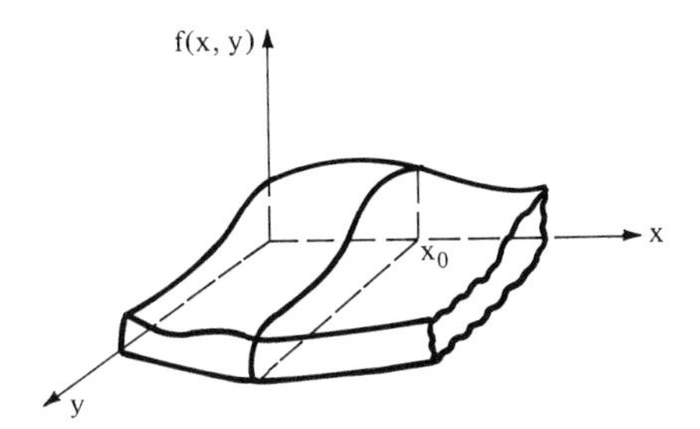

In particular, if X and Y are random variables with joint density function $f(x, y)$ and A is the event $Y \le y_0$, one obtains

$$P\{Y \le y_0 | X = x_0\} \overset{\text{def}}{=} \lim_{\varepsilon \to 0} \frac{\int_{x_0-\varepsilon}^{x_0+\varepsilon} \{\int_{-\infty}^{y_0} f(x, y)\, dy\}\, dx}{\underbrace{\int_{x_0-\varepsilon}^{x_0+\varepsilon} \{\int_{-\infty}^{\infty} f(x, y)\, dy\}\, dx}}. \tag{B1}$$

Note that the term in braces in the denominator of (B1) is just $f_X(x)$. If this, as well as the expression in braces in the numerator, is continuous and positive in x at x_0, (B1) becomes

$$\lim_{\varepsilon \to 0} \frac{[\text{approx. } 2\varepsilon]\{\int_{-\infty}^{y_0} f(x_0, y)\, dy\}}{[\text{approx. } 2\varepsilon]\{f_X(x_0)\}} = \int_{-\infty}^{y_0} \frac{f(x_0, y)}{f_X(x_0)}\, dy. \tag{B2}$$

Recall that f_Y, the density function of Y, is that function for which, for each y_0, $P\{Y \le y_0\} = \int_{-\infty}^{y_0} f_Y(y)\, dy$. Analogously, the conditional density of Y given that $X = x_0$ can be defined as that function $f_{Y|X}(y|x_0)$ for which, for each y_0,

$$P\{Y \le y_0 | X = x_0\} = \int_{-\infty}^{y_0} f_{Y|X}(y|x_0)\, dy. \tag{B3}$$

Comparing (B3) with (B2), we see that $f_{Y|X}(y|x_0) = f(x_0, y)/f_X(x_0)$. Pictorially (see drawing), slicing the graph of f with the plane $x = x_0$ yields the graph of a function of y which, when multiplied by the constant $1/f_X(x_0)$ so as to yield unit area under it, is the graph of $f_{Y|X}(y|x_0)$. Note the formal analogy of all this to the discrete case where $\int$ is replaced by $\sum$.

[For those who encounter conditional probability presented in a more formal fashion the preceding paragraph is replaced by defining $P\{A|X = x_0\} = g_A(x_0)$ (say) by saying that g_A is any function for which $E\{g_A(X)\} = P\{A\}$. It is not hard to see that, in "nice cases," the developments of the previous paragraph yield the same g_A. *Don't worry about this.*]

(iii) There are four cases in Bayes examples with $\Omega = \{\theta : a \leq \theta \leq b\}$: (1) X discrete and (a) θ with discrete law under ξ, (b) θ with continuous density under ξ; (2) X with continuous density function and (a) and (b) as previously. Case (1) (a) is literally what is presented in the discussion preceding Example 4.2. Case (2) (b) is obtained by the same formal development as (1) (a) with the interpretation of the previous paragraph ((ii)) where Y is replaced by θ. Thus, $f_{\theta;X}(x)f_\xi(\theta)$ is the joint density of X and θ, and $f_{\theta;X}(x_0)f_\xi(\theta)/\int_{-\infty}^{\infty} f_{\theta;X}(x_0)f_\xi(\theta)\,d\theta$ is the conditional density of θ given that $X = x_0$.

(1) (b) and (2) (a) are "mixed cases" not usually considered in books like Parzen's: One random variable is discrete, but the other is continuous. It is not hard to see how these can arise in practice, as in Example 4.2. The development preceding that example is now carried through in case (1) (b) as in the text; the probability that the parameter value is $\leq \theta_0$, given that $X = x_0$, is $\int_{-\infty}^{\theta_0} p_{\theta;X}(x_0)f_\xi(\theta)\,d\theta/\int_{-\infty}^{\infty}$ (same), by using the definition of $P\{A|B\}$ when A and B have positive probability; and differentiating this with respect to θ_0 yields $p_{\theta;X}(x_0)f_\xi(\theta)/\int_{-\infty}^{\infty}$ (same) $d\theta$ for the density function of θ given that $X = x_0$. Similarly, the definition (just as in (i)) of $P\{A|X = x_0\}$, with A the event $\{\theta = \theta_0\}$, yields for case (2) (a) $f_{\theta;X}(x_0)p_\xi(\theta)\Big/\sum_\theta$ (same) for the conditional (discrete) probability function of θ given that $X = x_0$; there is only a discrete set of θ-values for which this is positive.

So all four cases are covered in the development given in Chapter 4, allowing for slight changes of notation in cases other than (1) (a).

APPENDIX C

Some Inequalities and Some Minimization Methods

Estimation of the unknown parameters of the general linear model is an example of a statistical problem for which, within a certain class of procedures, i.e., linear unbiased estimators, best procedures exist (loss measured by square error). The problem of finding such best procedures is illustrated by several examples in Sections 5.1 and 5.2 and is treated abstractly in Section 5.4. From the standpoint of decision theory the problem of determining best procedures is the problem of minimizing the risk. Thus, in Section 5.1 we encounter a relatively easy problem, to minimize $\sum_{i=1}^{n} a_i^2$ subject to the restriction $\sum_{i=1}^{n} a_i = 1$. In Section 5.2 we deal with a somewhat more difficult problem, to minimize $\sum_{i=1}^{n} a_i^2$ subject to $\sum_{i=1}^{n} a_i b_i = 1$. In other places we encounter other questions of minimization. In this appendix we develop some useful inequalities. The use of some of these inequalities in this book is very limited, but most of the following results have great applicability. After a discussion of inequalities we use the two problems mentioned here to illustrate several methods of attacking minimization problems.

C.1. Inequalities

C.1.1. Convex Functions

The function $f(x) = x^2$ is a convex function. So are the following functions: $f(x) = |x|$, $f(x) = e^x$, and $f(x) = x^4 + x^2 + 1$. If you draw a rough graph of each function you will see that each picture has the following characteristic. Take any two points $(x_1, f(x_1))$ and $(x_2, f(x_2))$ on the graph. The line segment joining these points lies completely on or above the curve but never below the curve.

We may express this geometric idea as follows. Suppose $x_1 < x_2$ and $x_1 \le x \le x_2$. Then there is a number α satisfying $0 \le \alpha \le 1$ and $x = \alpha x_1 + (1 - \alpha)x_2$. The point in the plane having coordinates $(\alpha x_1 + (1 - \alpha)x_2, \alpha f(x_1) + (1 - \alpha)f(x_2))$ is the point a fraction α of the way along the line segment joining $(x_1, f(x_1))$ to $(x_2, f(x_2))$. Our geometric description of convexity says

$$f(\alpha x_1 + (1 - \alpha)x_2) \le \alpha f(x_1) + (1 - \alpha)f(x_2). \tag{C1}$$

A function having the property that for all x_1, x_2, and α, $0 \le \alpha \le 1$, (C1) holds, is said to be a *convex function.* When the graph of a function contains no linear segments (x^2 contains none, but $|x|$ does), the function is called *strictly convex.* In this case for all x_1, x_2, and α such that $x_1 < x_2$, $0 < \alpha < 1$,

$$f(\alpha x_1 + (1 - \alpha)x_2) < \alpha f(x_1) + (1 - \alpha)f(x_2). \tag{C2}$$

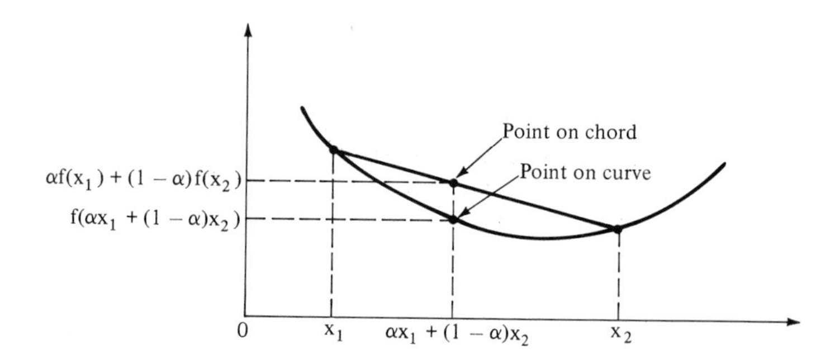

It is easy to show that many common functions are convex. Suppose at each x the function $f(x)$ has a second derivative $f''(x)$ and that $f''(x) \ge 0$. Then $f(x)$ is a convex function. If $f''(x) > 0$ for all x then $f(x)$ is a strictly convex function. For example, $f(x) = x^2$ is a strictly convex function since $f''(x) = 2 > 0$ for all x. This criterion does not apply to $f(x) = |x|$ since $f''(0)$ does not exist. If $f(x) = e^{ax}$ then $f''(x) = a^2 e^{ax} > 0$ for all x unless $a = 0$. Therefore unless $a = 0$, e^{ax} is a strictly convex function.

The assertion about second derivatives may be proved by several applications of the mean value theorem. Suppose $x < y$, and $0 \le \alpha \le 1$, $\beta = 1 - \alpha$. Then

$$\begin{aligned}
&\alpha f(x) + \beta f(y) - f(\alpha x + \beta y) \\
&= \alpha(f(x) - f(\alpha x + \beta y)) + \beta(f(y) - f(\alpha x + \beta y)) \\
&= \alpha(x - \alpha x - \beta y)f'(x^*) + \beta(y - \alpha x - \beta y)f'(y^*) \\
&= \alpha\beta(x - y)f'(x^*) + \alpha\beta(y - x)f'(y^*) \\
&= \alpha\beta(y - x)(f'(y^*) - f'(x^*)) \\
&= \alpha\beta(y - x)(y^* - x^*)f''(x^{**}) \ge 0.
\end{aligned}$$

In the first applications of the mean value theorem numbers x^* and y^* are found satisfying

$$x < x^* < \alpha x + \beta y \quad \text{and} \quad \alpha x + \beta y < y^* < y.$$

Therefore $x^* < y^*$. In the second application of the mean value theorem a number x^{**} satisfying $x^* < x^{**} < y^*$ is found. Observe that if $f''(x^{**}) > 0$ then $\alpha f(x) + \beta f(y) > f(\alpha x + \beta y)$ follows; i.e., strict convexity follows.

Suppose $f(x)$ is a convex function, $x_1, \ldots, x_n$ are n real numbers, and $\gamma_1, \ldots, \gamma_n$ are n nonnegative real numbers satisfying $\gamma_1 + \cdots + \gamma_n = 1$. Then

$$f(\gamma_1 x_1 + \cdots + \gamma_n x_n) \leq \gamma_1 f(x_1) + \cdots + \gamma_n f(x_n).$$

This relationship is proved by induction. If $n = 2$ we have the definition of a convex function. If $n = m + 1$ let

$$y = \frac{\gamma_1 x_1 + \cdots + \gamma_m x_m}{\gamma_1 + \cdots + \gamma_m}.$$

Note that $1 - \gamma_n = \gamma_1 + \cdots + \gamma_m$. Then $\gamma_1 x_1 + \cdots + \gamma_m x_m + \gamma_{m+1} x_{m+1} = (1 - \gamma_n)y + \gamma_n x_n$. Using the definition of a convex function

$$\begin{aligned}
&f(\gamma_1 x_1 + \cdots + \gamma_m x_m + \gamma_{m+1} x_{m+1}) \\
&\quad = f((1 - \gamma_n)y + \gamma_n x_n) \\
&\quad \leq (1 - \gamma_n) f(y) + \gamma_n f(x_n) \\
&\quad = (1 - \gamma_n) f\left(x_1 \gamma_1 \Big/ \sum_1^m \gamma_i + \cdots + x_m \gamma_m \Big/ \sum_1^m \gamma_i \right) + \gamma_n f(x_n) \\
&\quad \leq (1 - \gamma_n) \sum_{i=1}^m \frac{\gamma_i}{1 - \gamma_n} f(x_i) + \gamma_n f(x_n) \\
&\quad = \gamma_1 f(x_1) + \cdots + \gamma_n f(x_n).
\end{aligned}$$

In the next to last step we have used the inductive hypothesis that the assertion is true of any set of m numbers.

If we think of a random variable X satisfying $P(X = x_i) = \gamma_i$, $i = 1, \ldots, n$, then the assertion just proved states

$$f(EX) \leq Ef(X).$$

Whenever $f(x)$ is a convex function this assertion, known as *Jensen's inequality*, is true of all real-valued random variables X. A more general proof is given in Section 6.5.

We give a few simple illustrations of this very important result. For example the function $f(x) = x^2$ is a convex function. By Jensen's inequality $(EX)^2 \leq EX^2$ or $0 \leq EX^2 - (EX)^2 = \operatorname{var} X$. This result we already knew. More generally, suppose $r \geq 1$ is a number and $\alpha > 0$ is a number. The function $f(x) = |x|^r$ is a convex function. Therefore, by Jensen's inequality, $|EY|^r \leq E(|Y|^r)$. Choose for the random variable $Y = |X|^\alpha$. Then

$$(E|X|^\alpha)^r \leq E|X|^{\alpha r}.$$

If we take the αr root of both sides we obtain

$$(E|X|^{\alpha})^{1/\alpha} \le (E|X|^{\beta})^{1/\beta}$$

where $\alpha < \beta = \alpha r$ and β is otherwise arbitrary.

All the preceding remarks have an analogue for functions of n variables. A function $f(x_1, \ldots, x_n)$ of n variables is a convex function if the chord joining any two points of its graph lies on or above the graph. The condition (C1) can be interpreted as the analytic expression of this geometric definition for n variables. Simply interpret x_1 and x_2 as n-dimensional vectors.

More generally if $x_i = (x_{i1}, \ldots, x_{in})$ and $\gamma_1, \ldots, \gamma_n$ are nonnegative satisfying $1 = \gamma_1 + \cdots + \gamma_n$ then

$$f\left(\sum \gamma_i x_{i1}, \ldots, \sum \gamma_i x_{in}\right) \le \sum_i \gamma_i f(x_{i1}, \ldots, x_{in}).$$

If in addition strict inequality holds whenever at least two γ_i's are positive and the points $x_1, \ldots, x_n$ are distinct, then the function f is strictly convex.

There is a test, using second derivatives, for whether a function of n variables is convex. The description of this test requires the idea of a positive definite quadratic form. Suppose we are given a set of numbers a_{ij}, $i, j = 1, 2, \ldots, n$. If it is true that for every vector $(x_1, \ldots, x_n)$ having at least one nonzero coordinate $\sum_{i=1}^{n} \sum_{j=1}^{n} a_{ij} x_i x_j > 0$, then we say the quadratic form $\sum_1^n \sum_1^n a_{ij} x_i x_j$ is positive definite. Now suppose $f(x_1, \ldots, x_n)$ is given and $a_{ij} = \left.\dfrac{\partial^2 f}{\partial x_i \partial x_j}\right|_{(x_1, \ldots, x_n)}$. If for each point $(x_1, \ldots, x_n)$ these numbers are the coefficients of a positive definite quadratic form, then $f(x_1, \ldots, x_n)$ is a strictly convex function.

Suppose $f(x_1, \ldots, x_n) = x_1^2 + x_2^2 + \cdots + x_n^2$. Then $a_{ij} = 0$ if $i \ne j$ and $a_{ij} = 1$ if $i = j$. The associated quadratic form is therefore $x_1^2 + x_2^2 + \cdots + x_n^2 > 0$ unless $x_1 = x_2 = \cdots = x_n = 0$. Therefore $f(x_1, \ldots, x_n) = x_1^2 + \cdots + x_n^2$ is a convex function.

C.1.2. Schwarz's Inequality

Schwarz's inequality states that if U and V are random variables having finite second moments then the random variable $|UV|$ has a finite expectation and

$$(EUV)^2 \le EU^2 EV^2.$$

Further, if $(EUV)^2 = EU^2 EV^2$ then there are numbers a, b not both zero such that

$$0 = E(aU + bV)^2;$$

that is, $aU + bV$ is zero with probability one.

To verify the inequality, let $c = \sqrt{EU^2}$ and $d = \sqrt{EV^2}$ and note that if $c = 0$, then $U = 0$ with probability one and both sides of the inequality equal zero (similarly true if $d = 0$). If neither c nor d is zero, then

$$0 \leq (U/c - V/d)^2 = U^2/c^2 + V^2/d^2 - 2UV/cd$$

and, taking expectations,

$$0 \leq 1 + 1 - 2EUV/cd.$$

Thus $EUV \leq cd$. Similarly, using $0 \leq (U/c + V/d)^2$ leads to $EUV \geq -cd$, and the Schwarz inequality, $(EUV)^2 \leq c^2 d^2$, follows.

A famous special case of this arises when the sample space is $\{1, 2, \ldots, n\}$ and the $P\{1\} = P\{2\} = \cdots = P\{n\} = 1/n$. Suppose the random variables U and V are defined by

$$U(i) = a_i, \qquad V(i) = b_i, \qquad i = 1, \ldots, n.$$

Then $EUV = \sum a_i b_i / n$, $EU^2 = \sum a_i^2 / n$ and $EV^2 = \sum b_i^2 / n$. Therefore

$$|\textstyle\sum a_i b_i|^2 \leq \sum a_i^2 \sum b_i^2.$$

Further if $|\sum a_i b_i|^2 = \sum a_i^2 \sum b_i^2$ then there are numbers a and b not both zero such that $aa_i + bb_i = 0$, $i = 1, 2, \ldots, n$.

Another special case of Schwarz's inequality is obtained by putting $V = 1$. We then obtain

$$(EU)^2 \leq EU^2.$$

C.1.3. An Inequality

Suppose it is known of a random variable U that there are positive numbers m and M such that $-m \leq U \leq M$. Then $0 \leq m + U$ and $0 \leq M - U$. Therefore $0 \leq (m + U)(M - U)$. Taking expectations $0 \leq mM + MEU - mEU - EU^2$. If we suppose in addition that $EU = 0$, then it follows that $EU^2 \leq mM$. Further, if $EU^2 = mM$ then $0 = E(m + U)(M - U)$ and since $m + U \geq 0$, $M - U \geq 0$, it follows that with probability one $0 = (m + U)(M - U)$. Therefore the only values U can take are $-m$ or M.

C.2. Methods of Minimization

In many problems one knows on theoretical grounds that there is a minimum. This is true of the problems discussed in Chapter 5. It is sometimes true that one can guess the location of the minimum. The problem is then merely a problem of showing one's guess is correct. For this purpose, methods using inequalities like Jensen's or Schwarz's inequality are very useful in verifying the correctness of one's guess. But it is likewise true that sometimes one cannot guess the location of a minimum. In this case a very useful technique is to compute derivatives which are set equal to zero. We warn the reader that it is *not* generally true that the minimum of a function f can be found by setting its derivative equal to zero (e.g., minimize $f(x) = x$ on the set $0 \le x \le 1$), nor that every function attains its minimum (e.g., $f(x) = e^{-x^2}$ on $-\infty < x < \infty$). These phenomena do not arise in the present problem but may in others.

C.2.1. Direct Analytic Minimization of $\sum_1^n a_i^2$ with $\sum_1^n a_i = 1$

One must be cautious not just to try to minimize $\sum_1^n a_i^2$ without any attention to the restriction. The unrestricted minimum of $\sum a_i^2$, which quantity must be ≥ 0, is zero, which is achieved only when $a_1 = a_2 = \cdots = a_n = 0$; but these values of the a_i's do not satisfy $\sum a_i = 1$, so this is not a solution to the problem. There are two main ways of absorbing the restriction $\sum a_i = 1$ into the problem:

(i) Substitute $a_n = 1 - a_1 - \cdots - a_{n-1}$, minimizing $g(a_1, \ldots, a_{n-1}) = a_1^2 + a_2^2 + \cdots + a_{n-1}^2 + (1 - a_1 - \cdots - a_{n-1})^2$ with respect to $a_1, \ldots, a_{n-1}$ (with *no* restriction on these quantities), and then setting $a_n = 1 - \sum_1^{n-1} a_i$. This minimization can be achieved by solving the equations

$$\frac{\partial g(a_1, \ldots, a_{n-1})}{\partial a_i} = 0, \qquad i = 1, \ldots, n-1,$$

and making sure that the extremum so obtained is a minimum. For example, if $n = 2$, we would have the single equation $2a_1 - 2(1 - a_1) = 0$, with solution $a_1 = 1/2$ (hence, $a_2 = 1 - a_1 = 1/2$), and this gives a minimum since $d^2 g(a_1)/da_1^2 = 4 > 0$. Similarly, for general n we obtain the $n-1$ simultaneous linear equations

$$2a_i - 2(1 - a_1 - \cdots - a_{n-1}) = 0, \qquad 1 \le i \le n-1.$$

These equations say that each a_i is equal to $1 - a_1 - \cdots - a_{n-1}$; thus, all a_i's are equal. Substituting a_1 for each a_i in any of the equations, we then obtain $a_1 = \cdots = a_{n-1} = 1/n \left(\text{hence, } a_n = 1 - \sum_1^{n-1} a_i = 1/n\right)$. The condition that an extremum is actually a minimum is harder to write down and verify in problems involving many variables than in problems involving a single variable. In the present problem there is no difficulty, however, since we have obtained the *one* extremum of g, and since g grows arbitrarily large as one or more of the a_i's grows large in absolute value, this extremum can only be a minimum (if it were an extremum other than a minimum, there would have to be a local minimum at some other point, since g grows large as the point $(a_1, \ldots, a_{n-1})$ gets far from $(0, \ldots, 0)$; but that is impossible, since g was found to have only one extremum).

(ii) Use the method of Lagrange multipliers, which is a standard technique in calculus.

In Section 5.1, the problem of minimizing $\sum_1^n h_i a_i^2$ subject to $\sum_1^n a_i = 1$ is considered, where the h_i are given positive values. The method of (i) amounts to minimizing $\sum_1^{n-1} h_i a_i^2 + h_n(1 - a_1 - \cdots - a_{n-1})^2$. Setting the derivative with respect to a_i equal to 0, we obtain $0 = 2a_i h_i - 2(1 - a_1 - \cdots - a_{n-1})h_n = 2a_i h_i - 2a_n h_n$ for $1 \leq i \leq n-1$. Consequently, $a_i h_i = a_j h_j$ for all i and j. Writing this as $a_j^{-1} a_i = h_j h_i^{-1}$ and summing both sides over i, we obtain $\left(\text{since } \sum_1^n a_i = 1\right) a_j^{-1} = h_j\left(\sum_1^n h_i^{-1}\right)$, the same solution $a_j = h_j^{-1} \Big/ \sum_1^n h_i^{-1}$ as obtained by another method in Section 5.1.

C.2.2. A Geometric Argument

This is best visualized when $n = 2$ or $n = 3$ but applies for arbitrary n. Consider in the n-dimensional space of points $(a_1, a_2, \ldots, a_n)$ the family of spherical surfaces $S_c = \left\{(a_1, \ldots, a_n) : \sum_1^n a_i^2 = c^2\right\}$; for each constant $c > 0$ we obtain such a surface. The set $P = \{(a_1, \ldots, a_n) | \sum a_i = 1\}$ is a *hyperplane* (line when $n = 2$, plane when $n = 3$). The square of the distance of any point $(a_1, \ldots, a_n)$ of P from the origin is $\sum_1^n a_i^2$. The problem is to find the point q of P which is closest to the origin; the line from the origin to q will be perpendicular to P. The intersection of any sphere S_c with P is either a sphere of one lower dimension (check this when $n = 3$, where it is a circle), a point (if S_c is tangent to P), or is empty (if c is small enough). The value c for which S_c is just tangent to P will yield a point q (the point of tangency) of P which lies on an S_c of smallest possible radius c, i.e., which is closest to the origin. The point of tangency will be on the line $a_1 = a_2 = \cdots = a_n$, since P and all S_c's are symmetric about this line. It is now easy to see that the point $(1/n, \ldots, 1/n)$, with $c^2 = 1/n$, does the job.

C.2.3. Convexity and Symmetry: An Analytic Counterpart of the Argument of C.2.2.

The function $h(a_1, \ldots, a_n) = \sum_1^n a_i^2$ was shown to be strictly convex in the discussion of convex functions. Moreover, h is a *symmetric* function of its arguments in that permuting the order of particular values $a_1, \ldots, a_n$ of the arguments does not change the value of h. It is easy to prove that the minimum of h is attained. Suppose (this will lead to a contradiction) that $(a_1^*, a_2^*, \ldots, a_n^*)$ is a collection of numbers which are not all the same and which have unit sum, and that $(a_1^*, a_2^*, \ldots, a_n^*)$ minimizes h. Then, because of the symmetry, so does any permutation of these values; in particular, $p_1^* = (a_1^*, a_2^*, \ldots, a_n^*)$, $p_2^* = (a_n^*, a_1^*, a_2^*, \ldots, a_{n-1}^*)$, $p_3^* = (a_{n-1}^*, a_n^*, a_1^*, \ldots, a_{n-2}^*)$, $\ldots$, $p_n^* = (a_2^*, a_3^*, \ldots, a_n^*, a_1^*)$ are n different such minimizing vectors. Now, the strict convexity of h says that the average of h at these n points is at least as great as the value of h at the center of gravity of these n points (with weights $\gamma_i = 1/n$). The ith coordinate of this center of gravity is

$$n^{-1}(a_i^* + a_{i+1}^* + \cdots + a_n^* + a_1^* + \cdots + a_{i-1}^*) = 1/n, \quad \text{since} \sum a_i^* = 1.$$

Hence,

$$h(a_1^*, \ldots, a_n^*) = \sum_i \frac{1}{n} h(p_i^*) > h(1/n, 1/n, \ldots, 1/n),$$

which contradicts the supposition that $h(a_1^*, \ldots, a_n^*)$ was a minimum. Hence, the supposition that the a_i^* were not all equal is false, and $(1/n, \ldots, 1/n)$ is the *only* minimizing argument of $h(a_1, \ldots, a_n)$.

C.2.4. Use of Schwarz's Inequality

The argument used in C.2.3 is a very general argument applicable to many problems in mathematics. The simple problem we have been using for illustrative purposes can be treated with simpler methods as follows. To this end, consider a probability function which assigns probability $1/n$ to each of the values $a_1, \ldots, a_n$. The first moment of this probability function is $n^{-1} \sum_{i=1}^{n} a_i$, and the second moment is $n^{-1} \sum a_i^2$. Hence, assuming $\sum a_i = 1$, the variance of this probability function is $n^{-1} \sum a_i^2 - (n^{-1} \sum a_i)^2 = n^{-1}(\sum a_i^2 - n^{-1})$. This being a variance, it is ≥ 0; that is,

$$\sum a_i^2 \geq n^{-1}.$$

Moreover, equality can hold in this last inequality only if the above "variance" is zero, i.e., only if all a_i are actually equal (so that the probability is all concentrated at one point). Thus, the minimum value n^{-1} of $\sum a_i^2$ subject to $\sum a_i = 1$ is attained only if $a_1 = a_2 = \cdots = a_n = n^{-1}$.

We consider now the slightly more complicated problem, minimize $\sum_1^n a_i^2$ subject to $\sum_1^n a_i b_i = 1$. If we apply the Schwarz inequality we find

$$1 = \left| \sum_1^n a_i b_i \right|^2 \leq \sum_1^n a_i^2 \sum_1^n b_i^2.$$

Equality will hold exactly when there is a constant c such that $a_1 = cb_1, \ldots, a_n = cb_n$, and in this case $1 = \sum_1^n a_i b_i = c \sum_1^n b_i^2$ so that $a_i = b_i \Big/ \sum_1^n b_i^2$, $i = 1, 2, \ldots, n$. For the simpler problem $b_1 = \cdots = b_n = 1$ we obtain at once $a_i = 1/n$.

C.2.5. A Short Method of Presenting the Preceding Proof

Once we suspect (e.g., by looking at the case $n = 2$) what the solution is, we can write out a succinct proof of the result, as follows. If $\sum a_i = 1$, we have

$$\begin{aligned}\sum_1^n a_i^2 &= \sum_1^n [(a_i - n^{-1}) + n^{-1}]^2 \\ &= \sum_1^n (a_i - n^{-1})^2 + n^{-1} + 2n^{-1}\sum_1^n (a_i - n^{-1}) \\ &= n^{-1} + \sum_1^n (a_i - n^{-1})^2,\end{aligned}$$

since $\sum_1^n (a_i - n^{-1})$ vanishes because $\sum a_i = 1$. Now, $\sum_1^n (a_i - n^{-1})^2$ is nonnegative and equals zero if and only if each a_i is n^{-1}. Hence, $\sum_1^n a_i^2 \geq n^{-1}$, with equality if and only if each a_i is n^{-1}. This proof is really not much different from the previous one; it is given to indicate, as a method of attacking such problems, the idea of trying to break up the expression to be minimized into a constant term and a term which is nonnegative and which vanishes when the argument equals the suspected solution. To achieve this breakup, we grouped each a_i with a term n^{-1}, the suspected solution. This method of proof is used in Section 5.1 in the next more complicated case of linear unbiased estimation.

C.2.6. An Inductive Proof

Suspecting (e.g., from the result when $n = 2$, or by intuition) what the result is, we can prove it inductively, as follows: We want to prove the proposition P_n: "subject to $\sum_1^n a_i = L$, the sum $\sum_1^n a_i^2$ is minimized only by the choice $a_1 = \cdots = a_n = n^{-1}L$." To prove that P_n is true for all $n \geq 1$, we must prove (i) that it is true for $n = 1$ (which in the present problem is trivial), and (ii) that, for any $n \geq 1$, assuming P_n is true for all $n \leq N$, it follows that P_{N+1} is true. To prove the latter, we break up $\sum_1^{N+1} a_i^2$ into two parts, $\sum_1^N a_i^2$ and a_{N+1}^2, and use the inductive hypothesis (that P_N is true) on the first of these. We do this by noting that, for any fixed value h of a_{N+1}, the minimum of $\sum_1^N a_i^2$ subject to $\sum_1^N a_i = L - h$ is, by P_N, equal to $(L - h)^2/N$, attained only when $a_1 = a_2 = \cdots = a_N = N^{-1}(L - h)$. Hence, for any $a_1, \ldots, a_{N+1}$ with $\sum_1^{N+1} a_i = L$,

$$\sum_{1}^{N+1} a_i^2 \geq N^{-1}(L - a_{N+1})^2 + a_{N+1}^2,$$

with equality if and only if $a_1 = \cdots = a_N = N^{-1}(L - a_{N+1})$. The minimum of the right side with respect to a_{N+1} is easily seen (e.g., by setting the derivative equal to zero) to be attained if and only if $a_{N+1} = L/(N + 1)$, the right side becoming $L^2/(N + 1)$ in this case. Hence, if $\sum_{1}^{N+1} a_i = L$, we have

$$\sum_{1}^{N+1} a_i^2 \geq L^2/(N + 1),$$

with equality if and only if $a_1 = \cdots = a_{N+1} = L/(N + 1)$. This proves P_{N+1}.

In writing out this inductive proof, we chose the form of P_n to involve minimization of $\sum_{1}^{n} a_i^2$ subject to $\sum_{1}^{n} a_i = L$ (where L is any real number) rather than subject to $\sum_{1}^{n} a_i = 1$. This was done merely for convenience, since the two forms are closely related. If P_n had instead been stated with the restriction $\sum_{1}^{n} a_i = 1$, we would have proceeded by proving (assuming P_N) that, if $\sum_{1}^{N+1} a_i = 1$, then $\sum_{1}^{N} a_i^2 \geq N^{-1}(1 - a_{N+1})^2$ (with equality if and only if $a_1 = \cdots = a_N = (1 - a_{N+1})/N$) by noting that this is trivial if $1 - a_{N+1} = 0$, and that, if $1 - a_{N+1} \neq 0$, we can write $b_i = a_i/(1 - a_{N+1})$ for $1 \leq i \leq N$ and have $\sum_{1}^{N} b_i = 1$. Applying P_N to the b_i's, we would then obtain $\sum_{1}^{N} a_i^2/(1 - a_{N+1})^2 \geq N^{-1}$ and would then proceed as before, but with $L = 1$.

References

Anderson, T. W. (1958). *Introduction to Multivariate Statistical Analysis*. John Wiley, New York.
Cramér, H. (1946). *Mathematical Methods of Statistics*. Princeton University Press, Princeton, New Jersey.
Fisher, R. A. and Yates, F. (1957). *Statistical Tables for Biological, Agricultural, and Medical Research*, 5th Edition. Oliver and Boyd, Edinburgh.
Lehmann, E. L. (1959). *Testing Statistical Hypotheses*. John Wiley, New York.
Lehmann, E. L. (1983). *Theory of Point Estimation*. John Wiley, New York.
Mood, A. and Graybill, F. (1963). *Introduction to the Theory of Statistics*. McGraw-Hill, New York.
Parzen, E. (1962). *Stochastic Processes*. Holden-Day, San Francisco.
Pitman, E. J. G. (1939). "The estimation of the location and scale parameters of a continuous population of any given form." *Biometrika* **30**, 391–421.
Scheffé, H. (1959). *The Analysis of Variance*. John Wiley, New York.
Siegmund, D. (1985). *Sequential Analysis: Tests and Confidence Intervals*. Springer-Verlag, New York.
Stein, C. (1945). "A two sample test for a linear hypothesis whose power is independent of the variance." *Ann. Math. Statist.* **16**, 243–258.
Wald, A. (1947). *Sequential Analysis*. John Wiley, New York.
Wald, A. (1950). *Statistical Decision Functions*. John Wiley, New York.
Weiss, L. and Wolfowitz, J. (1974). *Maximum Probability Estimators and Related Topics*. Springer-Verlag, Berlin.
Wilks, S. S. (1962). *Mathematical Statistics*. John Wiley, New York.

Index